Spintronics

Puja Dey · Jitendra Nath Roy

Spintronics

Fundamentals and Applications

Springer

Puja Dey
Department of Physics and Centre for Organic Spintronics and Optoelectronics Devices (COSOD)
Kazi Nazrul University
Asansol, West Bengal, India

Jitendra Nath Roy
Department of Physics and Centre for Organic Spintronics and Optoelectronics Devices (COSOD)
Kazi Nazrul University
Asansol, West Bengal, India

ISBN 978-981-16-0068-5 ISBN 978-981-16-0069-2 (eBook)
https://doi.org/10.1007/978-981-16-0069-2

This Springer imprint is published by the registered company Springer Nature Singapore Pte Ltd.
The registered company address is: 152 Beach Road, #21-01/04 Gateway East, Singapore 189721, Singapore

Preface

Aim and Purpose

The philosophy, science and technology have given a new dimension to the field of information processing technology in the last few decades. Very large-scale integration (VLSI) technology has revolutionized the electronics industry and has established the 20th century as the computer age. Computers have enhanced human life to a great extent and its applications have penetrated into all areas of human society. The future generation of human civilization needs more energy-efficient devices, better healthcare, quality of living through integrating information and communication technologies (ICT), smart sensing technology, ubiquitous computing, big data analytics and intelligent decision-making system. Furthermore, future information processing schemes also have the potential for weather forecasting, earthquake prediction, artificial intelligence, space flight, antimissile system, telemedicine, multimedia etc. along with optical communication and networking technology. Therefore, dramatic technological solutions to these problems are needed, and unless we gear our thoughts towards a totally different pathway, we will not be able to further improve our information processing performance for the future.

Electron contains two fundamental degrees of freedom: the charge and the spin. Over the past years, only the charge of an electron was used in conventional electronic devices to store and control information. Such circuits are based on non-magnetic semiconductors, in which the electrons' spin does not play a role. But charge-based electronics have almost attained its saturation in the limits of storage density due to the remarkable increase of power consumption as a result of scaling-related development. An innovative idea that could offer a way out is thus the most wanted. The scientific world observed a most important breakthrough in the information technologies in 1988, when A. Fert and P. Grünberg discovered independently the giant magnetoresistance. This discovery rolls the wheel of the spintronics field, which relies not only on electrons' charge but also on electrons' spin, offering perspectives for a new generation of devices. Spintronic devices, on the other hand, use the spin degree of freedom to generate and control charge currents as well as to interconvert electrical and magnetic signals. By combining processing, storage, sensing and logic within a single integrated platform, spintronics can complement and, in some cases, outperform semiconductor-based electronics. Indeed, the seamless integration of these functions can bring enormous benefits in terms of scaling, power consumption and data processing speed of integrated circuits. The first application of spintronics in the realm of information processing was spin valves sensors, which are, for example, used in hard disk drive read heads and later in magnetic random-access memories (MRAMs).

It is also to be noted that we are in the era of daily explosions in the development of spintronic components for computing and other applications. The business of spintronics is booming worldwide. The field of spintronics is progressing rapidly and shows many dramatic opportunities for overcoming the limitations of electronics. The techniques of computing with electron spin may provide a way out of the limitations of computational speed and complexity inherent in charge-based electronic computing. The rapid development of femtomagnetism, nano-photonics, nonlinear optics and plasmonics has also enriched the field of spintronics and has opened up a new paradigm of computing and information processing. With inter-disciplinary researches into spintronics new concepts are coming up. The scope of works in this field has widened to cover a vast expanse—from micrometre to nanometre scales and from classical to quantum. This has brightened its future prospects of spintronics. Hence, a fundamental knowledge of spintronics is essential for the up-gradation and development of existing technology. Almost each and every month, new research findings are coming out in the field of spintronics. Therefore, the idea of different basic phenomena associated with spintronics that plays the pivotal role has been discussed in detail in this book. Some of the present and future applications that are based on spintronic devices are also discussed in this book. **It is expected that this book will serve two important purposes**:

1. To formally introduce senior-level undergraduate and graduate science and engineering students to emerging field of spintronics.
2. This book aims to provide the necessary foundation in spintronics which prepares the students for an intensive study of advanced topics at later stage.

Organization of the Book

The contents of this book are divided into ten chapters. Chapter 1 is the overview of spintronics. Chapters 2–6 are fundamentals of spintronics and should be treated as essential. Chapters 7–9 are little bit advanced, and Chap. 10 discusses the various present and future application aspect of spintronics. Although there are many more applications which are not included in this book, we feel we have covered the most important ones.

- Chapter 1 gives the overview and evolution of spintronics. It discusses the relative merits and demerits of spintronics over conventional electronics. Spin-dependent band gap in ferromagnetic materials is analysed. It also gives the outline of different emerging branches of spintronics.

- Chapter 2 lays the foundation of spintronics. Almost all the basic phenomena contributing to fascinating field of spintronics are discussed. These include some important notions, such as spin polarization, spin filter effect, spin injection, spin accumulation, spin relaxation, spin extraction etc. *Passive* spintronic devices, such as spin valves, are described in this chapter. This chapter also presents different kinds of spin relaxation mechanisms, such as the D'yakonov-Perel', the Elliott-Yafet, the Bir-Aronov-Pikus and hyperfine interactions with nuclear spins. *Field and heat-driven spintronics effect*, i.e., spin Hall effect, and Seebeck effect are also presented.
- Chapter 3 is devoted to discuss the most fundamental effect in spintronics, i.e., giant magnetoresistance (GMR). Different kinds of magnetoresistance and qualitative explanation of physical origin of GMR are explained. Spin-dependent and spin-flip scattering of electrons are described. Magnetoresistance theory based on *resistor network theory of GMR* is also discussed.
- Chapter 4 deals with the concept of tunnelling magnetoresistance, magnetic junctions and magnetic tunnel junction (MTJ). Quantum mechanical tunnelling of conduction electrons, which is at the origin of MTJ, is discussed in the light of *A* transfer matrix model. The Jullière formula is also explained elaborately.
- Chapter 5 emphasizes on *spin transfer torque* (STT). This chapter talks about spin transfer torque-driven magnetization dynamics and the possible applications of spin transfer torques in spintronic devices.
- Chapter 6 describes magnetic domain walls (DW) motion. Ratchet effect in magnetic DW motion is discussed. This chapter also gives the idea of current-driven DW motion.
- Chapter 7 deals with the emerging field of spintronics, i.e., optospintronics. The ultrafast manipulation of magnetic order by femtosecond lasers, as external stimuli, is discussed in this chapter. Special emphasis is given on*ultrafast* optical controlling of magnetic states of antiferromagnet, i.e., antiferromagnetic optospintronics. Spin–photon interaction, Faraday effect and inverse Faraday effect are presented. Outline of different types of all-optical spintronic switching is also given in this chapter.
- Chapter 8 presents one more promising branch of spintronics, i.e., terahertz spintronics, which bears novel application in THz range. Principle of operation of spintronic terahertz emitter and choice of materials for spintronic emitters are discussed in this chapter. Terahertz writing of an antiferromagnetic magnetic memory device is also discussed.
- Chapter 9 gives brief outline of one emerging branch of spintronics i.e., Semiconductor spintronics.
- Chapter 10 emphasizes on application side of spin-based devices. This chapter discusses several*modern spintronics devices that include GMR read head of modern hard disk drive, MRAM, position sensor, biosensor, magnetic field sensor, three terminal magnetic memory devices, spin FET etc.*

Target Audience

Immense care has been taken to keep the text as simple as possible. The students can read the book with minimum guidance from his/her instructor. The book begins with a thorough analysis of the subject in an easy-to-read style. The conventional approach appears to stress on the derivations, with a lesser attention paid to the physical concepts. As a result, the student feels the subject to be monotonous, detached and far from exciting. An attempt has been made in this book to balance the requirements of students taking a course and at the same time to make the subject interesting to them. Therefore, the mathematics is kept at the necessary minimum level and concepts are given priority over the derivations. Efforts are made to present the subject matter in a simple and lucid style. Basically, it is a student-friendly book written in a very simple language.

Request Feedback

Readers are requested to send their suggestions and comments for further improvement of this book. Please mention the title and the authors name in the subject line.

Asansol, India
2020

Dr. Puja Dey
Prof. Jitendra Nath Roy

Acknowledgements

It is a great pleasure to have the opportunity to thank all the people who contributed one way or another to our book.

First and foremost, we would like to express our sincere gratitude to Prof. Arghya Taraphder, Professor, Department of Physics, IIT Kharagpur, India, for guidance, support, sharing his knowledge and ideas in our every academic endeavours. This has indeed motivated us to carry out this book project. A special thank goes to Prof. Kushalendu Goswami, former professor of Department of Physics, Jadavpur University for his constant motivation and support till the completion of the book. We are grateful to Dr. S. M. Yusuf, Head, Solid State Physics Division, Bhabha Atomic Research Centre, Mumbai, India for always being available to scientific discussions. His motivating words and patronization have played a crucial role to venture into this book project. Great support and precious advice from Dr. Alok Banerjee, UGC-DAE CSR Indore is gratefully acknowledged.

One of the authors, Dr. Puja Dey would like to express her heart-felt gratitude to Prof. (Dr.) Wolfgang Weber, Professor, Dept. of Physics, University of Strasbourg, France for introducing Dr. Dey into the fascinating field of spintronics during her post-doctoral research at IPCMS, CNRS, Strasbourg, France.

We are whole-heartedly thankful to our Ph.D. research scholars: Mr. Apurba Pal of Kazi Nazrul University, Asansol, Mr. Debojit Deb and Ms. Debarati Nath of National Institute of Technology Agartala, Mr. Nitish Ghosh of Kazi Nazrul University, Asansol, who helped a lot in drawing figures, and thereby have contributed indirectly to this book. Their devotions and co-operations are also appreciatively acknowledged. We would also like to thank our M.Sc. project students, Mr. Ratan Das and Mr. Uttam Kumar Das, for their help.

We record our indebtedness to the authors of those original works from where a large number of information are liberally drawn while preparing the present book.

A *big thank you* goes to the publisher (Springer Science+Business Media, LLC, 233 Spring Street, New York, NY 10013, USA), for promptness and cordiality they provide to their authors.

Finally, we wish to deeply thank our parents, for the encouragement and love with which they have surrounded us during all these years.

Asansol, India
2020

Dr. Puja Dey
Prof. Jitendra Nath Roy

Contents

About the Authors

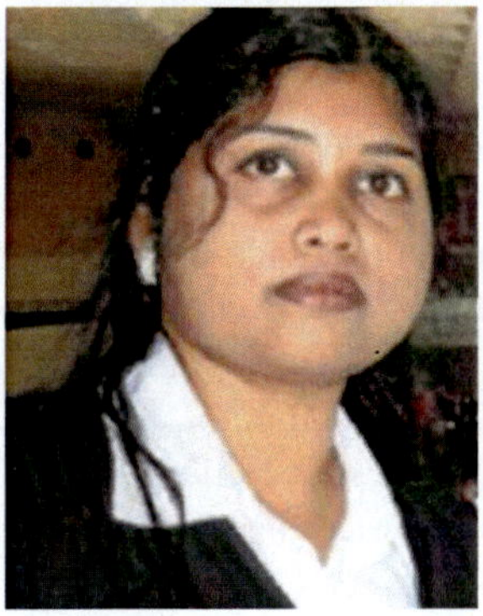

Dr. Puja Dey Associate Professor, Department of Physics and ***Joint Coordinator, Centre for Organic Spintronics and Optoelectronic Devices***, Kazi Nazrul University, C H Kalla, Asansol, WB, India.

She received M.Sc. and Ph.D. degree in Physics from University of North Bengal, India and Indian Institute of Technology Kharagpur, India, respectively. Dr. Dey received Institute Research Scholarship from I.I.T. Kharagpur and **Senior Research Fellowship** from *Council of Scientific and Industrial Research* (CSIR), Government of India. Dr. Dey has also received *Centre National de la Recherche Scientifique* (CNRS) Post-Doctoral Fellowship from CNRS, Strasbourg, France. Dr. Dey has overall 12 years of experience in teaching, research and administration. Presently, Dr. Dey is working as an Associate Professor in the Department of Physics and acting as ***Joint-Coordinator of Centre for Organic Spintronics and Optoelectronic Devices*** at Kazi Nazrul University, Asansol, WB, India. She has already published more than ***50 research papers*** in peer-reviewed journals and conference proceedings. She is co-author of ***two books***. Dr. Dey received two research grants, one from ***Department of Science and Technology*** **(DST) India**, another from **Board of Research in Nuclear Sciences (BRNS)**, ***Department of Atomic Energy, Govt of India***. Dr. Dey is the reviewer of many reputed journals. Dr. Dey is actively involved in research activities in different topics, such as spintronics, multi-functional materials, nanomaterials, optoelectronics and optospintronics and giving Ph.D. guidance to 6 ***Ph.D. students.***

Dr. Jitendra Nath Roy Professor, Department of Physics and ***Coordinator, Centre for Organic Spintronics and Optoelectronic Devices***, Kazi Nazrul University, C H Kalla, Asansol, WB, India.

He received M.Sc. and Ph.D. degree in Physics from Vidyasagar University, India. He is the recipient of ***University Silver Medal***. Dr. Roy received **Senior Research Fellowship** from Council of Scientific and Industrial Research (Government of India). Prof. Roy is also the recipient of ***International Sardar Patel Award*** from Sardar Vallabhbhai Patel Foundation, India, for his significant contribution in physical science. Dr. Roy has 20 years of experience in teaching research and administration. Presently, Dr. Roy is working as a Professor in the Department of Physics and acting as ***Coordinator of Centre for Organic Spintronics and Optoelectronic Devices*** at Kazi Nazrul University, Asansol, WB, India. He has already published more than ***139 research papers*** in peer-reviewed journals and conference proceedings. He is co-author of ***two books*** and ***one book chapter***. Dr. Roy received research grant from *AICTE, DST, BRNS (Govt of India)*. Dr. Roy is member of many International Advisory Committees, Technical Program Committees in various countries, acted as Panel Editor, Reviewer for reputed journals. He is a ***Fellow*** member of ***Optical Society of India.*** Prof. Roy has produced ***9 Ph.D. students*** in different topics of applied optics, photonics and optoelectronics.

Chapter 1
An Overview of Spintronics

1.1 Introduction

Spintronics, a portmanteau of spin-based electronics, is a new paradigm of electronics based on spin degree of freedom of the charge carriers. Under the scope of functionality of spintronics, both charge and spin properties of electrons can be utilized simultaneously. In this direction, production of devices with new functionality is a fascinating and promising field of research and has potential to revolutionize the world of electronics. Today, most of the technology are based on electrons and its flow throughout the devices or circuits. This is the way how *electronics devices* consisting of diodes, transistors, FETs, resistors etc. works. Up to this point of time, devices based on the principle of electronics have been realized by precisely controlling the charge of the electrons. For a long time, people did not exploit the fact that, apart from charge, every electron is also having spin, much like the earth precesses around its axis. On the question of the present day implementation of various microelectronic devices, we never consider the fact that nature has attributed electrons, along with charge, a spin property. Of course, all well-known ferromagnetic phenomena are ultimately the mere consequence of the diversified interplay and arrangement of electron spin. For a long time, in the field of research, spin-dependent electronic properties of ferromagnetic material and different micromagnetic phenomena have been supposed as two completely different cases. Although a huge research effort had been concentrated to elucidate the micromagnetic phenomena, research activities on the spin-dependent electronic properties of ferromagnetic systems were far less invasive. This is because the application of ferromagnets in industry was mostly limited so far, based on their bulk magnetization phenomenon only.

Furthermore, from the point of view of *information technology*, its central theme is the processing and storing of binary data. In solid-state systems, such operations were regularly implemented by controlling the charge of carriers, i.e., either of electrons or holes. Two physically distinguishable states are required to realize classical binary logic bits, i.e., ‘0’ and ‘1’. In earlier times, in computer memories those two states were defined by two distinctly different amounts of charge stored in a capacitor.

P. Dey and J. N. Roy, *Spintronics*,
https://doi.org/10.1007/978-981-16-0069-2_1

Alternatively, those two states were also realized by keeping two distinct voltage levels at some circuit. Such charge-based logic bits could be processed by incorporating switching devices, like metal oxide semiconductor field effect transistors (MOSFETs) in a circuit. In contrast, spintronics intends to exploit the spin property of charge carriers rather than charge to generate the desired outcome, namely, core information processing and storage functionalities. Both the approaches, i.e., (i) addition of spin degree of freedom to conventional charge-based electronic devices and (ii) sole incorporation of the spin property, have certain potential advantages like ***non-volatility, enhanced data processing speed, reduced consumption of electrical power*** **and much improved integration densities**, compared to conventional semiconductor devices. From the late twentieth century, such inclusion of spin, alone or in conjunction with charge, has been extensively exploited to ***process and store digital information*** encoded by the binary bits 0 and 1. The branch of spintronics in which there is no direct role of charge and encoding of information is done completely in the spin polarization, i.e., spin-up and spin-down states of electrons, is called as monolithic spintronics. In effect, spintronics, also known as 'magnetoelectronics', has become an emergent technology for various applications such as storing, encoding, accessing, processing and transmitting of information in some manner.

1.1.1 What Is Spintronics?

- *Spintronics is an emerging technology that exploits both the intrinsic spin of the electron and its associated magnetic moment.*
- *Spin is the intrinsic angular momentum of the negatively charge electron.*
- *Depending on the direction of spinning of electron, either in clockwise and anticlockwise direction, we may obtain two orientations of the associated spin-magnetic moment, given by the magnitudes* $\pm\frac{\hbar}{2}$. *Thus, spins of the electron exist in one of the two states, namely spin-up and spin-down, with spins either positive half or negative half.*
- *The two possible spin states naturally represent '0' and '1'. This bit of information is called* ***qubit***.
- *Motion of electron spin in a directional and coherent way set up spin current, circulation of which is supposed to transmit information in a spintronics device.*

1.1.2 Why Do We Need Spintronics?

This is mainly motivated by the failure of Moore's law. According to Moore's law, in electronics devices the number of transistors on a silicon chip roughly gets doubled every 18 months. However, different components, including transistors, in electronics devices have already reached nanoscopic dimensions. As per general consensus, further reduction of their size may result in the following consequences:

1. Production of intense heat might make the electronic circuit inoperable;
2. In the limit of nanoscale dimensions, quantum effects come into play instead of classical ones.

These restricted further size reductions of electronic component, such as transistors and others in electronic devices.

1.2 Comparative Merit–Demerit of Spintronics and Electronics

First, let us try to figure out the following.

1.2.1 What Are the Disadvantages of Electronics*?*

- *Consumption of high power.*
- *Higher degree dissipation of heat.*
- *Volatile electronic memory.*
- *Compactness is less, i.e., larger occupation of space on chip.*
- *Poor read and write speed because of inferior movement and controlling of electron.*

1.2.2 What Are then the Advantages of Spintronics Over Electronics?

- One of the prime advantages of spintronics stems from the spontaneous magnetization of ferromagnetic material by virtue of which ferromagnets tend to remain magnetized even after the withdrawal of any external magnetic field. This, in fact, created sparking interest in the computer hardware industry for the replacement of semiconductor-based components in computer hardware by the magnetic ones. Such effort initiated with the transformation of random access memory (RAM), which in turn leads to the evolution of magnetic random access memory (MRAM). MRAM is a non-volatile memory, like FLASH. Interestingly, MRAM does not require any electric current to retain the information, already written in electron's spin. This means spins would not change when electrical power is turned off.
- Such *all-magnetic RAM*, i.e., MRAM-based computer could retain all the information put into it. Significantly, when power is turned on, this special kind of computer does not require any '*boot-up waiting period*'.

- Another advantageous convenience of spintronics is that the usage of unique and specialized semiconductor is not required. Rather, incorporation of common metals like Cu, Al, Ag is enough to yield the desired functionality.
- Like ferromagnetic materials, antiferromagnets also bear a good number of properties that make them suitable for spintronic applications. Antiferromagnetic states are intrinsically non-volatile and additionally robust to external magnetic field. Most importantly, antiferromagnets are abundant in nature. Many ferromagnets, like iron and cobalt, become antiferromagnetic when oxidized, and they are good insulators also.
- Spin orientation of conduction electrons survives for relatively long time, on the order of nanoseconds. This makes spintronics devices promising for potential application in memory, storage and magnetic sensors.
- In order to operate, spintronics involves less power compared to that of conventional electronics. This is because the energy required to alter spin orientation is a minute fraction of the energy needed to make the electronics charge flow all around in the circuit.
- Spin is supposed to be more steady and 'reliable' than that of charge, when subject to the external perturbations like temperature, pressure or radiation. Hence, spintronics-based devices are expected to be better functioning in high temperature or radiation environments than that of electronics devices. Conceptually, spintronic devices should be more miniature, faster and more robust than electronic ones.

1.2.3 Advantages of Spintronics

To summarize, the advantages of spintronics are:

- *Consumption of low power.*
- *Smaller degree dissipation of heat.*
- *Non-volatile electronic memory.*
- *Compactness is more, i.e., lesser occupation of space on chip.*
- *Greater read and write speed because of superior and fast manipulation and controlling of electron spin.*
- *Spintronics uses very common metals like Cu, Al, Ag, instead of engineered semiconductor structure.*

1.3 Research Efforts Put on Spintronics

Nowadays, we are conversant with the idea that '*electron spin*' explicitly participates industry and has become an emergent technology for various applications such as storing, encoding, accessing, processing and transmitting of information in some manner. In this direction, spin relaxation and transport in both metallic and semiconducting samples are subject of intense research interest, not only for issues related to fundamental solid-state physics but also for their potential in electronic technology.

Designing and manufacturing of spintronic devices are executed by two different approaches. ***The first approach*** is to achieve improvement in the existing giant-magnetoresistive (GMR)-based technology. In this direction, a very preliminary attempt is to explore new materials having even larger electron spin polarization. Another attempt includes improving upon the existing GMR devices in order to obtain better spin filtering. In GMR-based industry, the already existing operational prototype device is the read head and memory storage cell. This is basically a GMR sandwich structure, consisting of alternating ferromagnetic and non-magnetic metallic layers. It is the relative magnetizations alignment in the adjacent ferromagnetic layers that decide the device resistance. For instance, device resistance is small for parallel alignment of magnetizations, whereas it is large for antiparallel alignment. Such change in resistance, referred to as magnetoresistance, is utilized to sense variations in magnetic fields. In recent efforts, magnetic tunnel junction-based devices have also been involved in GMR technology. In magnetic tunnel junctions, tunnelling current depends on the orientations of magnetizations of the electrodes. ***The second approach*** is rather more radical, where initiatives have been taken to find out novel avenues, both for production and application of spin-polarized currents. In this direction, research effort has been focussed on the spin transport in semiconductors to explore whether semiconductors can operate both as spin polarizers and spin valves. The significance of this attempt is that *amplification of signal* can be obtained in semiconductor-based spintronic devices, which could in principle function as multifunctional devices. Although the existing metal-based devices are successful as switches or valves, they do not amplify signals. A more important point is that such semiconductor-based devices are expected to be much easily integrated with conventional semiconductor industry. In order to effectively include spins into existing semiconductor-based devices, the technical challenges to achieve are efficient injection of electron spins followed by their transport into the device; finally controlling, manipulation and detection of this spin-polarized currents.

In this direction, spintronics has also extended its attention in the field of semiconductor devices, such as spin-FET, spin transistor, magnetic semiconductor devices etc. Very recently, advancement in the field of materials engineering presents a promising scenario to realize spintronic devices based on ***optical spin manipulation.*** Furthermore, application of electron and nuclear spins for quantum information processing and quantum computation has invoked tremendous ambition to the researcher. In fact, among many proposed hardware of quantum computer, the ones based on electron and nuclear spins have gained appreciable attentions. Evidently, spins of electrons and spin-1/2 nuclei could be perfect candidates for realization of quantum bits (qubits) given that their Hilbert spaces are well-defined and their decoherence is relatively slow.

1.4 Evolution of Spintronics

Let us go back into the history of spintronics. It emerged from the discoveries of the spin-dependent electron transport phenomena in solid-state devices in 1980s. Such discovery is about the injection of spin-polarized electron from a ferromagnetic to a normal metal by Johnson and Silsbee (1985) and the giant magnetoresistance by Albert Fert (Baibich et al. 1988) and Peter Grünberg (1988) (Baibich et al. 1989), independently. The Nobel Prize in Physics 2007 was awarded jointly to Albert Fert and Peter Grünberg for this discovery of giant magnetoresistance. Furthermore, beginning of spintronics can be tracked back from the pioneering ferromagnet/superconductor tunnelling experiments, carried out by Meservey and Tedrow and also from the early experiments performed on magnetic tunnel junctions by Julliere in 1970s (Julliere 1975). The concept of incorporation of semiconductors for spintronics applications was introduced through the theoretical proposal of a spin field effect transistor by Datta and Das (1990). Subsequently in spintronics, a new field of research, merging the technology of semiconductor and magnetism, has been developed. In this attempt, spin information of the electrons is intended to be utilized for extending the functionalities of the common transistor. This leads to the development that lead to '*going beyond transistor-based circuits*', hence prompting a model shift from the electronically driven to an entirely magnetically controlled digital logic. Based on magnetoresistive elements, the first design concepts for magnetic logic AND and OR gates have been proposed. This constituted the basic building blocks for the magnetic random access memory. Noteworthy, the non-volatile logic output of magnetic elements poses an immense advantage in comparison to that of conventional semiconductor-based electronics technology. This advent of technology is expected to decrease the consumption of power by several orders of magnitude. In 2012, IBM scientists observed and mapped the creation of persistent spin helices of synchronized electrons. Such observation of spin helices persisting for more than a nanosecond is actually a 30-fold increase from the previously observed results (Walser et al. 2012). This is even longer than the duration of a modern processor clock cycle. This finding opens new avenues of research for employing electron spins in information processing.

1.4.1 History of Spin

Stern–Gerlach Experiment

In 1925, Ralph De LaerKronig, George Uhlenbeck and Samuel Goudsmit, based on the anomalous Zeeman effect, postulated that along with orbital angular momentum, an electron possesses an additional angular momentum, arising out of the spinning motion about its own axis. Such spinning motion of electron is much like earth

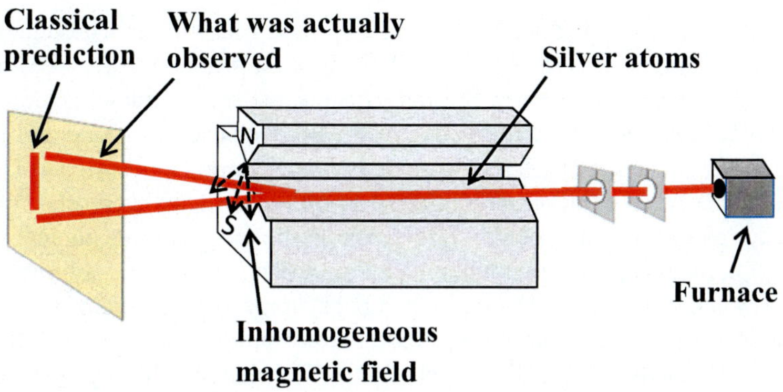

Fig. 1.1 Schematic diagram of Stern–Gerlach experimental setup

performing precession motion about its own axis and the magnitude of the corresponding angular momentum of self-rotation of electron is $\hbar/2$. The fixed magnitude of angular momentum associated with the spinning motion of electron suggested a physical interpretation of particles spinning around their own axis. The mathematical theory was worked out in depth by Pauli in 1927. Relativistic quantum mechanics, derived by Paul Dirac in 1928, included electron spin as an integral vital part of it.

However, the electron spin was already accidently evidenced in the very well-known Stern–Gerlach experiment, which is considered to be the turning point event in the history of spin. An experiment of atom's deflection named after Otto Stern and Walther Gerlach (Gerlach and Stern 1922; Stern 1921) performed in Frankfurt, Germany in 1920, was the first experiment to show the existence of an intrinsic property of the electron called 'spin'. Although their experimental result was not used to proof the existence of the electron spin, nowadays it is widely used to illustrate the existence of the spin and its quantization properties. Stern and Gerlach directed a silver beam through an inhomogeneous magnetic field. Then the beam hits a screen and shows how the Ag atoms are deflected after interacting with the inhomogeneous magnetic field (Fig. 1.1). In order to suppress the effect of Lorentz force, the silver beam was neutrally charged in the experiment.

Classically, the Ag atom is considered as spinning magnetic dipole. In the presence of a homogeneous magnetic field the dipole will precess due to the torque exerted by the magnetic field on it. If the magnetic field is inhomogeneous, the traversing dipole through the magnetic field will be deflected depending on its orientation. According to the dipole-magnetic field interaction, $F = \nabla(m \cdot H)$, where m is the dipole and H is the inhomogeneous magnetic field. Thus, one expects to see on the screen a smooth distribution of the Ag atoms. On the other hand, Bohr–Sommerfeld predicted that an atom of angular moment $L = 1$ would have a quantized magnetic moment with two equal sizes and of opposite directions. The aim of Stern–Gerlach experiment was to test the validity of this hypothesis (Stern 1921). Their result confirmed Bohr–Sommerfeld hypothesis for they observed two spots on the detector screen relative to two opposite magnetic moments. Later in 1927, a similar experiment using hydrogen atom, whose L = 0, was done by T. E. Phipps, and J. B. Taylor reproduced the two spots effect (Phipps and Taylor 1927). This posed a problem to Bohr–Sommerfeld hypothesis. The interpretation of Stern–Gerlach's results, nowadays, is referred to the electrons having a magnetic moment called spin. However, the concept of electron spin was first proposed in 1925 by Ralph De LaerKronig, Goudsmit and Uhlenbeck in order to explain the fine structures in the atomic spectra in the presence of external magnetic field known as Zeeman effect. While the quantum mechanics with three quantum numbers *n, l* and *m* could not explain the fine structures, a fourth quantum number was needed. Goudsmit and Uhlenbeck suggested the idea of spinning electron, which gives rise to an angular momentum in addition to the orbital angular momentum (Goudsmit and Uhlenbeck 1926). The idea of spinning electron did not convince Wolfgang Pauli, who argued that the electron is so small that it needs to rotate around itself with the speed of light in order to give rise to the measured angular momentum.

1.5 Quantum Mechanics of Spin

Theory of quantum mechanics, consisting of Heisenberg's matrix mechanics and Schrödinger's wave mechanics, predicts energy quantization and suggests a way to find out the energy difference between the levels. Moreover, it also allows one to calculate the probability of transition between the energy states having different quantization level. In wave mechanics, the wavefunction evolution with time and space for a single particle can be written by Schrödinger's equation as follows:

$$i\hbar \frac{d\psi(\vec{r})}{dt} = H_0 \cdot \psi(\vec{r}). \tag{1.1}$$

Neglecting spin, we may write

$$\begin{aligned} H_0 &= \frac{|\vec{p}|^2}{2m} + V(\vec{r}) \\ \vec{p} &= p_x\hat{x} + p_y\hat{y} + p_z\hat{z} = -i\hbar\frac{d}{dx}\hat{x} - i\hbar\frac{d}{dy}\hat{y} - i\hbar\frac{d}{dz}\hat{z} \\ \vec{r} &= \left[x\hat{x} + y\hat{y} + z\hat{z}\right], t \end{aligned} \tag{1.2}$$

where 'hats' indicates unit vectors along the axes of coordinates. It is noteworthy that Eq. (1.1) does not include the spin part. Hence the question raises, how to include the 'spin' part?

1.5.1 Pauli Spin Matrices

Inclusion of spin part was done by Wolfgang Pauli, who derived an equation to replace Eq. (1.1). This equation is known as the Pauli equation. It is well known that any physical observable is correlated with an operator in quantum mechanics. The operator should be linear in case of Schrödinger formalism, whereas it would be in matrix form in Heisenberg formalism. Now, eigenvalues of those linear operators are actually the expectation values of their corresponding physical observables. More clearly, those expectation values are expected to appear if measurements are carried out on those physical quantities in experiments. Likewise, spin is a physical observable since its associated angular momentum is a measurable quantity (discussed in Stern–Gerlach experiment). Therefore, a quantum mechanical operator must be associated with the spin. Such quantum mechanical operators have been derived by Pauli for the spin components along three orthogonal axes S_x, S_y and S_z. Those were come out to be three 2×2 matrices, which are known as ***Pauli spin matrices***. The approach of Pauli was based on the following facts:

(i) Upon measuring the component of spin angular momentum for an electron along any of the coordinate axes, we obtain the result as $+\hbar/2$ or $-\hbar/2$;
(ii) Similar to orbital angular momentum, the operators associated with the components of spin angular momentum should obey commutation rules under operations along three mutually orthogonal axes. Let us briefly discuss the commutation relations satisfied by the operators of the orbital angular momentum as given below:

$$\begin{aligned} L_x L_y - L_y L_x &= i\hbar L_z \\ L_y L_z - L_z L_y &= i\hbar L_x \\ L_z L_x - L_x L_z &= i\hbar L_y. \end{aligned} \tag{1.3}$$

These equations express that the operators associated with the components of orbital angular momentum along any two mutually orthogonal axes could not be measured simultaneously and that with absolute precision, except the component associated with the third axis disappears. Similar commutation relations were adopted by Pauli for the operators associated with the spin angular momentum components, i.e., S_x, S_y and S_z along three mutually orthogonal axes. These are given by

$$\begin{aligned} S_x S_y - S_y S_x &= i\hbar S_z \\ S_y S_z - S_z S_y &= i\hbar S_x \\ S_z S_x - S_x S_z &= i\hbar S_y. \end{aligned} \tag{1.4}$$

In Stern–Gerlach experiment, z-axis is assumed as the axis joining the South to North Pole of the magnet. Two traces have been obtained on the photographic plate.

Such observations were interpreted as being caused by spin angular momentum S, having two values $\pm\hbar/2$ of its z components. Hence, the matrix operator S_z should be (i) 2×2 matrix and (ii) eigenvalues must be $\pm\hbar/2$. We understand that a 2×2 matrix having eigenvalues of $\pm\hbar/2$ would be the matrix of the form:

$$M_{2\times 2} = \frac{\hbar}{2}\begin{pmatrix} 1 & 0 \\ 0 & -1 \end{pmatrix}. \tag{1.5}$$

In addition, Pauli also defined the first three dimensionless matrices σ_x, σ_y and σ_z such that

$$S_x = \frac{\hbar}{2}\sigma_x;\ S_y = \frac{\hbar}{2}\sigma_y;\ S_z = \frac{\hbar}{2}\sigma_z. \tag{1.6}$$

Since, S_x, S_y and S_z must have eigenvalues of $\pm\hbar/2$, σ matrices have eigenvalues of ± 1. Furthermore, Eq. (1.4) mandates that

$$\begin{aligned} \sigma_x\sigma_y - \sigma_y\sigma_x &= 2i\sigma_z \\ \sigma_y\sigma_z - \sigma_z\sigma_y &= 2i\sigma_x \\ \sigma_z\sigma_x - \sigma_x\sigma_z &= 2i\sigma_y. \end{aligned} \tag{1.7}$$

According to Eqs. (1.5) and (1.6)

$$\sigma_z = \begin{pmatrix} 1 & 0 \\ 0 & -1 \end{pmatrix}. \tag{1.8}$$

Hence, the other two matrices, which have eigenvalues of ± 1 and obey Eq. (1.7), are

$$\begin{aligned} \sigma_x &= \begin{pmatrix} 0 & a \\ a^* & 0 \end{pmatrix} = \begin{pmatrix} 0 & 1 \\ 1 & 0 \end{pmatrix} \\ \sigma_y &= \begin{pmatrix} 0 & b \\ b^* & 0 \end{pmatrix} = \begin{pmatrix} 0 & -i \\ i & 0 \end{pmatrix}. \end{aligned} \tag{1.9}$$

These are the famous Pauli spin matrices, which according to Eq. (1.6) act as operators for the corresponding spin components. Additionally, square of each of the Pauli spin matrices is the 2×2 unit matrix [I]. Thus,

$$\begin{aligned} |S|^2 &= |S_x|^2 + \left|S_y\right|^2 + |S_z|^2 \\ |S|^2 &= \left(\frac{\hbar}{2}\right)^2[I] + \left(\frac{\hbar}{2}\right)^2[I] + \left(\frac{\hbar}{2}\right)^2[I] \\ |S|^2 &= 3\left(\frac{\hbar}{2}\right)^2[I] + \bar{s}(\bar{s}+1)\hbar^2[I] \end{aligned}$$

$$\bar{s}(\bar{s}+1) = \frac{3}{4} \Rightarrow \bar{s} = \frac{1}{2}. \tag{1.10}$$

1.5.2 Eigenvectors of the Pauli Matrices: SPINORS

In quantum mechanics, the state of any physical system is identified with a wavefunction (in a complex separable Hilbert space) or by a point (projective Hilbert space). Each vector in the wavefunction is called 'ket' $|\psi\rangle$. The eigenvalues of the Pauli spin matrices are ± 1. We denote the corresponding eigenvectors as $|\pm\rangle$.

Matrix $\boldsymbol{\sigma}_z$

The eigenvectors of $\boldsymbol{\sigma}_z$ should satisfy eigenvalue equation as shown below:

$$\sigma_z|\pm\rangle_z = \pm 1|\pm\rangle_z. \tag{1.11}$$

These eigenvectors are given by

$$|+\rangle_z = \begin{pmatrix} 1 \\ 0 \end{pmatrix}$$
$$|-\rangle_z = \begin{pmatrix} 0 \\ 1 \end{pmatrix}. \tag{1.12}$$

Matrix $\boldsymbol{\sigma}_x$

The eigenvectors of $\boldsymbol{\sigma}_x$ should satisfy eigenvalue equation as shown below:

$$\sigma_x|\pm\rangle_x = \pm 1|\pm\rangle_x. \tag{1.13}$$

These eigenvectors are given by

$$|+\rangle_x = \frac{1}{\sqrt{2}}\begin{pmatrix} 1 \\ 1 \end{pmatrix}$$
$$|-\rangle_x = \frac{1}{\sqrt{2}}\begin{pmatrix} 1 \\ -1 \end{pmatrix}. \tag{1.14}$$

These eigenvectors are orthogonal and can be expressed as follows:

$$|\pm\rangle_x = \frac{1}{\sqrt{2}}\left[|+\rangle_z \pm |-\rangle_z\right]. \tag{1.15}$$

Matrix $\boldsymbol{\sigma}_y$

The eigenvectors of $\boldsymbol{\sigma_y}$ should satisfy eigenvalue equation as shown below:

$$\sigma_y|\pm\rangle_y = \pm 1|\pm\rangle_y. \tag{1.16}$$

These eigenvectors are given by

$$|+\rangle_y = \frac{1}{\sqrt{2}}\begin{pmatrix} 1 \\ i \end{pmatrix}$$
$$|-\rangle_y = \frac{1}{\sqrt{2}}\begin{pmatrix} 1 \\ -i \end{pmatrix}. \tag{1.17}$$

These eigenvectors are orthogonal and can be written as follows:

$$|\pm\rangle_y = \frac{1}{\sqrt{2}}\left[|+\rangle_z \pm i|-\rangle_z\right]. \tag{1.18}$$

These eigenvectors of Pauli spin matrices are the examples of SPINORS. As we have found, these are basically 2 × 1 column vectors, representing the spin state of an electron. If the SPINORS are known, then electron's spin orientation of a given state can be easily deduced.

1.6 Dynamics of Magnetic Moments: *Landau-Lifshitz-Gilbert Equation*

It is well known that the basis of all magnetic phenomena is the interactions between magnetic moments and magnetic fields. On the one hand, as knowledge of such interactions is indeed important to understand several magnetic phenomena, on the other hand, they may be applied to derive diversified functionalities in many ways. Magnetic moment of a ***homogeneously*** magnetized materials for a given volume $\boldsymbol{V}$ is given by $m = VM$, where $\boldsymbol{M}$ is the magnetization. Straightforwardly, we may say that if $\boldsymbol{V}$ denotes the atomic volume, then $\boldsymbol{m}$ is the magnetic moment per atom; similarly, if $\boldsymbol{V}$ is the volume of the magnetic solid, then m corresponds to the total magnetic moment of the solid. The latter case is often referred to as the 'macrospin approximation'. Furthermore, considering ***inhomogeneous magnetized materials***, conceptually the magnetic solid can be subdivided into small regions. Magnetization of those small regions may be assumed to be homogeneous. However, the dimension of those regions is large enough that, in general, the magnetization dynamics can be explained classically.

It is well known that in the absence of any damping effect, the precessional motion of a magnetic moment is described by the torque equation. Now, following quantum mechanics, the angular momentum ($\boldsymbol{L}$) associated with a magnetic moment $\boldsymbol{m}$ is

$$\vec{L} = \vec{m}\Big/\gamma, \tag{1.19}$$

where γ is the gyromagnetic ratio. Application of magnetic field, $\boldsymbol{H}$ exerts torque on the magnetic moment $\boldsymbol{m}$ given by

$$\vec{\tau} = \vec{m} \times \vec{\mathrm{H}}. \tag{1.20}$$

Again, the variation in angular momentum with time corresponds to the torque:

$$\frac{d\vec{L}}{dt} = \frac{d}{dt}\left(\frac{\vec{m}}{\gamma}\right) = \vec{m} \times \vec{\mathrm{H}}. \tag{1.21}$$

Now, if the spins are not only subjected to the external magnetic field, but several factors like magnetocrystalline anisotropy, shape anisotropy, magnetic dipole interaction etc. are also affecting the spins, then the situation would become much more complicated. These factors are also expected to contribute to the thermodynamical potential, Φ. The collective effect and consequences, arising out of these contributions, can be approximated as an effective magnetic field:

$$H^{eff} = -\frac{\partial}{\partial M}. \tag{1.22}$$

Therefore, following Eq. (1.21), the motion of the magnetization vector can be written as the following equation:

$$d\vec{m}\Big/dt = \gamma\vec{m} \times \vec{H}^{eff}. \tag{1.23}$$

This equation is named after Landau and Lifshitz. It illustrates the precession of the magnetic moment around the effective field $\vec{H}^{eff}$. As already mentioned, $\vec{H}^{eff}$ has many contributions and hence it can be written as follows:

$$\vec{H}^{eff} = \vec{H}_{ext} + \vec{H}_{ani} + \vec{H}_{dem} + \cdots \tag{1.24}$$

where $\vec{H}_{ext}$ is the external applied field, $\vec{H}_{ani}$ is the anisotropy field and $\vec{H}_{dem}$ is the demagnetization field. It is noteworthy that, apart from $\vec{H}_{ext}$, the other contributions to $\vec{H}^{eff}$, i.e., $\vec{H}_{ani}$, $\vec{H}_{dem}$ etc. are material-dependent. Now, if a magnetic material is exposed to optical excitation, there may be some optically induced modifications in the material-dependent components of fields, as mentioned above. This in turn causes change in $\vec{H}^{eff}$ and thereby giving rise to ***optically induced magnetization dynamics***. At equilibrium, the time variation of angular momentum is zero and consequently, the torque is zero. Moreover, the motion of a precessing magnetic moment towards equilibrium can be understood by including a viscous damping term. In this direction, a dissipative term ($-\frac{\partial \vec{m}}{\partial t}$), proportional to the generalized velocity, is included with the effective magnetic field. This dissipative term decelerates the motion of the magnetic moment and finally brings the magnetic moment $\boldsymbol{m}$ parallel to $\vec{H}^{eff}$. This ultimately provides Landau–Lifshitz–Gilbert (LLG) equation of motion:

$$\partial \vec{m} \Big/ \partial t = \gamma \vec{m} \times \vec{H}^{eff} + \frac{\alpha}{\left|\vec{m}\right|} \vec{m} \times \partial \vec{m} \Big/ \partial t, \tag{1.25}$$

where α is the dimensionless phenomenological Gilbert damping constant.

The LLG equation is extensively employed to investigate and explain the switching dynamics of small magnetic particles. For sufficiently small particles, magnetization may be supposed to remain constant during its reversal process. In this case, the only contributions to $\vec{H}^{eff}$ comes from the anisotropy field, demagnetizing field and the applied external magnetic field. On the other hand, for larger samples and in case of inhomogeneous magnetization dynamics, the magnetic moment becomes a function of spatial coordinates. Noteworthy, in this case, exchange interaction also contributes to $\vec{H}^{eff}$. The LLG equation also provides us an opportunity to calculate the evolution of the spin system in the atomistic limit using Langevin dynamics. In fact, this has proved to be a powerful and efficient route to model ultrafast magnetization processes.

A limitation of the LLG equation includes that for a very short time scale, even shorter than the spin–orbit coupling of the order of 20 fs, the description with a single gyromagnetic ratio fails. In this case, spin and orbital contributions must be considered separately.

1.7 Spin-Dependent Band Gap in Ferromagnetic Materials

A spin-polarized current can be obtained by injection of unpolarized one into ferromagnetic materials. This is due to the spin-dependent band structure found in this material. This ferromagnetism property originates from the tendency of the electron spin to align in the same direction due to Pauli exclusion principle. Pauli stated that two electrons with the same spin cannot be in the same position. Thus, the complete wavefunction of the two electrons would be *antisymmetric* given that the two electrons interchange their positions. Therefore, the electrons will feel an additional effective repulsion due to the Pauli principle in addition to the Coulomb one.

The difference in energy between two electron systems with a symmetric or *antisymmetric* spin part of the wavefunction is referred to as the exchange energy E_{ex}. In 1928, Heisenberg introduced a microscopic origin of the exchange energy by constructing a new effective Hamiltonian which aligns the spin parallel or antiparallel depending on the sign of the exchange term. For example, in iron (Fe, Z = 26), cobalt (Co, Z = 28) and nickel (Ni, Z = 29) atoms, the exchange term has a negative sign resulting in a lower total energy that favour the alignment of the spin to be parallel in the 3d shell. However, in the period table, the subsequent element copper (Cu) with Z = 29 does not exhibit any signature of ferromagnetism. This is because of completely filled 3d shell of Cu, in which there is no flexibility of the electronic arrangement that is requisite to show ferromagnetism. The Stoner criteria determine if a 3d transition metal is stable against the formation of a ferromagnetic or paramagnetic state. This criterion is schematically illustrated by the spin-resolved 3d energy band structure, where the up-spin and down-spin 3d energy bands are represented at the two sides of the energy axis (Fig. 1.2). According to Stoner model, there are formation of energy bands where the 3d-spin sub-bands, i.e., 3d up-spin and down-spin bands are shifted with respect to each other along the vertical energy axis owing to the presence of exchange interaction. As a result of this *shift* between the 3d-spin sub-bands, if the increase in exchange energy supersedes the increase in the kinetic energy, ferromagnetic ordering occurs. Thanks to their spin-dependent states, the flow of current inside a ferromagnet is affected by the direction of the electron spin, leading to two transport channels, spin-up channel and spin-down channel; thus ferromagnetic material can be considered as two different materials simultaneously occupying the same space. Those 'two materials' have different elec-

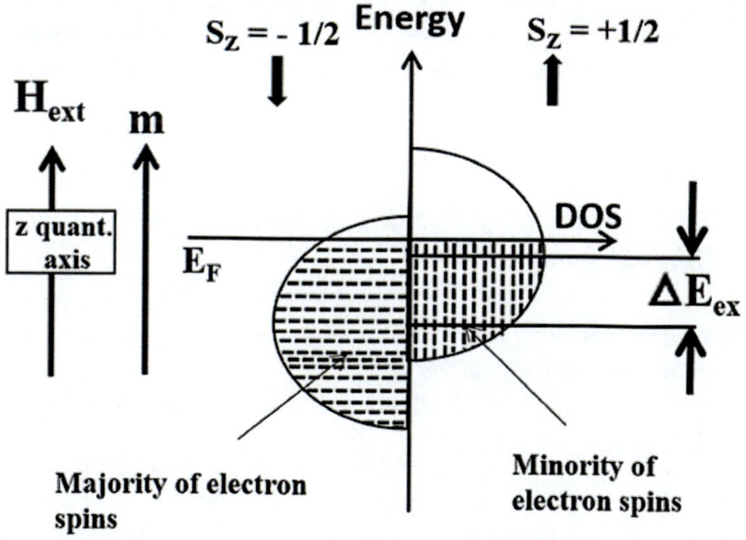

Fig. 1.2 Schematic representation of Stoner model of ferromagnetic transition metals illustrated for 3*d* shell. Spin states with the largest number of electrons are called the majority spins, and those with the smallest number of electrons are called the minority spins. Centre of the majority and minority spin bands are separated by the exchange splitting E_{ex}

trical properties and different electric resistance. This leads to one of the interesting applications of ferromagnetic material, that is to read and store information using this magnetoresistance effect.

1.8 Functionality of 'Spin' in Spintronics

Electrons have much more functionality than merely causing the current flow. Successful utilization of their spins may open up several new possibilities. It is well known that the spin is a quantum-mechanical property, which is depicted by a rotation, either clockwise or anticlockwise, around its own axis. This in turn gives rise to a magnetic moment. Therefore, the electron may be regarded as a minute magnet in which either the magnetic North or South Pole are supposed to points upwards, which gives rise to spin-up or spin-down condition of electrons. Consequently, electron spins could be controlled by applying an external magnetic field in a systematic manner. Electronic spins also play vital role in determining the magnetic properties of a material. Utilizing this phenomenon in spintronics, the goal is to control and manipulate spin-polarized transport through ferromagnetic contacts for information processing and other applications. The proposed usage of electronic states, representing only 0 or 1, within a semiconductor is purely binary. In that case, the range of eight bits can represent every number between 0 and 255, *but only one number at a time.* In a different approach, the 'spin-up' and 'spin-down' states can be represented as superposition of 0 or 1, which are referred to as spintronics quantum bits, and are also known as qubits. Qubits can represent every number between 0 and 255 *simultaneously.*

1.8.1 Basic Principle of Working of All Spintronic Devices (Simple Scheme)

- The information is written and stored in the particular spin orientations, i.e., either in up- or in down-spin directions.
- Information is transferred along the wire through spins attached to mobile electrons.
- Finally, information is captured and read at a terminal where mobile electrons are being collected. Most importantly, spin polarization of conduction electrons survives in the nanoseconds time scale. This lifetime of electron spin orientation is relatively long compared to tens of femtoseconds, during which momentum of the conduction electron decays. This makes spintronic devices potential candidates for memory storage and magnetic sensors applications, and, also particularly advantageous for quantum computing where electron spin would represent 'qubit' of information.

Using this spintronics technology, many spintronic devices are making in progress that stems from combining the advantages of magnetism and semiconductors. Those devices are supposed to be non-volatile, versatile, fast and capable of storing and processing data simultaneously. Furthermore, in order to operate, those devices consume less energy. In this context, it should be mentioned that spintronic devices are either revolutionizing or having potential to revolutionize high-density data storage, microelectronics, sensors, quantum computing, bio-medical applications etc. Some of the spintronic devices are magnetic read head, magnetoresistive random access memory (MRAM), spin transistor, spin torque oscillators etc.

1.9 Different Branches of Spintronics

Spintronics has been extensively researched worldwide in the last few decades. Even today it is one of the most attractive and interesting research fields due to its huge prospect in industrial applications. Extensive research in this field has opened up new avenues in the implementation of innovative and improved spintronic devices. Developments have been observed in both the front (i) choice of material and (ii) method of manipulation of the magnetic state.

1.9.1 Branching Based on Choice of Materials

Spintronics started its journey with ferromagnetic materials and has proved its potential towards realization of spintronic devices. But, nowadays, more versatile materials have drawn widespread attention and have shown spintronic phenomena. The front runner is antiferromagnetic materials. They are greatly more common and are allowed in each magnetic symmetry group, in contrast to ferromagnets (FMs). Antiferromagnets (AFMs) can be insulators, metals, semimetals, semiconductors or superconductors, whereas FMs are primarily metals. Rigorous research leads to material based classification of spintronics, like metallic spintronics, semiconductor spintronics, molecular spintronics etc. (Fig. 1.3).

Metallic spintronics

The simplest approach of producing spin-polarized current is to flow electronic current through a ferromagnetic material having spontaneous magnetization. Discovery of giant magnetoresistance (GMR) device is the result of the most direct application of this effect. Typical GMR device comprises at least two ferromagnetic layers and a non-magnetic metallic spacer layer in between them. During the parallel alignment of the magnetization vectors of those two ferromagnetic layers, electrical resistance exhibits low value, which means a higher current yield at constant voltage. However, during antiparallel orientation of magnetization of the ferromagnetic layers, the electrical resistance will be low pretty high. This constitutes a

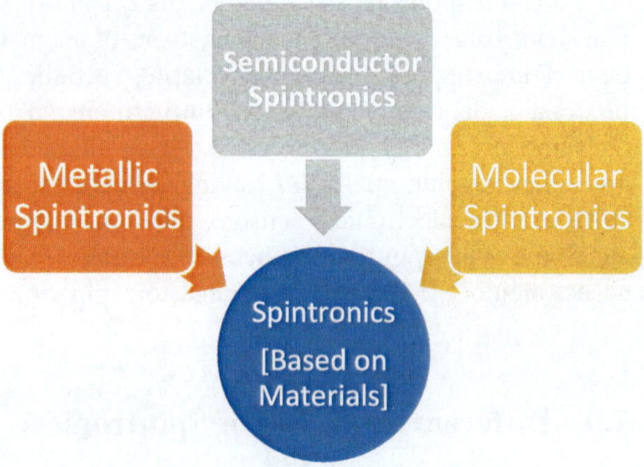

Fig. 1.3 Material-based spintronic division

magnetic field sensor. Depending on construction, two kinds of GMR could be realized in devices: (1) current-in-plane (CIP), in which electric current flows parallel to the layers, and (2) current-perpendicular-to-plane (CPP), in which electric current flows perpendicular to the layers.

Semiconductor spintronics

Whereas the above narration of spintronics is that of metallic spintronics, spintronics using inorganic semiconductors has also been potentially explored. Materials investigated for this purpose are GaAs, Si and many more *diluted magnetic* (ferromagnetic) *semiconductor* materials. Among these inorganic semiconductors, GaAs exhibits strong spin–orbit interaction. Applied gate voltage is competent to rotate the injected spins in GaAs and be used to realize spin transistors as proposed by Das and Datta. However, spin MOS field effect transistors (FETs) are likely to be realized in Si as it is a light element and possesses lattice inversion symmetry. Of late, several research groups are investigating spin injection and spin transport rigorously. This field is acknowledged as the second pillar of spintronics, namely, semiconductor spintronics (Holub et al. 2007).

Molecular spintronics

Since 1999, molecular spintronics, a promising branch of spintronics, catches the attention of the people in spintronics and in molecular electronics, that is, molecular spintronics. Molecular spintronics is considered to be a promising research direction in a field of spintronics, next to metallic spintronics and inorganic semiconductor spintronics. It has been observed that a molecule shows a comparably lesser spin–orbit interaction. A spin–orbit interaction is responsible for loss of spin coherence. Therefore, realization of quantum computation needs materials having a smaller spin–orbit interaction and a so-called Sugahara-Tanaka-type spin MOSFETs (Matsuno et al. 2004). Recently, widespread research has been started in nano-carbonaceous molecules (graphene, carbon nanotube and fullerene) and organic molecules for the advancement in this research field.

1.9.2 Branching Based on Magnetic Manipulation

Efficient manipulation of magnetic state of a material is most important for the development of spintronics devices. In this regard, four main methods (magnetic, strain, electrical and optical) have been reported to mediate the magnetic states (Fig. 1.4). Strong magnetic field, exchange bias and field cooling are the key to magnetic control, whereas magnetic anisotropy effect and meta-magnetic transition are the basics of strain control. Optical control comprises ultrafast laser pulse, thermal and electronic excitation, an inertia-driven mechanism. Electric control involves both the electric field and electric current. Magnetic manipulation is usually used in FM spintronics, particularly with a view to exchange bias, as the pinning layer to alter ferromagnetic moments. But, with the quick progress of AFM spintronics, exchange bias has become more and more important to control the magnetic structure of AFMs by itself or in combination with other fields. However, the creation of exchange bias requires ferromagnets, signifying that it is straightforward to be modulated by outer perturbation. Nevertheless, the velocity of exchange bias-controlled spin reorientation is much slower. Hence, some alternative powerful techniques are needed to resolve these issues. The important aspect of strain-induced magnetic anisotropy or meta-magnetic transition is that no ferromagnets are required to change the spin configuration at room temperature. On the other hand, electrical control also does not need any ferromagnets, magnetic field or field cooling to perform fast switching.

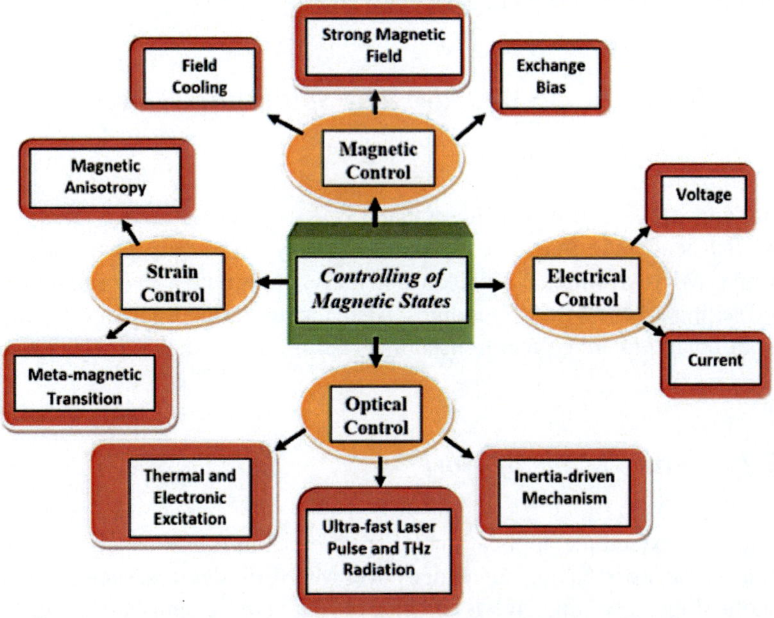

Fig. 1.4 Various methods adopted for manipulation of magnetic states in spintronics

These outstanding characteristics show prospect for coming days of spintronics devices like memory resistors. On comparison, the strain and electric methods are limited so far to a small range of materials and require a typical crystal structure. As far as the optical manipulation is concerned, spin dynamics are being widely studied, on a femtosecond scale. This has been discussed at length in Chap. 7 of this book. They demonstrate huge prospect in designing of ultrafast information processing and recording devices.

1.10 Applications

The most prominent applications of spintronics are the magnetic read heads for the magnetic data storage device and MRAM. Read heads of modern hard disk drives are based on GMR or TMR effect. GMR effect has also been used to construct magnetic sensors. GMR sensors have a variety of applications:

1. In the field of precision engineering and robotics, fast and accurate sensing of position and motion of several mechanical components.
2. Guiding missile motion.
3. In computer video games, sensing of position and motion of several electrical and mechanical components.
4. Acting as *automotive sensors* in diversified fields, such as in fuel handling system, speed controlling, navigation etc.

1.10.1 Presently, Do We Have Any Spintronics-Powered Products?

In this context, we may first mention that 'GMR-based spin valves' are used in hard disks read heads and magnetic sensors, nowadays. The very next name we may mention is MRAM, which is advancing steadily. Importantly, *EverSpin* have already started selling 4-Mbit MRAM modules. Also, the feasibility for spin valve transistors and spin-polarized field effect transistors is being investigated.

1.10.2 Future Scope of Work

We intend to extend the applications of spintronics in future through the advent of spin-based transistor having advantages over MOSFET devices, such as steeper sub-threshold slope. The objective is to achieve *non-volatile* spin-based logic devices, which might enable scaling beyond 2025. Extensive research has been focussed on spin-transfer torque-based logic devices at *Intel*. Such devices are now part of the International Technology Roadmap for Semiconductors (ITRS) and are potential candidates for inclusion in future computers. Crocus and NEC are about to develop logic-in memory applications.

1.11 Conclusions

In conclusion, solely spin or in combination with charge can be utilized to derive the requisite functionality for information processing and storing digital information. These are the fundamental issues of spintronics. Technology related to spintronics has already revolutionized the *information storing*, i.e., storage density of hard drives by the advent of GMR-based read head of magnetic hard disk drive. Consequently, universal memory and low-power computation can potentially be realized in modern technology.

Initial successes in this field include

- Development of GMR read heads to sense magnetic storage, i.e., magnetic hard disk drive,
- Non-volatile magnetic random access memory (MRAM),
- Programmable spintronic logic devices,
- Rotational speed control system,
- Positioning control devices in robotics,
- Perimeter defence systems,
- Magnetometers, etc.

Implementation of spintronics in the field of ***information processing*** is a novel attempt. This is motivated by the expectations that spintronics may offer a more *energy-efficient* route in comparison to traditional transistor-based paradigm.

Recent research activities include

- Implementation and involvement of spintronics in the still growing field of information technology, particularly for ***computing*** and **signal processing**.
- ***Spin-based quantum computers***, which operate on the basis of reversible quantum dynamics. As a result, such computers do not dissipate any energy at all to complete a logic operation. Therefore, understanding the science and technology of spintronics has become imperative for the students of basic science and engineering disciplines.

1.12 Exercises

1. Define exchange splitting of 3d spin-up and spin-down bands in ferromagnet with relevant band diagram.
2. What will happen to the outcoming beams, if the non-uniform magnetic field is changed into uniform magnetic field in Stern–Gerlach experimental setup?

3. Define SPINORS. How SPINORS can be used to deduce the electron's spin orientation?
4. In quantum mechanics, how a physical quantity is expressed in Heisenberg's mechanics and Schrodinger's wave mechanics?
5. Show that the spin angular momentum operators satisfy the commutation relation?
6. Derive Landau–Lifshitz–Gilbert (LLG) equation of motion.
7. Discuss different branches of spintronics.

References

M.N. Baibich, J.M. Broto, A. Fert, F.N. VanDau Nguyen, F. Petroff, P. Etienne, G. Creuzet, A. Friederich, J. Chazelas, Giant magnetoresistance of (001)Fe/(001)Cr magnetic superlattices. Phys. Rev. Lett. **61**(21), 2472–2475 (1988)

G. Binasch, P. Grünberg, F. Saurenbach, W. Zinn, Enhanced magnetoresistance in layered magnetic structures with antiferromagnetic interlayer exchange. Phys. Rev. B **39**(7), 4828 (1989)

S. Datta, B. Das, Electronic analog of the electro-optic modulator. Appl. Phys. Lett. **56**(7), 665–667 (1990)

W. Gerlach, O. Stern, Der experimentelle Nachweis der Richtungsquantelung im Magnetfeld. Zeitschriftfür Physik **9**, 353–355 (1922)

S. Goudsmit, G.E. Uhlenbeck, Over Het Roteerende Electron En de Structuur der Spectra. Physica **6**, 273 (1926)

M. Holub, J. Shin, D. Saha, P. Bhattacharya, Electrical spin injection and threshold reduction in a semiconductor laser. Phys. Rev. Lett. **98**(14), 146603 (2007)

M. Johnson, R.H. Silsbee, Interfacial charge-spin coupling: injection and detection of spin magnetization in metals. Phys. Rev. Lett. **55**(17), 1790–1793 (1985)

M. Julliere, Tunneling between ferromagnetic films. Phys. Lett. A **54**(3), 225–226 (1975)

T. Matsuno, S. Sugahara, M. Tanaka, Novel reconfigurable logic gates using spin metal–oxide–semiconductor field-effect transistors. Jpn. J. Appl. Phys. **43**, 6032 (2004)

T.E. Phipps, J.B. Taylor, The magnetic moment of the hydrogen atom. Phys. Rev. **29**, 309 (1927)

O. Stern, Ein Weg zur experimentellen Prüfung der Richtungsquantelung im Magnetfeld. Zeitschriftfür Physik **7**, 249–253 (1921)

M. Walser, C. Reichl, W. Wegscheider, G. Salis, Direct mapping of the formation of a persistent spin helix. Nat. Phys. **8**(10), 757 (2012)

Chapter 2
Basic Elements of Spintronics

In this chapter, we will discuss several basic phenomena that give rise to spintronics properties (Bandyopadhyay and Cahay 2008; Zutic et al. 2004).

2.1 Spin Polarization

An ensemble of electrons is said to be polarized if the electron spins have a preferential orientation. If in an ensemble of N electrons, N_1 electrons have up-spin and N_2 electrons have down-spin orientations, then spin polarization could be defined as

$$P = \frac{N_1 - N_2}{N_1 + N_2}\%, \text{ where } N_1 + N_2 = N \tag{2.1}$$

(a) ***Unpolarized electron beam***—If the number of electrons having spin-up orientations and that having spin-down orientations are the same, i.e., $N_1 = N_2$, then $P = 0$, thus the ensemble becomes unpolarized. This kind of ensembles is said to be unpolarized electron beam.

(b) ***Polarized electron beam***—If the number of electrons having spin-up orientations and that having spin-down orientations are not the same, i.e., $N_1 \neq N_2$, P should have some finite value, thus the ensemble should have some spin polarization. This kind of ensembles is said to be polarized.

   (i) **Fully Polarized Electron**: In this case, all the electron spins have same orientations (either spin-up or spin-down) and all electron spins can be described by a single spin function. This kind of ensemble of electrons is said to be fully spin polarized. In this case, if N be the total number of electrons in the ensemble, then either $N_1 = N$ and $N_2 = 0$ or $N_2 = N$ and $N_1 = 0$, thus $P = 100\%$, hence fully spin-polarized electron beam.

P. Dey and J. N. Roy, *Spintronics*,
https://doi.org/10.1007/978-981-16-0069-2_2

Spin polarization P is less than one for metallic ferromagnetic elements. For instance, in case of Fe, Co and Ni elemental ferromagnet P is 0.44, 0.34 and 0.11, respectively. However, in some materials spin polarization has been found to be unity. Examples are CrO_2, Fe_3O_4 and some of the manganites. This feature originates from the fact that in these materials one spin-split band is totally empty, which results in the metallicity of the electrons of one spin band and insulating of those of another spin band. Hence, these materials are named as *half-metallic or half-insulating ferromagnets.*

2.2 Spin Filter Effect

Spin-dependent scattering, suffered by electrons in a ferromagnet because of its spin-dependent band structure (*discussed above*), leads to spin filtering effect (Ohno et al. 1999; Jedema et al. 2001; Appelbaum et al. 2007; Saikin 2004; Yu et al. 2012; Suzuki et al. 2009; Tang et al. 2002; Schmidt et al. 2000; Wu et al. 2010). Spin-dependent scattering mechanism will be discussed in detail in Chap. 3. Let us consider a passage of electronic current through a ferromagnet, which is magnetized to saturation. Now, magnetization is defined as: $M \approx n_\uparrow - n_\downarrow$, where $n_\uparrow$ and $n_\downarrow$ are the total number of up-spin and down-spin electrons, respectively. It can be clearly seen from the filled states of the band structure, shown by striped portion, that the number of up-spin and down-spin electrons are different in case of ferromagnet (Fig. 2.1a), whereas they are equal in case of paramagnet (Fig. 2.1b). Therefore, it is quite natural that such imbalance in the number of up-spin and down-spin electrons in ferromagnets could possibly give rise to a net magnetization, whereas paramagnet could not have such net magnetization. Furthermore, direction of that magnetization in ferromagnets follows the spin direction, i.e., up-spin or down-spin directions of the electrons that

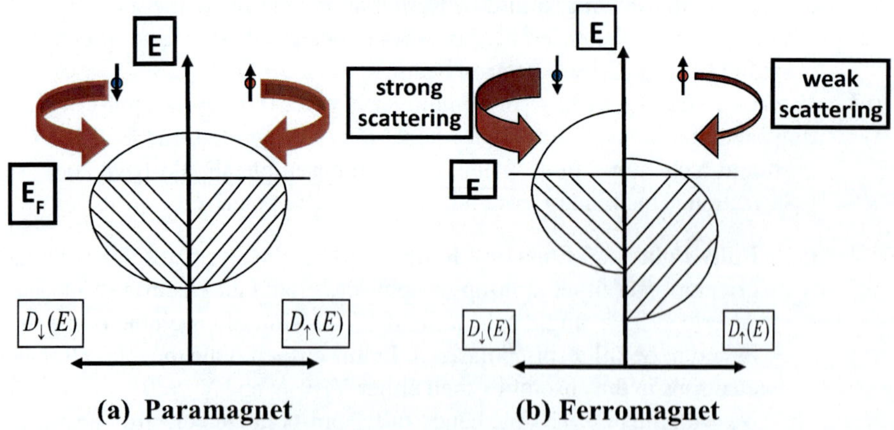

Fig. 2.1 Spin-resolved energy band structure for **a** paramagnet and **b** ferromagnet

are having majority population. Therefore, we define electrons having their spin magnetic moment parallel to magnetization M as ***majority spin electrons***, whereas electrons having their spin magnetic moment antiparallel to magnetization M as ***minority spin electrons***.

2.2.1 What Is Spin Filter Effect?

It has been found that majority spin electrons are more easily transmitted through ferromagnet than minority spin electrons. This phenomenon is known as ***spin filter effect*** (Appelbaum et al. 2007) and is characterized by a parameter called *spin asymmetry* '***A***'. Now, spin filtering effect or spin asymmetry arises due to this spin-dependent scattering and is defined as

$$A = \left(\mathrm{I}^{\uparrow} - \mathrm{I}^{\downarrow}\right) / \left(\mathrm{I}^{\uparrow} + \mathrm{I}^{\downarrow}\right) \tag{2.2}$$

where current contributed by the electrons having spin magnetic moment parallel/antiparallel to M is denoted by $\mathrm{I}^{\uparrow}/\mathrm{I}^{\downarrow}$.

2.2.2 Physical Interpretation of Spin Asymmetry 'A'

In an attempt to give physical interpretation of A considering only the intensity of the transmitted electrons, let us note that the empty states available for minority spin electrons for scattering into are much more than that for majority spin electrons (Fig. 2.1a). Thus, minority spin electrons possess a larger absorption coefficient than that of majority spin electrons. Such difference in absorption coefficients between majority and minority spin electrons gives rise to spin filtering or $\boldsymbol{A}$ as two branches of electrons, majority and minority spin electrons, traveling through the ferromagnetic film.

To summarize, during a passage of electronic current through a ferromagnet, because of its spin-dependent scattering, both the absorption coefficients and phases for majority and minority spin electrons within the ferromagnet would be different. This gives rise to spin filter effect, which plays a very critical role in exhibiting spintronics effect.

2.2.3 Spin Detection Efficiency

In this context, a quantity named 'spin detection efficiency' has been defined. Such quantity is a measure of how efficiently a ferromagnet can filter spin, i.e., how selectively it can transmit electrons of a particular spin orientation impinging on it from the paramagnet. It should be noted that if a ferromagnetic detector transmits only one kind of spin, then the spin detection efficiency is 100%. In analogy to optics, sometimes spin detectors are referred to as 'spin analysers'.

2.3 Spin Generation and Injection

In spintronics, generation of spin-polarized electrons plays a pivotal role for making spintronic devices (Ohno et al. 1999; Jedema et al. 2001; Appelbaum et al. 2007; Saikin 2004; Yu et al. 2012; Suzuki et al. 2009; Tang et al. 2002; Schmidt et al. 2000; Wu and Jiang 2010). Several methods are taken up for generation of spin-polarized electron in non-magnetic (NM) materials (see Fig. 2.2) (Hirohataa et al. 2020). They are (i) spin injection from a ferromagnetic (FM), (ii) a magnetic field, (iii) an electric field, (iv) electromagnetic wave introduction, (v) Zeeman splitting, (vi) spin motive force, (vii) a thermal gradient and (viii) mechanical rotation. The frequently used method is spin injection from a ferromagnetic material. Here, conventional ferromagnetic metals, half-metallic ferromagnets and dilute magnetic semiconductors (DMS) are attached to a non-magnetic metal or semiconductor through an ohmic contact or a tunnel barrier. Though a stray field at the edge of a FM can induce a population difference in spin-polarized electrons in a non-magnetic material, yet it is not easy to control due to its well-defined edge shape. An electric field can initiate the movement of spin-polarized electrons in a NM material towards a desired direction based on spin Hall effects. Circularly polarized light stimulates spin-polarized electrons in a semiconductor. Through reverse effect, circularly polarized light can be produced by a spin-polarized electron current. This can also be extended for spin generation by electromagnetic waves. Thermal gradient-driven spin Seebeck and Nernst effects can also generate spin-polarized carrier. In a DMS, spin imbalance can be induced at the Fermi level through Zeeman splitting.

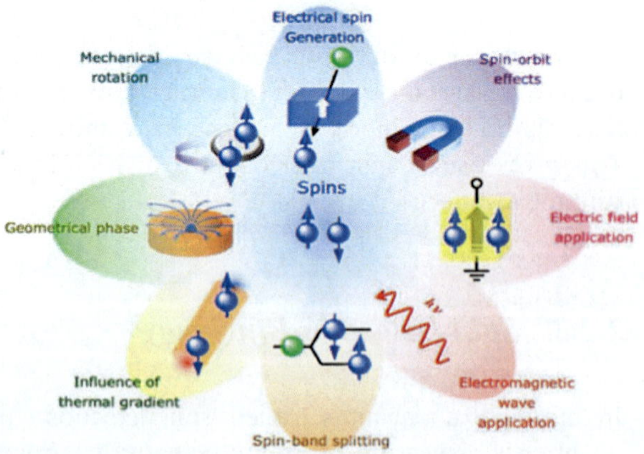

Fig. 2.2 Different methods for generation of spin-polarized electron in non-magnetic (NM) materials (Taken from Hirohata et al. 2020)

2.3.1 *What Is Spin Injection?*

In general, when electrons move in a material, it carries both charge ($-e$) and spin degrees of freedom ($\hbar/2$). It is straightforward to state that in case of paramagnets, conduction electrons do not intrinsically have any net spin polarization. This means that in a paramagnet the population of two species of electrons, i.e., up-spin and down-spin electrons, is equal under equilibrium. On the other hand, ferromagnets have a non-zero spin polarization, i.e., a net spin magnetic moment of electrons under equilibrium conditions. In order to attain spin transport in a device, as is requisite in spintronics, the primary condition is to achieve an imbalance in the number of two species of the spin carriers so that the net spin magnetic moment becomes finite.

It is noteworthy that in spintronics device, often we have to use paramagnetic metallic or semiconducting component as its part. Thus following the prerequisite condition of achieving spin transport in such paramagnetic components of spintronics devices, net spin polarization **P** [given by Eq. (2.1)] can be generated in a paramagnet either through electrical spin injection of charge carriers from a ferromagnet. Creation of such a non-equilibrium situation in a material in terms of spin magnetic moment of charge carriers is generally termed as 'spin injection'.

2.3.2 *Transport Method*

The transport method, i.e., *electrical injection* of spins, is the most suitable method for electrical device applications. It requires successfully injecting an imbalance of spin from a ferromagnet into the paramagnet in the form of a current. It is well known that the magnetization (M) of the ferromagnet is $M \propto n_{\uparrow}-n_{\downarrow}$, where $n_{\uparrow}(n_{\downarrow})$ are the populations of the majority (minority) spin electrons. $n_{\uparrow}(n_{\downarrow})$ can be found by taking integration over the filled states of the up-spin (down-spin) energy band. Accordingly, in a ferromagnet the electrons at the Fermi level (E_F) possess certain spin polarization. In a half-metallic ferromagnet the charge carriers are highly spin polarized ($\boldsymbol{P} \sim$ 100%). Thus, in the ferromagnetic material the electric current constitutes a net flow of spins. This spin-polarized electron current in a ferromagnetic material is generally referred to as 'spin current'.

Now, let us consider a spin transport experiment, which is *all-electrical*. In this case, spin injection into the paramagnet can be obtained by employing contacts, commonly described as 'spin injectors'. This ferromagnetic contact could be transition metals, half-metals or diluted magnetic semiconductors. In a paramagnetic/non-magnetic material (NM), at equilibrium, the spin magnetic moments are aligned randomly in space. Thus, inside the device the average spin magnetic moment of the electron spin that ensembles at any position and time would be zero. This in turn implies that the conduction electrons in a non-magnetic material are unpolarized. Application of an external bias voltage induces a flow of this unpolarized ensemble of charge carriers and consequently results in charge current. However, from spintronics point of view, we need to achieve 'spin transport', rather than mere charge transport. In this attempt, the basic principle is to create a net *non-zero* spin moment inside the device through the spin injection process. This is the fundamental idea of *electrical spin injection,* which is shown in Fig. 2.3.

If we do suitable biasing so that net electron flow takes place from a half-metallic ferromagnet to a paramagnetic material as shown in Fig. 2.4, then subsequently the spins of the ferromagnetic material are injected into the paramagnetic material through the interface. Thereby, the *spin injection* process could be realized in this case. As per general consensus, these injected spins suffer spin relaxation process mainly owing to spin–orbit interaction. Moreover, because of paramagnetic material, the spin polarization of the injected spins also decreases during their passage away from the interface, as shown in Fig. 2.4.

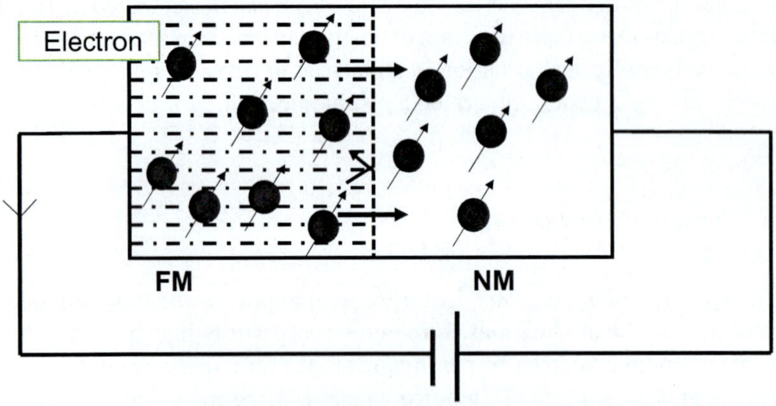

Fig. 2.3 Illustration of electrical spin injection

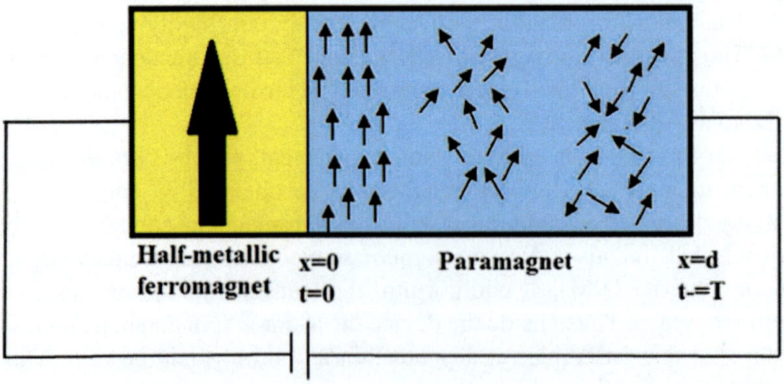

Fig. 2.4 Spin injection across an interface of a half-metallic ferromagnetic and a paramagnetic material

2.3.3 *Knowledge of Some Essential Parameters of Injector Ferromagnet*

In order to understand the spin injection process, knowledge of the following quantities of an injector ferromagnet is essential as discussed below:

(a) **Spin polarization of the density of states (DOS), P_{DOS}:**

This is defined by

$$P_{DOS} = \left| \frac{\rho_\uparrow - \rho_\downarrow}{\rho_\uparrow + \rho_\downarrow} \right| \tag{2.3}$$

where $\rho_\uparrow$ ($\rho_\downarrow$) represents DOS at E_F of charge carriers having spin moments parallel (antiparallel) to the direction of magnetization of the ferromagnet. It is noteworthy that M and P_{DOS} do not essentially have either the same magnitude or the sign. Interestingly, for Co and Ni, contribution of minority spins is dominant at E_F, thus resulting in negative P_{DOS}.

(b) **Conductivity polarization ($P_{\sigma F}$)**: This term arises from the fact that the two spin species in a ferromagnet may also have different mobilities and is defined as follows:

$$P_{\sigma F} = \left| \frac{\sigma_\uparrow - \sigma_\downarrow}{\sigma_\uparrow + \sigma_\downarrow} \right| \tag{2.4}$$

where $\sigma_\uparrow$ ($\sigma_\downarrow$) represents conductivity of charge carriers having spin moments parallel (antiparallel) to the direction of magnetization of the ferromagnet. Counterintuitively, $P_{\sigma F}$ is not straightforwardly identical to P_{DOS} and they might even possess different signs. For instance, in case of Co and Ni, $P_{\sigma F} > 0$; on the other hand, in case of Fe, $P_{DOS} > 0$ whereas $P_{\sigma F} < 0$.

(c) **Spin injection efficiency (η)**: Spin polarization of the electron ensemble or current injected from the ferromagnet to a paramagnet is termed as the spin injection efficiency, η. This is defined by

$$\eta = \left| \frac{J_\uparrow - J_\downarrow}{J_\uparrow + J_\downarrow} \right| \tag{2.5}$$

where $J_\uparrow$ ($J_\downarrow$) is the density of current corresponding to the majority (minority) spin species, obtained just after the spin injection at the interface. It can be understood that in the ferromagnet, $J_\uparrow \neq J_\downarrow$ since $n_\uparrow \neq n_\downarrow$.

2.3.4 Sustainability of Spin Polarization in Paramagnet

Now, the question is whether such inequality in the number of electrons having up-spin and down-spin orientations in the ferromagnetic material can possibly be sustained in the paramagnet as well, after electron spin injection from ferromagnet to paramagnet? To answer this question, let us recall the current continuity equation, which in steady-state mandates, $\nabla \cdot J = 0$.

Therefore,

$$J^{\uparrow}_{ferromagnetic} + J^{\downarrow}_{ferromagnetic} = J^{\uparrow}_{paramagnetic} + J^{\downarrow}_{paramagnetic} \tag{2.6}$$

According to general consensus, the inequality of $J^{\uparrow}_{paramagnetic}$ and $J^{\downarrow}_{paramagnetic}$, i.e., $J^{\uparrow}_{paramagnetic} \neq J^{\downarrow}_{paramagnetic}$ is not required in the above equation. This means normally there will be no spin-polarized current in the paramagnet. It is noteworthy that such inequality of $J^{\uparrow}_{paramagnetic}$ and $J^{\downarrow}_{paramagnetic}$, hence spin-polarized current, would appear in the paramagnet if there is 'resistivity matching', i.e., the resistance of the ferromagnet and paramagnet should be about equal, which is the case when both ferromagnet and paramagnet happen to be metal. However, 'resistivity matching' would be impossible to attain when the ferromagnet happens to be a metal (e.g., cobalt or iron) and the paramagnet happens to be a semiconductor (e.g., GaAs or Si) or an insulator (e.g., Al_2O_3) (Ohno et al. 1999). In this scenario, one of the solutions is the introduction of a tunnel barrier in between the ferromagnetic and paramagnetic material/layer, which can circumvent the resistivity mismatch problem. However, this topic has been dealt with in detail in the subsequent discussions.

In the context of spintronics device application, the discussion on spin injection efficiency at the ferromagnet/paramagnet interface has attracted intense focus on it. Spin injection efficiency is a measure of how efficiently a ferromagnet, which is in good electrical contact with a paramagnet, can inject spin into it. It should also be noted that if a ferromagnet injector injects only one kind of spin—either the majority or the minority spin exclusively—into a paramagnet, then it is an ideal spin injector. In this case, the spin injection efficiency is 100%. However, transition metal ferromagnets typically inject/transmit both majority and minority spins, albeit not equally. Hence, in this case both spin injection and detection efficiencies should be less than 100%. In analogy to optics, sometimes spin injectors are also referred to as 'spin polarizers'. High efficiency (ideally 100%) of electrical spin injection/detection across a ferromagnet/paramagnet interface is the requirement for the operation of the spin-valve, GMR devices and the spin field effect transistors (discussed in the subsequent chapters).

2.3.5 Discussion of Spin Injection Process in Two Cases

In this direction, we will discuss spin injection process in two distinctively different cases of electron transfer through interface: (*i*) from a ferromagnetic (FM) material to a non-magnetic (NM) one, and (*ii*) from one ferromagnetic material (FM1) to another ferromagnetic FM (FM2) material.

(i) ***Spin injection from a ferromagnetic (FM) to a non-magnetic (NM) material***

In order to assess the spin injection process from a ferromagnetic to a non-magnetic/paramagnetic material through an interfacial barrier, the associated spin injection efficiency (η) can be calculated using the following equation:

$$\eta = \frac{r_F P_{\sigma F} + r_i P_{\sigma i} + r_N P_{\sigma N}}{r_F + r_N + r_i} \tag{2.7}$$

where $P_{\sigma F}$, $P_{\sigma N}$ and $P_{\sigma i}$ are the conductivity polarizations of the bulk ferromagnet, bulk paramagnet and that of their interface, respectively (Zutic et al. 2004). η is calculated as a weighted average of $P_{\sigma F}$, $P_{\sigma i}$ and $P_{\sigma N}$ and the weights are proportional to the corresponding resistances. Effective resistance of the bulk ferromagnet, bulk paramagnet and the interface between them are denoted by r_F, r_N and r_i, respectively. It should be noted that it is the characteristics of charge injection from the ferromagnet that decide $P_{\sigma i}$. Obviously, $P_{\sigma N} = 0$ as per definition. Thus, Eq. (2.7) reduces to

$$\eta = \frac{r_F P_{\sigma F} + r_i P_{\sigma i}}{r_F + r_N + r_i} \tag{2.8}$$

The quantity η gives us a measure how *efficiently*, as the name suggests, spins are getting injected into the non-magnetic/paramagnetic material from the ferromagnetic one. Now, we will discuss the diversified cases based on the above-mentioned expression of η.

Case (i): Ohmic contact between metallic ferromagnetic and paramagnetic material

It can be understood that in this case, $r_i = 0$ and $r_F \approx r_N$, thus yielding $\eta \approx P_{\sigma F}$. As generally $P_{\sigma F}$ is recognized to be high, it thus implies significant spin injection in this case. In fact, experimental observation of spin injection has indeed been obtained almost in all-metal structures (Jedema et al. 2001).

Case (ii): Ohmic contact between metallic ferromagnetic and paramagnetic semiconductor: conductivity mismatch problem

It can be understood that in this case, $r_i = 0$ and $r_F \ll r_N$, thus yielding $\eta \ll P_{\sigma F}$, i.e., poor injection of spin. This in turn implies that spin injection process does not take place efficiently from a metallic ferromagnetic to a semiconducting paramagnetic material through an *Ohmic contact*. This phenomenon is known as *conductivity mismatch problem.* Such problem is quite well-known and has been attempted to overcome by incorporation of some novel spin-injecting materials, such as diluted magnetic semiconductors, half-metallic ferromagnets etc. These proposed spin injector materials offer a scenario, where $r_F \approx r_N$, thus yielding $\eta \approx P_{\sigma F}$, implying reasonably good spin injection.

Case (iii): Tunnel/Schottky barrier at metallic ferromagnetic and paramagnetic semiconductor interface
In this case, $r_i \gg r_F, r_N$, thus yielding $\eta \approx P_{\sigma i}$. For a tunnel barrier (discussed in Chap. 4), the wavefunctions and hence the transmission probabilities are different for spin-up and spin-down electrons at E_F of the injector ferromagnetic materials. Thus, at the tunnel barrier the conductivities corresponding to the two spin species are different, which means $P_{\sigma i} \neq 0$. This in turn again signifies efficient spin injection through the barrier (Rashba 2000). However, if spin-independent thermionic emission process results in carrier injection over the barrier, then $P_{\sigma i} \approx 0$.

Physical picture of spin injection from ferromagnetic injector to paramagnetic material
Whenever electronic current flows from a ferromagnetic injector into a paramagnetic material, there is a change in distribution of up-spin and down-spin carriers. We found similar phenomenon occurring across a superconductor and normal metal interface, where normal current converts into supercurrent. In that case, quasiparticles and Cooper pairs have a difference in electrochemical potential near the interface (Jedema et al. 2001). In order to analyse spin injection phenomenon considering difference in electrochemical potential between up-spin and down-spin electrons, let us consider a ferromagnetic and normal metal interface in the plane $x = 0$ assuming that the system is homogeneous in the y and z directions. As shown in Fig. 2.5, the ferromagnet stretches in the region x < 0 and the normal metal in the region x > 0. Thicknesses of both materials are such that those are greater than their respective spin diffusion lengths.

Let us suppose that the positive terminal of the voltage source (battery) is connected to the normal metal and the negative terminal to the ferromagnetic contact so that the current of electrons flows from the left to right in the x-direction, i.e., from the ferromagnet to normal metal only. Now, considering the situation that the rate of electron scattering events that do not flip spin is far larger than the spin-flip scattering rate at any arbitrary coordinate point x, we can define individual spatially varying electrochemical potential for the up-spin ($\mu_\uparrow$) and down-spin ($\mu_\downarrow$) electron channels. Specifically, at the ferromagnetic and normal metal interface these electrochemical potentials are expected to be quite different from each other and here lies the origin of spin injection at ferromagnet and normal metal interface. To quantify such difference, the conductivity and current density associated with the up-spin and down-spin electrons are expressed as follows:

$$\sigma^{\uparrow}(\mathrm{x}) = \alpha(\mathrm{x})\sigma, \qquad (2.9)$$

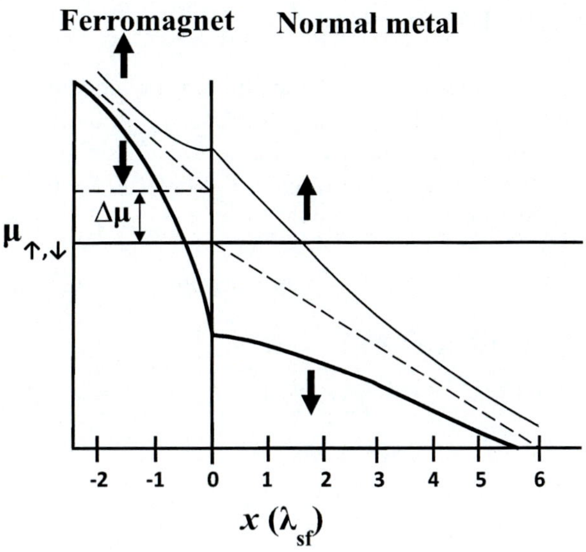

Fig. 2.5 Illustration of the electrochemical potential difference for up-spin and down-spin electrons near a ferromagnetic metal/non-magnetic metal interface when a current flowing through the junction (Figure adapted and redrawn from Bandyopadhyay and Cahay 2008)

$$\sigma^{\downarrow}(x) = (1 - \alpha(x))\sigma, \tag{2.10}$$

$$\text{and } J_{\uparrow}(x) = \beta(x)J, \tag{2.11}$$

$$J_{\downarrow}(x) = (1 - \beta(x))J, \tag{2.12}$$

where ***J*** denotes the total current density flowing across the interface and ***J*** has been found to be independent of *x* in steady state due to current continuity. Here, α(x) and β(x) are two variables.

Now, let us discuss what happens in case of ***(i)*** far from the interface and ***(ii)*** near the interface.

(i) ***At a distance far from the interface***

The two electrochemical potentials, $\mu_{\uparrow}(x)$ and $\mu_{\downarrow}(x)$, are expected to converge towards each other since the population of up-spin and down-spin electrons should be near their equilibrium values. This means

$$\beta(-\infty) = \alpha_F, \tag{2.13}$$

$$\text{and } \beta(+\infty) = \alpha_N = 0.5 \tag{2.14}$$

where the subscripts ‘F’ and ‘N’ stand for ferromagnet and normal metal (or non-magnet), respectively.

(ii) ***At the interface***

At the interface, due to the difference in conductivities of the two materials, $\alpha(x)$ is expected to change abruptly. However, in the absence of spin-flip scattering mechanisms at the interface, $\beta(x)$ must be continuous. Consequently, at and at the close vicinity of the interface, $d\beta/dx \neq 0$ and $\mu_\uparrow(x) \neq \mu_\downarrow(x)$. Now, introducing the weighted electrochemical potential

$$\mu_0 = \alpha(x)\mu_\uparrow(x) + (1-\alpha)\mu_\downarrow(x), \tag{2.15}$$

it can be shown that

$$\frac{d}{dx}\mu_0 = -\frac{e}{\sigma(x)J} \tag{2.16}$$

where $\sigma(x) = \sigma_\uparrow(x) + \sigma_\downarrow(x)$ and J denotes the total current density flowing through the interface. We understand that $\mu_0(x)$ is the electrochemical potential value when there is equilibrium distribution between the up-spin and down-spin electrons in the current. As discussed earlier, at and in the vicinity of the interface, the discontinuity of μ_0 leads to a boundary resistance defined as follows:

$$R_b = \frac{\Delta\mu}{eJ} = \frac{\mu_{\mathrm{F},0}(\mathrm{x}=0-) - \mu_{\mathrm{N},0}(\mathrm{x}=0+)}{eJ} \tag{2.17}$$

where

$$\begin{aligned}\mu_{\mathrm{F},0}(\mathrm{x} = 0-) &= \alpha_{\mathrm{F}}\,\mu_{\mathrm{F},\uparrow}(\mathrm{x} = 0-) + (1-\alpha_{\mathrm{F}})\,\mu_{\mathrm{F},\downarrow}(\mathrm{x} = 0-),\\ \mu_{\mathrm{N},0}(\mathrm{x} = 0+) &= \alpha_{\mathrm{N}}\,\mu_{\mathrm{N},\uparrow}(\mathrm{x} = 0+) + (1-\alpha_{\mathrm{N}})\,\mu_{\mathrm{N},\downarrow}(\mathrm{x} = 0+)\end{aligned} \tag{2.18}$$

This electrochemical potential difference is the driving force for the spin current conversion across the interface

In this way, we use the concept of electrochemical potentials for the up-spin and down-spin channels to calculate the amount of spin polarization of the current in a semiconductor in close contact with a ferromagnetic injector or in a spin valve consisting of a semiconductor sandwiched between two ferromagnetic contacts.

(ii) ***Spin injection from one ferromagnetic (FMI) to another ferromagnetic (FM2) material***

Although less discussed, we present here description of spin injection process from a ferromagnetic material to another. Let us do suitable biasing so that net electrons flow takes place from one ferromagnetic to another ferromagnetic material as shown in Fig. 2.6, then subsequently the spins of the first ferromagnet (FM1) are injected into the second one (FM2) through the interface. This is followed by an exchange interaction between injected electron spins with the spins of the electrons of the host

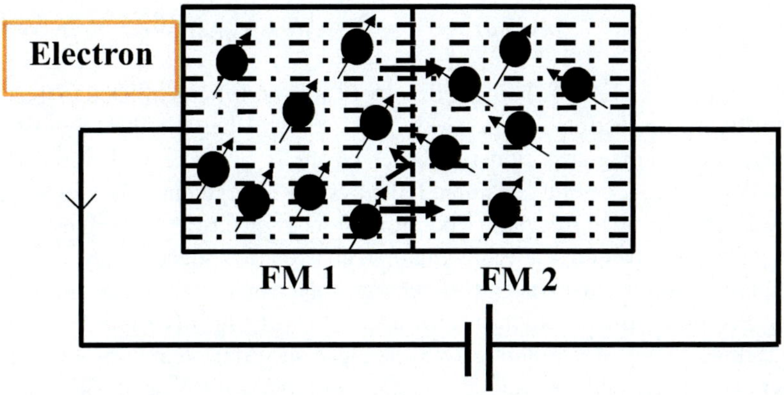

Fig. 2.6 Spin injection across an interface of a ferromagnetic material, FM1 to another FM2

FM2. This in turn excites a precession of the electron spins of the host and may consequently results in the switching of the magnetization of FM2. Since many electrons are interacting and as generally supposed that the electron spins are subjected to only exchange interaction; therefore, in this case we can consider the conservation of angular momentum of the electron system.

Hence, analogous to that of charge current density, density of spin angular momentum, i.e., spin current density ($\hat{J}^S$) can also be defined as below:

$$\hat{J}^S(\vec{x}, t) = \sum \vec{v_L}(t)\overrightarrow{S_L}(\vec{x}, t) + (exchange\,mediated\,term) \tag{2.19}$$

Here, $\vec{S}_L(\vec{x}, t)$ *and* $\vec{v}_t(t)$ are the density of electron spin and velocity of ith electron, respectively, and $\sum$ stands for summation over all the concerned electrons. The first term represents the spin current, realized by the flow of spin-polarized electrons, whereas the second term represents the exchange interaction-mediated transfer of spin angular momentum. Thus, conservation of spin angular momentum and charge can be stated as follows:

$$\frac{\partial \vec{s}}{\partial t} + div\,\hat{J}^s = 0;\ \frac{\partial \rho}{\partial t} + div\,\hat{J}^Q = 0 \tag{2.20}$$

where $\vec{s}$ is the density of spin angular momentum, ρ is the charge density and $\vec{J}^Q$ is the electric current density.

Now, let us consider a large ferromagnetic material subjected to the application of an electric field. In the ferromagnetic materials, larger density of majority spin electrons, n_+ than that of minority spin electrons, n_- results in non-zero $\vec{s}$. Thus we can express:

$$\vec{s} = \vec{s}_+ + \vec{s}_- = \frac{\hbar}{2}\vec{e}_{spin}(n_+ - n_-)$$

$$\rho = (\rho_+ + \rho_-) = (-e)(n_+ + n_-) \tag{2.21}$$

Here, $\vec{s}_+(\vec{s}_-)$ and $\rho_+(\rho_-)$ are the density of spin and charge related to the majority (minority) spin electrons, respectively. It is known that the direction of angular momentum and magnetization for the electrons is opposite with respect to each other. As generally can be considered, the system is large enough compared to that of the electronic mean free path. Therefore, flow of electrons takes place diffusively by undergoing repeated scattering and then acceleration successively.

Figure 2.7 shows that a negatively charged electron ($-e$) is under the application of an electric field $\vec{E}$, which in turn accelerates it towards right direction. After undertaking a short duration of travel, the electron suffers scattering by a scattering center and consequently, its velocity changes. Followed by this scattering, the electron again accelerates until it experiences the next collision. It is well known that the average distance, traversed by electrons, between two consecutive collisions is termed as the mean free path, which is typically 10 nm for metal at 300 K. Hence, a physical picture can be framed where in each spin sub-channel, diffusive transport of the majority and minority spin electrons takes place independently. It should be noted that in those two spin sub-channels, both the charge densities and drift velocities $(\vec{v}_+, \vec{v}_-)$ of electrons are different. Therefore, electric conductivities (σ_+, σ_-) (= (charge density) × (mobility)) of electrons are also different in those two spin sub-channels. Thus, under the application of an electric field, along with drift charge current density, $\vec{j}^{Q,Drift}$, we also obtain a drift spin current density $\hat{j}^{S,Drift}$.

$$\begin{aligned} \hat{j}^{S,Drift}* &= \vec{v}_+\vec{s}_+ + \vec{v}_-\vec{s}_- = \frac{\hbar}{2}\vec{e}_{spin}\left[\frac{(\sigma_+ - \sigma_-)}{-e}\vec{E}\right] \\ \vec{j}^{Q,Drift} &= \vec{v}_+\rho_+ + \vec{v}_-\rho_- = (\sigma_+ + \sigma_-)\vec{E} \end{aligned} \tag{2.22}$$

Here, we have neglected the exchange mediated term (i.e., second term) for simplicity.

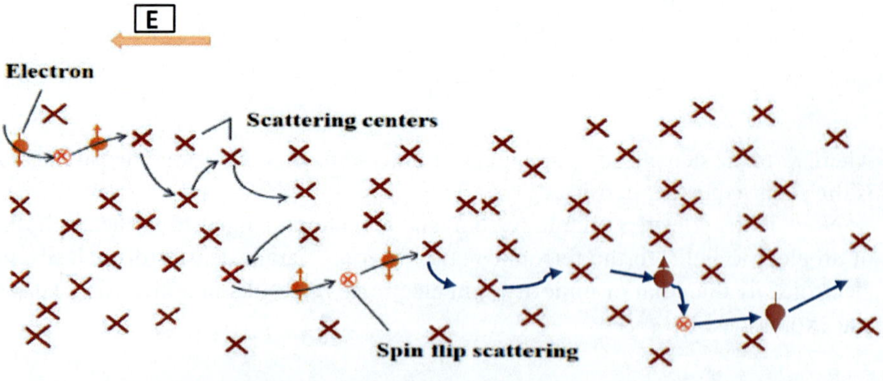

Fig. 2.7 Drift motion of free electron in conductive material

2.4 Spin Accumulation

2.4.1 *What Is Spin Accumulation?*

As a consequence of spin injection from ferromagnet to paramagnet in a ferromagnet/paramagnet heterostructure, a particular species of spin will accumulate in the paramagnet near the interface (Jedema et al. 2001). Describing the ferromagnet based on the Stoner–Wohlfarth model, let us assume that the up-spin electrons are the majority in the ferromagnet. Thus, the density of up-spin electrons in the ferromagnet is much larger than that of down-spin electrons and consequently, up-spin electrons become the major contributors of current injected by the ferromagnet. This results in an excess of up-spin electrons and a net magnetic moment of such spin-polarized ensemble within the paramagnet near the paramagnet/ferromagnet interface. This phenomenon is referred to as ***spin accumulation***. Since this is a non-equilibrium phenomenon, such accumulated spins are not expected to spread over the entire paramagnet forever because of *spin-flip scattering* events, which might convert some of the up-spin electrons of the ensemble into down-spin electrons. This in turn brings the non-equilibrium phenomenon into equilibrium one, where at the steady state the population of up-spin and down-spin electrons becomes the same far from the interface into the bulk of the paramagnet. Thus, spin accumulation is quite likely to gradually decay with distance away from the interface. In this context, we may define and extract a '***spin accumulation length***' which is the characteristic distance over which the accumulated spin decays to *1/e* times its magnitude at the interface.

In order to discuss spin accumulation and its decay, let us consider an interface between a ferromagnet (FM) and a non-magnetic material (NM) (Fig. 2.8a). In either a bulk metallic or semiconducting non-magnetic material, the energy dispersion relation could be simply considered as parabolic. As already discussed, if the material is non-magnetic, the two spin channels have the same mobility.

Thus, under the application of a small electric field, Fermi energy surface for the up-spin and down-spin channels on the (k_x, k_y) plane will be shifted by an equal amount, as shown in Fig. 2.9a. Denoting Δk as the shift in momentum space for both spin Fermi surfaces, it must satisfy the following equation:

$$F = -eE = \hbar\frac{dk}{dt} = \hbar\frac{\Delta k}{\tau_m}, \tag{2.23}$$

where $\boldsymbol{F}$ is the force acting on the electron, $\boldsymbol{E}$ is the applied electric field, $\boldsymbol{e}$ is the charge on the electron and τ_m is the scattering time of the electron. τ_m is related to the mobility of the carrier (μ) by

$$\mu = \frac{e\tau_m}{m^*} \tag{2.24}$$

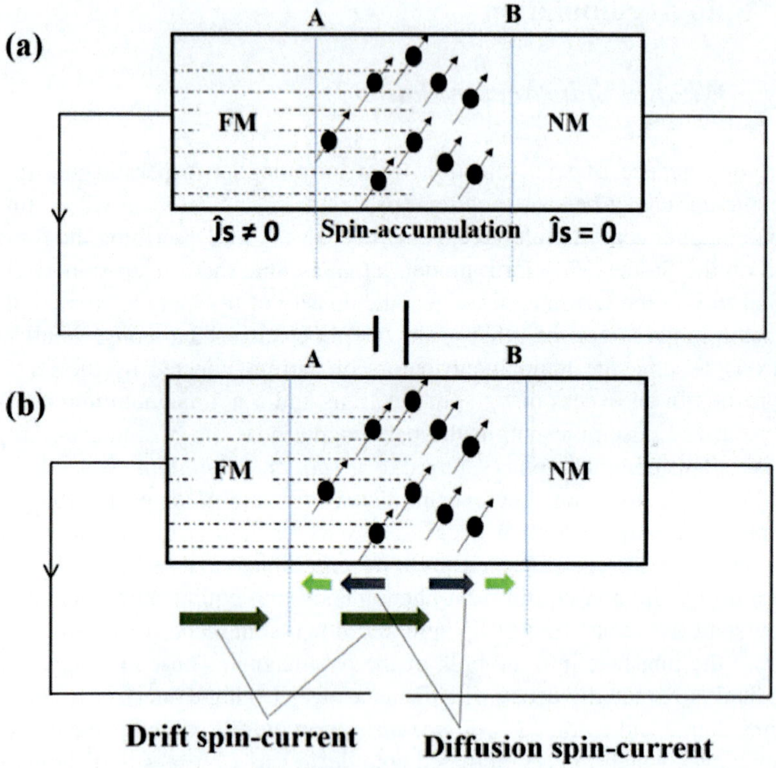

Fig. 2.8 Spin accumulation, diffusion and then relaxation during spin transport from FM to NM. **a** Spin current at section A has been found to be finite, whereas that at B is zero. **b** The number of accumulated spins has been found to decrease because of *spin-flip scattering* process occurring at the interface region

where m^* = electron effective mass. On the other hand, in case of a ferromagnetic metal, owing to the two spin channels having different mobilities, under the application of a small electric field, the shift of the Fermi surface of up- and down-spin channel will be different (Fig. 2.9b).

Now, let us consider two sections A and B in the ferromagnetic (FM) and non-magnetic (NM) layer, respectively, where A and B are so assumed that they are sufficiently far away from the interface (Fig. 2.8a). Consequently, there should be finite spin current at A but zero spin current at B. It follows from the law of conservation of the spin momentum [Eq. (2.20)] that spin accumulation will take place around the interface with time. As already mentioned, spin accumulation is not expected to sustain indefinitely at the interface. The accumulated spins either lose its identity through *spin-flipping* induced by ***spin–orbit interaction*** or they may give rise to diffusion spin current $\left(\hat{j}^{S,Diffusion}\right)$ flowing from the interface to the bulk of the non-magnetic material (Fig. 2.8b) governed by the equations

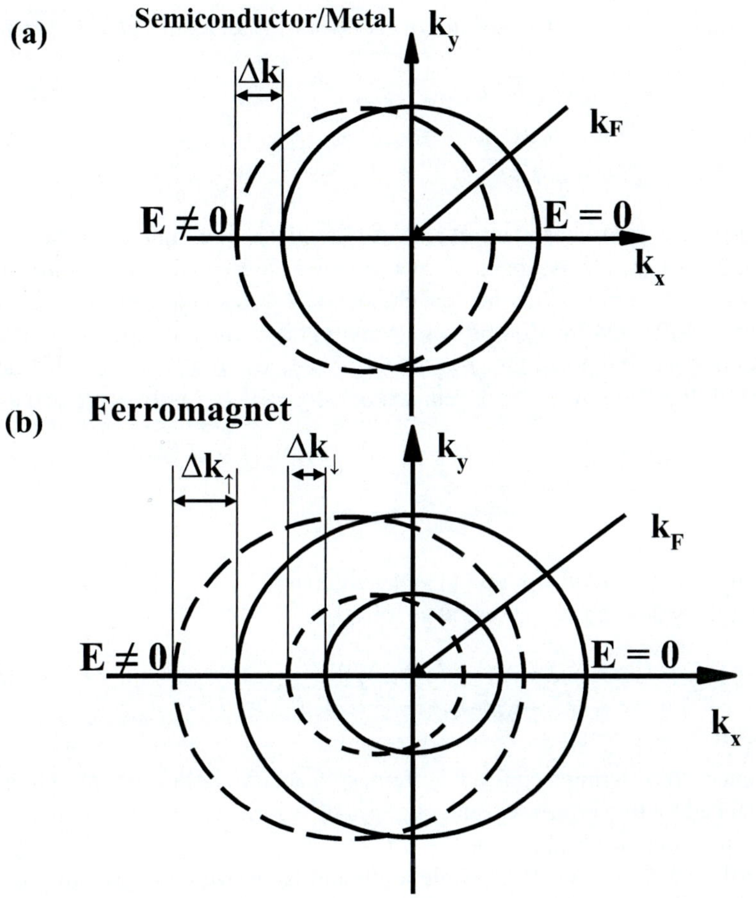

Fig. 2.9 **a** Shifting of Fermi surface in a paramagnetic metal or semiconductor in the presence of a constant external electric field. This in turn leads to current conduction. **b** As per the generalization of Mott model of the ferromagnet where up-spin and down-spin electrons have different mobilities (Figure adapted and redrawn from Bandyopadhyay and Cahay 2008)

$$\hat{j}^{S,Diffusion} = -D_\uparrow \nabla\vec{s}_\uparrow - D_\downarrow \vec{\nabla}\vec{s}_\downarrow$$
$$\vec{j}^{S,Diffusion} = -D_\uparrow \vec{\nabla}\rho_\uparrow - D_\downarrow \vec{\nabla}\rho_\downarrow \tag{2.25}$$

where $D_\uparrow/D_\downarrow$ is the diffusion constant of electrons in up (↑)/down (↓) spin sub-channels. Relation between diffusion constants and conductivities, as proposed by Einstein, for metals and semiconductors takes the form like:

$$\sigma_{\uparrow\downarrow} = N_{\uparrow\downarrow} e^2 D_{\uparrow\downarrow} : Metals$$
$$\sigma_{\uparrow\downarrow} = \frac{N_{\uparrow\downarrow} e^2 D_{\uparrow\downarrow}}{k_B T} : Non\text{-}degenerate\, semiconductor \tag{2.26}$$

Gradient of the electron density in each spin sub-channels can be expressed by

$$\vec{\nabla}_{n\uparrow} = \pm N_{\uparrow}\vec{\nabla}_{\mu\uparrow} : Metals$$
$$\vec{\nabla}_{n\uparrow} = \pm\frac{n_{\uparrow}\vec{\nabla}_{\mu\uparrow}}{k_B T} : Non-degenerate\, semiconductors \tag{2.27}$$

where $\mu_{\uparrow\downarrow}$ = spin-dependent chemical potentials. The sign $\pm$ in the second line in Eq. (2.27) may correspond to electrons or holes in the semiconductor. Equations (2.27) are only valid when the change in $\mu_{\uparrow\downarrow}$ is small. Considering simultaneously the drift and diffusion process (Saikin 2004; Yu et al. 2012), let us introduce an electrochemical potential, $\bar{\mu}_{\uparrow\downarrow} = \mu_{\uparrow\downarrow} - e\varphi$, where φ is the electrostatic scalar potential. Involving $\bar{\mu}_{\uparrow\downarrow}$, charge current density can be simply expressed as follows:

$$\vec{J}^{Q}_{\uparrow} = \vec{J}^{Q,Drift}_{\uparrow} + \vec{J}^{Q,Diffusion}_{\uparrow}$$
$$= \sigma_{\uparrow\downarrow}\vec{\nabla}\mu_{\uparrow\downarrow}/e \tag{2.28}$$

Equation 2.28 holds for up (↑) spin sub-channel. Thus, the total spin and charge current densities can be expressed as follows:

$$\vec{J}^{S} = \frac{\hbar}{2}\vec{e}_{spin}\frac{\vec{J}^{Q}_{\uparrow} - \vec{J}^{Q}_{\downarrow}}{-e}; \vec{J}^{Q} = \vec{J}^{Q}_{\uparrow} + \vec{J}^{Q}_{\downarrow} \tag{2.29}$$

Hence, considering a constant current, spin accumulation continues to grow till it is balanced by two processes, namely, spin diffusion and spin relaxation. After that, the system attains a steady state.

Furthermore, we adopt a simple argument to derive an expression for the ***spin accumulation length***, which is the length scale over which the spin accumulation fall offs away from the ferromagnet/non-magnet interface. As already mentioned, we expect the spin accumulation length decay exponentially away from the interface as $\sim e^{-x/\lambda_{sd}}$ where λ_{sd} is the spin accumulation length.

2.4.2 Estimation of Spin Accumulation Length by a Simple Method

Let us consider the spin injection mechanism for an up-spin electron injected from a ferromagnetic contact into a non-magnetic material. As generally recognized from previous discussions, the spins of the injected electrons would eventually be flipped after undergoing a random motion in the paramagnet. Over an ensemble of many electrons, the average distance from the interface over which spin flipping takes place can be roughly assumed as the spin diffusion length. At the onset let us assume that an up-spin electron injected from the ferromagnetic layer into the non-magnetic layer

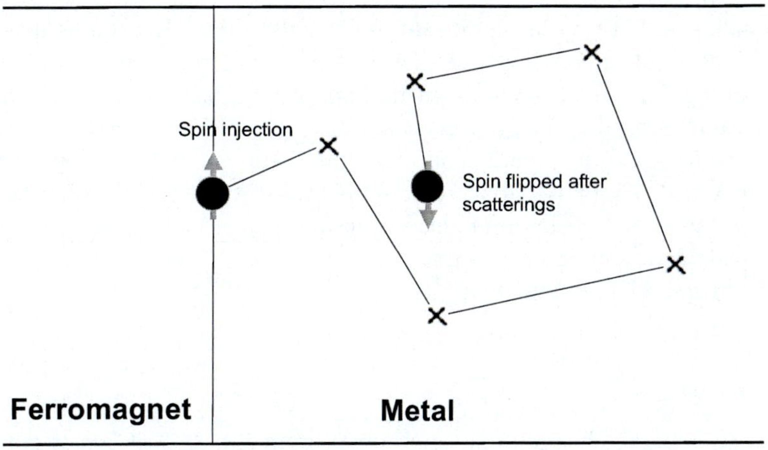

Fig. 2.10 Spin injection of an up-spin electron from a ferromagnetic contact into a paramagnet is demonstrated. Injected up-spin electron suffers N momentum relaxing scattering events (indicated by crosses) after it undergoes random walk in the paramagnet. Eventually, spin of the injected electron is flipped. Spin flip occurs at an average distance (over an ensemble of many injected carriers) of roughly the spin diffusion length from the interface (Figure adapted and redrawn from Bandyopadhyay and Cahay 2008)

will undergo *N* momentum-relaxing collisions before being flipped, as demonstrated in Fig. 2.10.

We denote the corresponding mean free path, i.e., the average distance between momentum scattering collisions by λ_f and the average *spin flip time* by $\tau_{\uparrow\downarrow}$. Now, the target up-spin electron is allowed to move randomly in three dimensions in equal amount. Then after suffering each collision, the average distance in a direction perpendicular to the interface that the electron penetrates into the non-magnetic material, i.e., spin accumulation length λ_{sd} is given by

$$\lambda_{sd} = \lambda_f \sqrt{N/3} \tag{2.30}$$

On the other hand, the total distance travelled by the injected up-spin electron is N * λ_f, which in turn equals the velocity of the injected electron at the *Fermi* level, i.e., Fermi velocity, v_{F} times the spin-flip time τ_{s} (=$\tau_{\uparrow\downarrow}$ =$\tau_{\downarrow\uparrow}$)

$$\text{i.e.,}\quad N * \lambda_f = v_{\mathrm{F}} * \tau_{\mathrm{s}} \tag{2.31}$$

Assuming elastic collisions here, the magnitude of the electron's velocity is invariant. Furthermore, we suppose that the carriers are injected with the Fermi velocity v_{F}. From Eqs. (2.30) and (2.31), we obtain the expression

$$\lambda_{sd} = \sqrt{\frac{\lambda_f v_F \tau_S}{3}} \tag{2.32}$$

Equation (2.32) is quite significant in the sense it clearly exhibits that λ_{sd} and λ_f are indeed related but not equal. For instance, by doping silver with increasing levels of a gold, i.e., by a non-magnetic impurity, λ_{sd} has been found to decrease. The reason for this drop in λ_{sd} originates from a decrease in the electronic mean free path λ_f with increasing impurity concentration. Furthermore, larger concentration of gold impurities is supposed to increase spin–orbit scattering of electrons because of heavy gold atoms. This in turn reduces spin-flip time τ_s and hence causes decrease in λ_{sd} with increasing levels of a gold.

From **Eq.** 2.31, it follows that

$$\frac{\tau_s}{\tau_m} = N \tag{2.33}$$

where τ_m is the momentum relaxation time. From **Eq.** 2.33, it comes out that the spin relaxation time τ_s is directly proportional to the momentum relaxation time τ_m.

Now, let us define the spin polarization α of the injected current as

$$\alpha(\mathrm{x}) = (\mathrm{J}\uparrow(\mathrm{x}) - \mathrm{J}\downarrow(\mathrm{x}))/\mathrm{J} \tag{2.34}$$

where J is the total current density $J{\uparrow}(x) + J{\downarrow}(x)$ injected across the interface, x being the direction of current flowing perpendicular to the interface. Net spin accumulation in the non-magnet/paramagnet for a current J flowing through the junction can be found in the following manner.

As can be generally understood, J does not depend on x; hence, we may equate the net spin injection across the interface given by

$$\left[\frac{d(\mathrm{n}\uparrow - \mathrm{n}\downarrow)}{dt}\right]_{x=0} = \frac{\mathrm{A}\alpha(0)\mathrm{J}}{e} \tag{2.35}$$

(where A = cross-sectional area at the ferromagnet/non-magnet junction) to the rate of decrease of the total spin concentration in the whole volume of the paramagnet. In this attempt, we obtain

$$\left[\frac{d(n\uparrow - n\downarrow)}{dt}\right]_{x=0} = \frac{A}{\tau_S}\int_0 (n_\uparrow - n_\downarrow)dx \tag{2.36}$$

Since the decay of the spin accumulation has been assumed to be exponential with characteristic decay constant or spin accumulation length λ_{sd}, therefore, we may write

$$n_\uparrow(x) - n_\downarrow(x) = n_0 e^{-x/\lambda_{sd}} \tag{2.37}$$

where $\mathrm{n}_0 = \mathrm{n}_\uparrow(0) - \mathrm{n}_\downarrow(0)$.

Thus from Eqs. 2.35 to 2.37, we get

$$n_0 = \frac{\alpha(0) J \tau_s}{e\lambda_{sd}} = \frac{3\alpha(0) J \lambda_{sd}}{e v_F \lambda_f} \tag{2.38}$$

For $\alpha(0) = 1$ (half-metallic contact), $v_F = 10^6$ m/s, $\lambda_f = 5$ nm, $\lambda_{sd} = 100$ nm, and a typical current density $J = 10^3$ A/cm^2, we get $n_0 = 4 \times 10^{22}$ m^{-3}. It is well known that in normal metals, the electron concentration is of the order of several 10^{28} m^{-3}. Hence, the net spin accumulation in a normal metal (paramagnet) is found to be typically small with only one part in 10^6 of the electrons being spin polarized. Correspondingly, the magnetic field B associated with this spin accumulation is given by

$$\mathrm{B} = \mu_0 \mathrm{M} = \mu_0 \mathrm{n}_0 \mu_\mathrm{B} = 10^{-9}\,\text{Tesla} \tag{2.39}$$

which is very small compared to the magnetic field due to the current flowing through the interface and generating the spin accumulation.

2.5 Spin Relaxation

2.5.1 What Is Spin Relaxation?

The concept of spin relaxation is very much significant in spintronics. Here, information is encoded by employing the spin polarization state of either single electron or that of an ensemble of electrons, where spin polarization hosts the information (Dresselhaus 1955; Bychkov and Rashba 1984; Jedema et al. 2002; Pramanik et al. 2006). In order to ensure the reliability of such process, random and spontaneous depolarization of spins, i.e., 'spin relaxation' must be prohibited.

In general, when an electron is introduced in a solid, the interaction between the electron and the environment affects its spin orientation. Actually, the environment may give rise to an effective magnetic field in a solid that interacts with the spin of the charge carriers and thereby causes the alteration of its orientation. Such magnetic field in a solid arises from ***(i)*** the spins of other electrons and holes existing in the solid; ***(ii***)nuclear spin; ***(iii)*** phonons or vibrating atoms giving rise to time-dependent magnetic field in some circumstances and ***(iv)*** spin–orbit interactions in the solid. Any effective magnetic field interacts with the spin magnetic moment of the electron with the interaction energy, $\boldsymbol{E}_{rel} = \boldsymbol{g}_0 \boldsymbol{\mu}_B \vec{\boldsymbol{s}}.\vec{\boldsymbol{B}}$. Such interaction can alter the electron's spin state.

Now, let us consider electrons encountering different effective magnetic field within the medium through which it is flowing:

(i) If the spin polarization of an electron is already parallel or antiparallel to that of effective magnetic field, then the spin of the electron will not change since the spin polarization is already an eigenstate [discussed in Chap. 1] and therefore stable. As can be understood, such situation is a bit rare.

(ii) Most of the time, electron's spin points along some arbitrary direction in space, and hence they are not either parallel or antiparallel to that effective magnetic field. As a result, such spin will undergo the familiar Larmor precession about the effective magnetic field, $\boldsymbol{B}_{\boldsymbol{eff}}$ with an angular frequency $\Omega_{eff} = {g\mu_B B_{eff}}/{\hbar}$, where g is the Landé g-factor of the medium. Time evolution of the spin, undergoing Larmor precession (which follows from time-dependent Pauli Equation) is described by the equation $\frac{d\vec{s}}{dt} = \vec{\Omega}_{eff} \times \vec{S}$, which present us a picture where the initial spin will not be stable and begin to change with time when the electrons interact with the effective magnetic field.

Furthermore, $\vec{B}_{eff}$ depends on the velocity of electrons $\vec{v}$ or wavevector $\vec{k}$ (will be discussed later in detail). Since different electrons suffer scattering differently, hence electron's velocity or wavevector changes randomly. This in turn leads to randomization of $\vec{B}_{eff}$. Consequently, the axis of precession of electron's spin changes its direction and the frequency of Larmor precession also changes. Since scattering events are random, such changes occur randomly in time, thereby resulting in the orientation of electron's spin changes randomly and gradually with time. This leads to spin relaxation. On the other hand, if the spin changes suddenly and discretely in time, then such phenomenon is referred to as '*spin flip*', which corresponds to an '*up-spin*' state becoming a '*down-spin*' state and vice versa. As discussed in Chap. 1, '*up-spin*' and '*down-spin*' states would not have any coupling between them since they are mutually orthogonal and the corresponding matrix element connecting these two states should be zero. It is the 'scatterer' that couples and causes transitions from one state to another, thereby resulting in a *spin flip. Noteworthy*, the scatterer must have an internal magnetic field of some sort.

2.5.2 What Is Spin–Orbit Interaction?

First, we should have knowledge on 'spin–orbit interactions' before going into details of spin relaxation mechanisms. The spin–orbit interaction or coupling is the interaction of a particle's spin with its motion. In order to understand the interaction process, let us take a simple example of an electron orbiting around a nucleus in an atom. In such case, 'spin–orbit interaction' is basically an electromagnetic interaction of *spin magnetic moment* of *electron* with the *magnetic field, which is produced due to the orbiting negatively charged electron around the positively charged nucleus*, hence have coined the name spin–orbit interaction. A well-known consequence is the shifting of the atomic energy levels of the electrons, thus causing splitting in atomic spectral lines. Similar effect can also be found in case of protons and neutrons moving inside the nucleus. In the arena of spintronics and many spin-based devices, spin–orbit interaction is one of the major mechanisms that determine the spin relaxation process; therefore, it is indeed necessary to learn the spin–orbit interaction in detail.

Here, we will describe spin–orbit interaction by simply employing electrodynamics (semi-classical) and quantum mechanics (non-relativistic) that agree quite well with the experimental observations. Even more precise results need meticulous derivation starting from Dirac equation and need to involve quantum electrodynamics.

Following the above discussions, it appears that in general, spin–orbit interaction is an interaction between spin magnetic moment of a spinning particle and the magnetic field, generated by the spinning particle's orbit itself in a strong electric field. Thus, in order to understand the interaction process, the first question is: *what is the energy of a magnetic moment in a magnetic field?*

Energy of a magnetic moment subject to a magnetic field is given by

$$E = -\vec{\mu} \cdot \vec{B} \tag{2.40}$$

where $\vec{\mu}$ is the magnetic moment of the particle and $\vec{B}$ is the magnetic field it experiences. Here, we will concentrate on the spin–orbit interaction experienced by an electron inside an atom. Let us now find out the expression for $\vec{B}$ and $\vec{\mu}$ in this case, one by one.

Magnetic field ($\vec{B}$)

Let us suppose that in an atom a negatively charged electron is orbiting around the nucleus with an orbiting velocity $\vec{v}$ and the radius of the orbit is $\vec{r}$ (Fig. 2.11). Thus, it feels an electric field due to the positively charged nucleus.

Obviously, no magnetic field could be realized in the rest frame of the nucleus; however, a magnetic field will appear in the rest frame of the electron as follows: An observer sitting on the electron and moving with it thinks that the electron is at rest and the nucleus is revolving around it with a velocity $-\vec{v}$. To the observer, the radius of the orbit will be $-\vec{r}$ and the nuclear charge will be $+Ze$, where Z = atomic

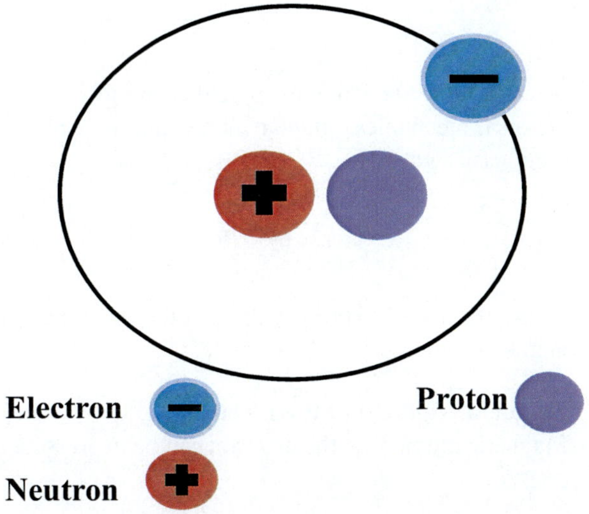

Fig. 2.11 Schematic drawing of electron orbiting around the nucleus

number and $e =$ elementary charge. Consequently, according to Biot–Savart's law, the magnetic flux density at the position of the electron will be given by

$$\vec{\boldsymbol{B}}_0 = Ze\frac{\vec{r} \times \vec{v}_{orbit}}{4\pi \in_0 c^2 r^3} \tag{2.41}$$

Now, the Coulomb electric field $\vec{E}$ seen by the orbiting electron due to the positively charged nucleus is

$$\vec{\boldsymbol{E}} = \frac{\boldsymbol{Ze\vec{r}}}{4\pi \in_0 \boldsymbol{r}^3}, \tag{2.42}$$

Hence, in the rest frame of the electron

$$\vec{\boldsymbol{B}} = -\frac{\vec{v} \times \vec{\boldsymbol{E}}}{c^2} \tag{2.43}$$

Since the force related to $\vec{E}$ is radial in nature, we can rewrite $\vec{\boldsymbol{E}} = \left|\frac{E}{r}\right|$ and momentum of the electron $\vec{\boldsymbol{\rho}} = m_e v$. Substituting $\vec{\boldsymbol{E}}$ and $\vec{\boldsymbol{v}}$ in Eq. 2.43 and altering the order of the cross product result in

$$\vec{\boldsymbol{B}} = \frac{\vec{\boldsymbol{r}} \times \vec{\boldsymbol{P}}}{m_e c^2}\left|\frac{E}{r}\right|.$$

$$\text{Again, } \vec{\mathbf{E}} = -\nabla V \tag{2.44}$$

According to central field approximation, electrostatic potential is spherically symmetric; therefore, V is only a function of radius r. Thus,

$$\left|\vec{E}\right| = \frac{\partial V}{\partial r} = \frac{1}{e}\frac{\partial U(r)}{\partial r} \tag{2.45}$$

where $\boldsymbol{U}(= Ve)$ denotes the potential energy of the electron in the central field. From classical mechanics, angular momentum of a particle is $\vec{\boldsymbol{L}} = \vec{\boldsymbol{r}} \times \vec{\boldsymbol{p}}$. Putting it all together we get

$$\vec{\boldsymbol{B}} = \frac{1}{m_e e c^2}\frac{1}{r}\frac{\partial U(r)}{\partial r}\vec{\boldsymbol{L}} \tag{2.46}$$

Equation 2.46 expresses the fact that the magnetic field is parallel to the orbital angular momentum of the particle.

Magnetic Moment of the Electron (μ)

Magnetic moment of the electron can be expressed as

$$\vec{\mu} = -g_s \mu_B \vec{\mathbf{S}}/\hbar \tag{2.47}$$

where $\boldsymbol{S}$ is the spin angular momentum vector, $\boldsymbol{\mu}_B$ is the Bohr magneton and $g_s \approx 2$ denotes the electron spin g-factor. It is obvious from Eq. 2.4 that $\boldsymbol{\mu}$ is antiparallel to the spin angular momentum.

In effect, spin–orbit interaction energy is made up of two parts. The Larmor part is related to the interaction, occurring between the magnetic moment of electron and the magnetic field produced by the nucleus. The Larmor interaction energy is

$$\Delta H_L = -\mu \cdot \boldsymbol{B} \tag{2.48}$$

By using the expressions of the magnetic moment and the magnetic field, we obtain from Eq. 2.4

$$\Delta H_L = \frac{2\mu_B}{\hbar m_e e c^2} \frac{1}{r} \frac{\partial U(r)}{\partial r} \boldsymbol{L}.\boldsymbol{S} \tag{2.49}$$

The second part is related to Thomas precession for the electron's curved trajectory (we would not go into details of this). Actually, the electron in a rotating orbit is constantly accelerating because the direction of the velocity is changing with time, even though the magnitude is not changing. Therefore, it is not enough to transform the laboratory frame to the rest frame using the electron's instantaneous velocity. This requires an additional correction factor ½, also known as Thomas correction factor, to be introduced when the above fact is taken into account.

Thus, the net spin–orbit interaction energy takes the form

$$\Delta H \equiv \Delta H_L + \Delta H_T = \frac{\mu_B}{\hbar m_e e c^2} \frac{1}{r} \frac{\partial U(r)}{\partial r} (\boldsymbol{L}.\boldsymbol{S}) \tag{2.50}$$

The net effect of Thomas precession comes from the reduction of the Larmor interaction energy by factor ½, which is familiar as Thomas half. It is obvious that the above energy depends on the scalar product $\vec{L}.\vec{S}$, where $\vec{L}$ and $\vec{S}$ are the orbital and the spin quantum number, respectively. Hence, the above equation reflects the interaction or coupling between the magnetic moment and the orbital angular momentum associated with an electron in the atom and thereby denotes spin–orbit interaction.

2.5.3 Spin Relaxation Process

Let us return back to our discussion on the spin relaxation process. In spintronics, after spin injection from the ferromagnetic contact, the spin-polarized carriers flow within the paramagnetic semiconductor under the application of a bias voltage, which in turn produces a transport-driving electric field inside the semiconductor. During the travel of spin-polarized carriers, spins undergo different interactions (spin–orbit, hyperfine, carrier–carrier interactions etc.) with their environments and their initial orientations get altered to different extent considerably. As a result, the spin polarization of the

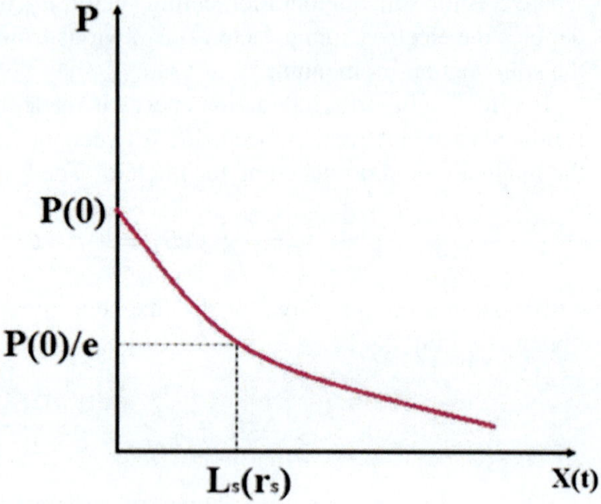

Fig. 2.12 Illustration of spin relaxation. The magnitude of spin polarization of the electron ensemble decreases exponentially with time and distance

spin ensemble decreases both with time (t) and distance (x), as estimated from the point of spin injection into the paramagnetic semiconductor (Fig. 2.12). Such gradual loss in the magnitude of the spin polarization of the injected electron ensemble in the paramagnetic medium is termed as spin relaxation phenomenon. The variation in the injected spin polarization with respect to distance/time in the paramagnetic semiconductor has been found to follow an exponential decay, as demonstrated in Fig. 2.12.

As is shown in Fig. 2.12, spin relaxation or diffusion length/time, i.e., L_S/τ_S is defined as the distance/duration over which the spin polarization reduces to *1/e* times of its initial value. When $x \gg L_S$ (or $t \gg \tau_S$), then spin polarization of the ensemble tends to zero, i.e., P → 0, implying complete loss of spin polarization. Thus, spin relaxation phenomenon in a paramagnet tends to bring the non-equilibrium population of spin back to the equilibrium unpolarized condition. Here lies the paramount importance of spin relaxation in spintronics since everyone is concerned with using spin polarization of either a single-charge carrier or the net spin polarization of ensemble of charge carriers to encode and decode information. If the given spins were to host the information reliably for considerable time, it must be protected against random or spontaneous depolarization caused by a various relaxation mechanism. Thus, with the objective to attain 'non-zero spin polarization', the continuous effort is to suppress spin relaxation process; in other words to improve spin relaxation length and time into the paramagnetic semiconductor.

Several mechanisms are there in solids and those are responsible for spin relaxation of electrons in the conduction band of a semiconductor. Among them, the four main spin relaxation processes are: (**a**) Elliott–Yafet mode of spin relaxation, (**b**) D'yakonov-Perel' mode of spin relaxation, (**c**) Bir-Aronov-Pikus mode of spin relaxation and (**d**) hyperfine interaction of electron spins with nuclear spins.. Spin–orbit interaction is the primary cause of the first two mechanisms, whereas the third one arises due to the exchange coupling between electron and hole spins. The last one, as already mentioned, is the familiar hyperfine interaction originating due to the interaction between carrier and nuclear spins.

2.6 Elliott–Yafet Mechanism

2.6.1 *What Is the Prerequisite Condition for Elliott–Yafet Mechanism?*

Elliott–Yafet spin relaxation mechanism in real crystals occurs owing to the presence of *spin–orbit interaction* in the solid (Pramanik et al. 2006). In fact, this mechanism arises owing to the fact that in a real crystal, Bloch states, i.e., the solutions of the Schrödinger equation in the periodic lattice potential, are not pure spin eigenstates ($|\uparrow\rangle$ or $|\downarrow\rangle$), but an admixture of both

$$u_k\left(\vec{r}\right) = a_k\left(\vec{r}\right)|\uparrow\rangle + b_k\left(\vec{r}\right)|\downarrow\rangle \tag{2.51}$$

This means that electron's spin in a crystal does not really have one of two fixed spin polarizations, i.e., spin-up or spin-down with a unique axis defined by spin quantization. Rather, the electron's spins are orientated either in pseudo-up or down directions. More subtle point is that the degree of admixture of up-spin or down-spin electrons, determined by the quantities a_k and b_k, is a function of the electronic wavevector k. This in turn implies that for electrons in a crystal, its spin orientation depends on the electron's wavevector k. This is evident from the schematic demonstration (Fig. 2.13), where for an arbitrary band, energy has been plotted as a function of the wavevector. Such plot explores that the spin orientations of electrons (shown by arrow) are different in different wavevector states. However, corresponding to each wavevector state there are two possible mutually antiparallel spin orientations.

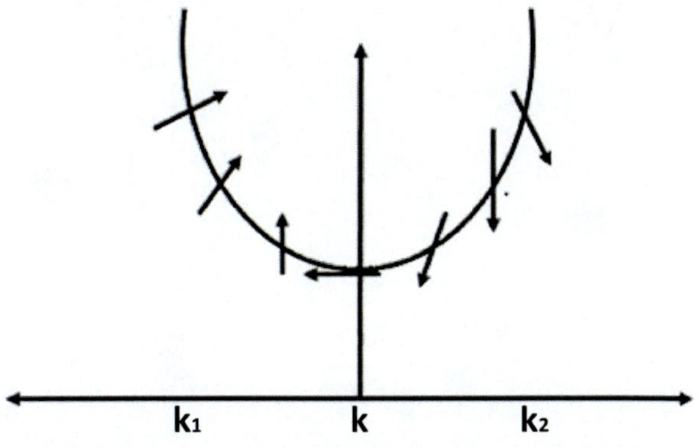

Fig. 2.13 Energy dispersion relation showing the spin polarizations at different wavevector states

Thus, it comes out that spin orientations associated with different wavevectors states (ks) can have arbitrary angle between them.

2.6.2 *Elliott–Yafet Mode of Spin Scattering Mechanism*

On this background, let us consider a collision event takes place between an electron and a non-magnetic scatterer. The scatterer may be a non-magnetic impurity, device boundary or phonon. Let us suppose k is the wavevector of the electron in the crystal. As a result of such scattering phenomenon, the momentum (p) and hence k of the electron gets changed given that $p = \hbar k$. This in turn causes the change in the spin orientation of that electron as well, since the spin orientations associated with the wavevector states, before and after scattering, say $k_{initial}$ and k_{final}, are never mutually parallel. Thus it comes out that in a real crystal, momentum-relaxing collision results in transitions between different wavevectors states of electrons, which in turn causes the change in spin orientation of that electrons and thereby resulting in *spin relaxation*. This constitutes the basis of the well-known Elliott–Yafet mode of spin relaxation process (Fig. 2.14).

Now, the extent of spin relaxation, i.e., change in such spin orientation of electrons is a function of how much the wavevector changes. For instance, collisions of electrons with charged impurities, generally do not change the wavevector appreciably. Hence, such collisions are not very effective in changing spin orientation or giving rise to spin relaxation. However, in case of collisions of electrons with certain types of acoustic phonons, scattering occurs preferably through large angles. Therefore, in

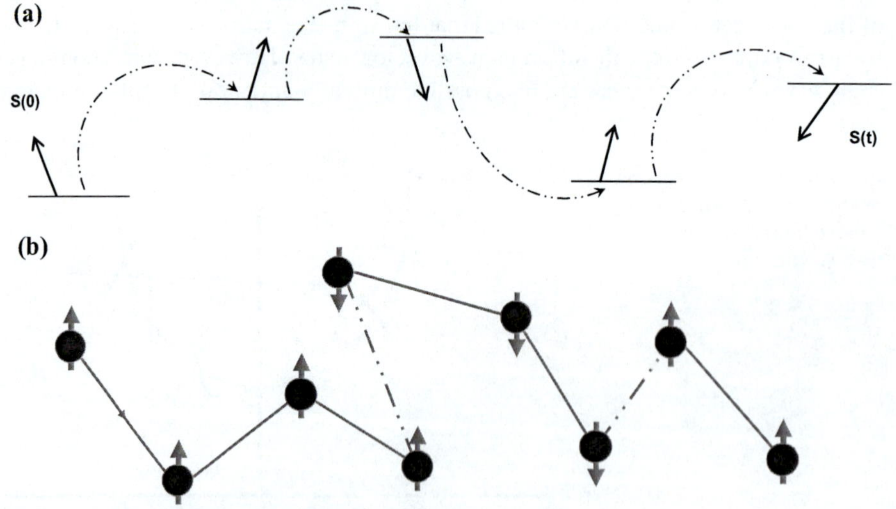

Fig. 2.14 Schematic description of Elliott–Yafet spin relaxation mechanisms

this type of scattering wavevector changes by a lot, and thereby will be very effective in relaxing spin.

2.6.3 *Where Does the Elliott–Yafet Mechanism of Spin Scattering Occur?*

At the onset, Elliott–Yafet relaxation process is considered to be primary mechanism for spin relaxation in low mobility semiconductors, e.g., organic materials. This is because in organics the momentum-relaxing scattering events are frequent.

To summarize, in this relaxation mechanism the spin relaxation does not occur due to the mere presence of spin–orbit interaction in the system. In this case, spin relaxation occurs only if the scattering of carriers takes place during transport. This implies higher is the spin scattering rate subject to the higher momentum scattering rate. Noteworthy, this spin relaxation mechanism must be associated with some degree of momentum relaxation because it is the change in wavevector that results in the change in spin.

2.7 D'yakonov-Perel' Mechanism

2.7.1 *What Is the Prerequisite Condition for D'yakonov-Perel' Mechanism?*

The origin of D'yakonov-Perel' spin relaxation mechanism is also spin-orbit coupling. It is the lack of inversion symmetry in some solids that causes this spin relaxation process to become the dominant relaxation mechanism in those systems. Let us segregate the discussion in two divisions:

(i) In some cases, for example in compound semiconductors, inversion symmetry becomes absent owing to its crystallographic structure. Examples are inorganic semiconductors like GaAs (Group III–V semiconductors), ZnSe (Group II–VI semiconductors). In these semiconductors, the very presence of two distinct atoms in the Bravais lattice breaks their inversion symmetry. This kind of crystallographic inversion asymmetry, also known as ***bulk inversion asymmetry***, leads to ***Dresselhaus spin–orbit interaction*** (Dresselhaus 1955). Furthermore, bulk inversion symmetry has also been found to be absent in disordered organic semiconductors.

(ii) In a different scenario, external electric field could be applied or there may be a built-in electric field in a solid. Such external or built-in electric fields can also break the inversion symmetry. As a result, conduction band energy profile becomes inversion asymmetric along that electric field direction. This kind of inversion asymmetry is called ***structural inversion asymmetry***, which gives rise to ***Rashba spin–orbit interaction*** (Bychkov and Rashba 1984). Now, the presence of structural inversion asymmetry in case of disordered organic semiconductors is typically due to the microscopic electric fields arising owing to the presence of charged impurities and surface states, i.e., dangling molecular bonds.

It comes out that both Dresselhaus and Rashba spin–orbit interactions lift off the degeneracy between the two spin states corresponding to any arbitrary wavevector. Therefore, those non-degenerate up- and down-spin states possess different energies corresponding to the same wavevector state. In effect, these spin–orbit interactions resemble effective magnetic fields in a sense that magnetic fields also transform the degenerate up- and down-spin states to non-degenerate one, for any given wavevector due to Zeeman interaction.

2.7.2 D'yakonov-Perel' Mode of Spin Scattering Mechanism

Thus, it comes out that both bulk and structural inversion asymmetries, i.e., Rashba and Dresselhaus interactions, give rise to effective electrostatic potential gradient. This in turn generates electric field experienced by charge carriers. Hence, electrons in this kind of solid, having inversion asymmetry, will experience strong spin–orbit interaction. Now, considering the rest frame of reference of a mobile charge carrier, such electric field may be considered to be Lorentz transformed to an effective magnetic field $\vec{B}\left(\vec{v}\right)$, which in turn is a function of the carrier's velocity $\vec{v}$. Consequently, spin of charge carriers, i.e., electrons in an inversion asymmetric solid is supposed to perform continuous Larmor precession about that $\vec{B}\left(\vec{v}\right)$ with the precession axis being collinear with the magnetic field.

Now, let us suppose a simplified picture of an ensemble of electrons drifting and diffusing in such an inversion asymmetric solid. Let us first consider the velocity $\vec{v}$ of all the electrons in the ensemble be the same and does not change with time. Then the magnetic field, $\vec{B}\left(\vec{v}\right)$, experienced by all the electrons will also be the same. Consequently, spin magnetic moment of every electron in the ensemble should perform precession about this constant magnetic field with a fixed frequency. Interestingly, this does not cause any spin relaxation at all! In order to understand this, let us consider that all the electrons in the ensemble are in the same spin polarization states just after spin injection into the solid, i.e., at the onset of their journey through the solid under the applied bias. Then after any arbitrary time interval, all of them must have undergone same degree of precession by exactly the same angle. Hence, the ensemble of electrons again has the same spin polarization. Although the direction of this spin polarization might be different from the initial one, but that does not matter. More significant point is that the magnitude of this ensemble averaged spin polarization does not change with time. Therefore, this scenario does not lead to any spin relaxation.

Now, let us consider a different scenario where the ensemble of electrons undergoes momentum-relaxing scattering events. This in turn results in the random change of their corresponding $\vec{v}$ with time. Accordingly, the magnitude of $\vec{B}\left(\vec{v}\right)$ would also get changed since $\vec{B}\left(\vec{v}\right)$ is proportional to the velocity $\vec{v}$ of the charge carrier. Thus, in this case $\vec{B}\left(\vec{v}\right)$ is expected to have a large distribution of values. Consequently, every electron in the ensemble should perform precession about the magnetic field, experienced by it, with different precession frequencies. In this scenario, after a certain interval of time after spin injection into the solid, different electrons would have precessed by different angles since their scattering histories are different. Even if considering an over-simplified picture of all the electrons in the ensemble were injected with the same spin polarization, after a given time from the spin injection into solid the orientation of the precessing spins would be different for various electrons in the ensemble. Consequently, spin polarization of the ensemble of electrons gradually becomes out of phase with respect to each other during the passage of carriers through the solid, and the ensemble averaged spin polarization decays with time. Finally, after a sufficiently long time, the average spin polarization of the ensemble will decay to zero. This is the conceptual representation of the D'yakonov-Perel' mode of spin relaxation (Fig. 2.15).

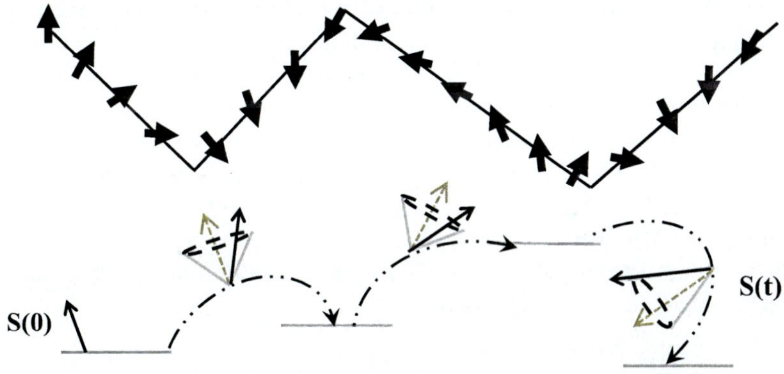

Fig. 2.15 Schematic description of D'yakonov-Perel' spin relaxation mechanisms

2.7.3 Is Frequent Momentum Scattering Actually Be Beneficial for Spin Longevity!!

It is the Dresselhaus and Rashba spin–orbit interactions that generated *velocity-dependent* magnetic field $\vec{B}\left(\vec{v}\right)$, which triggers D'yakonov-Perel' relaxation. Therefore, reduction in $\vec{B}\left(\vec{v}\right)$ might suppress D'yakonov-Perel' relaxation. Now, if the carrier experiences momentum relaxing scattering phenomenon quite *frequently*, i.e., carriers are having low mobility and small momentum relaxation time, then $\vec{v}$ is small implying $\vec{B}\left(\vec{v}\right)$ is small. This means that slower moving electrons experience a smaller $\vec{B}\left(\vec{v}\right)$. Thus, they should be less susceptible to D'yakonov-Perel' relaxation and consequently their spin lifetimes should be large. This phenomenon originally noted by D'yakonov and Perel' has come to be known as '***motional narrowing***'. It seems to imply that high carrier mobility or high saturation velocity of charge carriers might lead to the dephasing of their spin polarizations. In most of the cases this phenomenon has been found out to be indeed true. In a different scenario, very strong momentum-relaxing mechanism has been found to cause a larger spread in the velocity of the carriers, and thereby tend to put the spin polarizations of different electrons immediately out of phase with respect to one another. Thus, it cannot be asserted in absolute scale a priori whether momentum-relaxing collisions are beneficial or detrimental to spin lifetimes. The consequence may vary from system to system. Noteworthy, in case of Elliott–Yafet mechanism, spin relaxation rate is directly proportional to the momentum scattering rate.

Furthermore, as we recognize that D'yakonov-Perel' spin relaxation mechanism is less effective in case of low mobility materials than that of high mobility materials, thus in D'yakonov-Perel' process spin relaxation rate could be considered as inversely proportional to the momentum scattering rate. Significantly, this constitutes the basis to distinguish Elliott–Yafet spin relaxation process from D'yakonov-Perel' process. We emphasize that distinction between these two spin relaxation processes is based on the fact that the dependences of their spin relaxation rates on their momentum scattering rate, i.e., mobility is opposite.

2.7.4 Where Does D'yakonov-Perel' Mechanism Occur?

The primary spin relaxation process in ***high mobility semiconductors*** is usually the D'yakonov-Perel' mechanism.

2.8 Bir-Aronov-Pikus Mechanism

Bir-Aronov-Pikus (BAP) mechanism is a spin relaxation mechanism of electrons in some special kind of semiconductors that have simultaneous existence of both electrons and holes in significant concentration. Naturally, proximity of electrons and holes is quite close in this case. This, in turn results in overlapping of their wavefunctions, which would cause an exchange interaction between them.

Such exchange interaction, involving both electrons and holes, can be written in terms of Hamiltonian as

$$H = A\vec{S}.\vec{J}\delta\left(\vec{r}\right) \tag{2.52}$$

where ***A*** is proportional to exchange integral between conduction and valence states, $\vec{J}$ is the angular momentum operator for holes, $\vec{S}$ is the spin operator for electrons and $\vec{r}$ is the relative position of the electron and the hole. Owing to this exchange interaction, electron spins are supposed to perform precession motion along some magnetic field. Noteworthy, such effective magnetic field is determined by hole spins, $\vec{J}$. In this context, we should recall that hole spin and momentum relax very fast in the valence band because of its strong spin–orbit interaction. Such fluctuations in the total spins of holes produce a fluctuating effective magnetic field. In the limit of strong hole spin relaxation, there is random change in the effective magnetic field even before the electron spin completes a full precession, thereby causing the relaxation of electron spin. Thus, it comes out that flipping of hole's spin results in the flipping of the spin of electron as well due to electron–hole coupling (Fig. 2.16). This is the concept of Bir-Aronov-Pikus mode of spin relaxation of electrons.

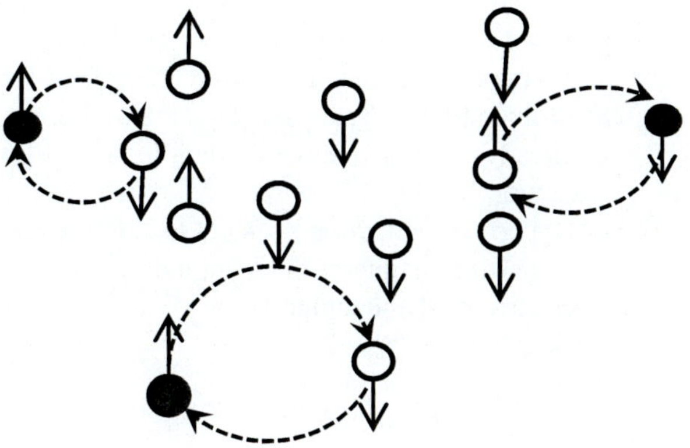

Fig. 2.16 Schematic drawing of BAP mechanism

2.8.1 *Where Does the Bir-Aronov-Pikus Mechanism of Spin Scattering Occur?*

Bir-Aronov-Pikus spin relaxation mechanism is dominant in bipolar semiconductors. As the mechanism suggests, this mode of spin relaxation is ineffective in case of unipolar transport, when current carriers are either electrons or holes, but not both simultaneously. It is noteworthy that in spin-based organic light-emitting diodes, this spin relaxation process plays a dominant role.

2.9 Hyperfine Interaction

In solids, nuclear spins may generate an effective magnetic field. Such effective magnetic field may interact with the carrier electron spins via *hyperfine interactions*. This results in spin relaxation of electron (Fig. 2.17). The Hamiltonian describing this interaction is given by

$$H_{nuclear} = \vec{S}.\sum_i A_i \vec{I}_i \tag{2.53}$$

where $\vec{I}_i$ = spin of the ith nucleus, $\vec{S}$ = electron spin, A = constant and A_i is the corresponding coupling coefficient. To summarize,

(i) Hyperfine interaction is the sort of magnetic interaction that takes place between the spin magnetic moments of the electrons and the nuclei and results in dephasing of electron spins.
(ii) Hyperfine magnetic field is created by an ensemble of nuclear spins.
(iii) Dominant for *quasi-static* carriers, which are strongly localized in space having no resultant momentum. Therefore, such carriers are virtually immune to both Elliott–Yafet and D'yakonov-Perel' spin relaxations processes. As already discussed above in detail, these relaxation processes require carrier motion.
(iv) Hyperfine interaction does not lead to the complete loss of spin polarization.
(v) Hyperfine magnetic field, though weak, is dependent on both the biasing electric field and temperature.

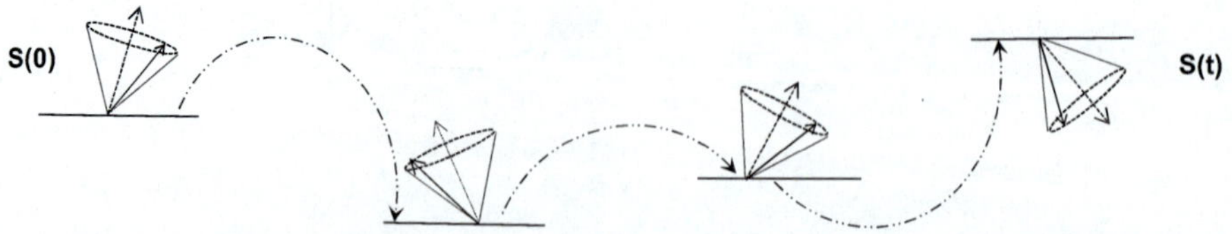

Fig. 2.17 Schematic description of hyperfine interaction-mediated spin relaxation mechanisms

Case I: In case of isotopes of majority of the ***inorganic semiconductors***, nuclear spins are ***non-zero***. For instance, all natural isotopes of Ga, In, Al, Sb and As have finite nuclear spins accompanied by substantial magnetic moments. Hence, hyperfine interaction is supposed to be particularly significant in technologically relevant III–V semiconductors, e.g., (Al)GaAs, In(Al)As etc.

Case II: Cd, Zn, S, Se and Te have isotopes, carrying nuclear spins, with less natural abundance.

- Hyperfine interaction in II–VI semiconductors is weaker compared to their III–V counterparts.
- Hyperfine interaction has been observed to be weak in case of elemental semiconductors, such as Si and Ge.

Case III: For organic semiconductors, two main constituent elements are carbon and hydrogen. In case of carbon atoms, the most abundant (98.89%) isotope 12C possesses zero nuclear spin. Thus, it is not supposed to contribute to the hyperfine interaction. The major contributions to hyperfine interaction in organics originate from the hydrogen atoms (1H, abundance > 99.98%). Negligible contributions to hyperfine interaction arise from 13C (natural abundance 1.109%) and other minor constituent elements.

2.10 Spin Valve

Spin valve is the basic building block of spintronics (Jedema et al. 2002; Pramanik et al. 2006). It is at the core of constructing diversified spintronics-based data storage devices, such as GMR read head sensors of computer hard disk drives, magnetic random access memory (MRAM) devices, spin-transfer torque-based devices, spin-polarized light emitting diodes etc. Furthermore, spin valve is also widely used to study and measure properties associated with spin transport in paramagnetic metals or semiconductors, like spin diffusion length, L_s and then spin relaxation time τ_s related by the equation

$$L_S = \sqrt{D_S \tau_S} \tag{2.54}$$

where $\boldsymbol{D_S}$ = spin diffusion coefficient. Therefore, spin valve is an important measurement device as well. Thus, we may consider spin valve as an extremely important device, both from the viewpoint of spin-physics and spintronics or optospintronics device applications.

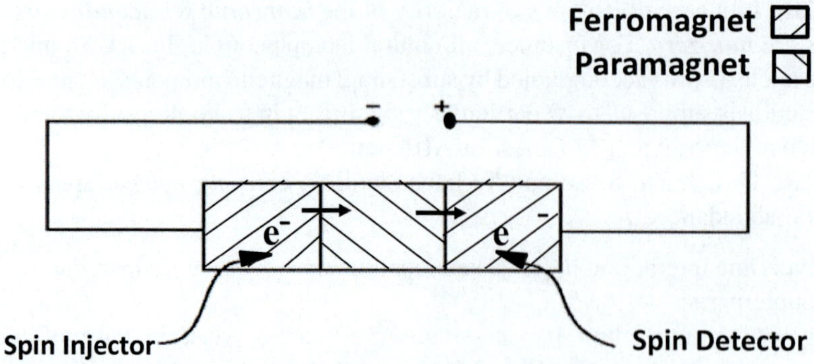

Fig. 2.18 Schematic depiction of a spin valve

2.10.1 What Is Spin Valve Device?

In its simplest form, a spin valve is a trilayered device, in which two ferromagnetic (FM) electrode layers of different coercivities are deposited sandwiching a paramagnetic material in between them Fig. 2.18. It should be noted that those ferromagnetic electrodes generally are not magnetically coupled with each other. Consequently, application of a global magnetic field can tune their magnetizations independently. Now, under the application of electrical bias, one of these ferromagnetic electrodes that injects spins from its *quasi*-Fermi level into that of paramagnet, as discussed before, acts as ***spin injector***.

The second ferromagnet, posing unequal spin-up and spin-down density of states (DOS) at the Fermi level, preferentially transmits spins of one particular orientation. Therefore, this ferromagnetic layer is called ***spin detector***.

2.10.2 Operation of Spin Valve

Depending on the construction and exact nature of the paramagnetic and ferromagnetic layers of the spin valves, their spin transport mechanism is quite different. Several models are used to understand the physics involved in the observed spin valve effect in this kind of three-layer system. We will present those models in relevant cases in the subsequent chapters. Here, we will briefly present the operation of spin valve.

Figure 2.19 exhibits the typical spin valve magnetoresistance trace of a spin valve. According to general measurement protocol, a very high magnetic field is applied on the structure which magnetizes both ferromagnetic electrodes in the direction of the applied field. This in turn brings the magnetizations of the two ferromagnetic electrodes parallel to each other. Generally, such configuration corresponds to very low device resistance, given the fact that the spin polarizations of the two ferromagnetic

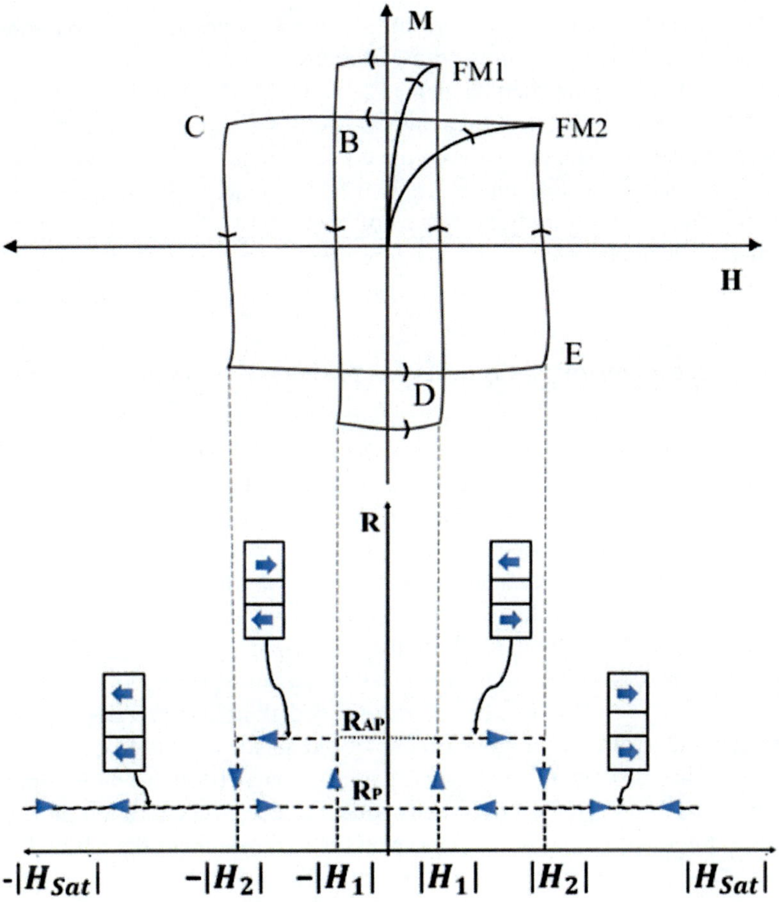

Fig. 2.19 Graphical representation of a spin valve response

electrodes of the device have the same sign (e.g., cobalt and nickel). Subsequently, the magnetic field is decreased and after sweeping past zero, its direction is made reversed, which is then represented by its negative sign. At the onset, we would like to mention that spin valve magnetoresistance versus magnetic field curve follow the magnetization hysteresis loops, i.e., magnetization versus magnetic field curves of the two ferromagnetic electrodes as a whole, as shown in Fig. 2.19. During scanning of the magnetic field when it attains the coercive field value, H_{c1} of one of the ferromagnetic electrodes, then that particular ferromagnetic electrode flips its magnetization, whereas the other ferromagnetic electrode having higher coercivity H_{c2} still retains its original direction of magnetization. As a result, the magnetizations of those two ferromagnetic electrodes become antiparallel and the corresponding device resistance becomes very high. By further increasing the magnetic field in the reverse direction, at some point the coercive field of the second ferromagnetic electrode, i.e., H_{c2} is reached and the magnetization of that ferromagnetic electrode flips too. Hence, the magnetizations of those two ferromagnetic electrodes are again parallel and the corresponding device resistance drops. Therefore, a resistance peak is supposed to be observed between the coercive fields H_{c1} and H_{c2}, where the height of this peak is $\Delta R = R_{AP} - R_P$. Likewise, if the magnetic field is scanned in the reverse direction, the same peak will reappear between the coercive fields. The background resistance is R_P. It is understandable that if the two ferromagnetic electrodes have opposite signs of spin polarization, then instead of a spin valve 'peak', spin valve 'trough' would have been observed.

2.10.3 Description of Spin Valve Device Experiments

In this experiment, resistance of the device is measured with the variation of the applied magnetic field (H). Figure 2.19 gives the pictorial demonstration of the spin valve response. Let us assume the effect associated with a regular spin valve in which a low resistance state is associated with parallel configuration, whereas a high resistance state is associated with that of antiparallel configuration. Let us also suppose that the coercive fields of those two ferromagnetic electrodes are given by $|H_1|$ and $|H_2|$ with $|H_1| < |H_2|$. Initially, a strong magnetic field (H), such that H ($= H_{sat}$) $\gg |H_2|$, is applied to the device. Consequently, as shown in Fig. 2.19, both the ferromagnetic electrodes become magnetized along the direction of this applied magnetic field. Corresponding resistance of the device is measured and denoted by R_P, where the subscript (P) indicates 'parallel' magnetization configuration. After that, the applied magnetic field value is decreased, swept through zero and then reversed. Interestingly, at the very moment when the value of the applied field, i.e., H exceeds $|H_1|$ in the reverse direction (i.e., $-|H_2| < \mathrm{H} < -|H_1|$), the ferromagnetic electrode having lower coercivity (i.e., $|H_1|$) flips its magnetization (Fig. 2.19). As a result, the magnetizations of those two ferromagnetic electrodes become antiparallel to each other. Corresponding resistance of the device is again measured at this step and denoted by R_{AP}, where the subscript (AP) indicates 'antiparallel' magnetization configuration. In case of a regular spin valve, as described above, the jump of the device resistance in increasing direction at $H = -|H_1|$ implies $R_{AP} > R_P$.

On further increasing the magnetic field in the same reverse direction, at some point of time it reaches the value of the coercive field of the second ferromagnetic electrode, i.e., $\mathrm{H} = -|H_2|$. Quite expectedly, at this point the second ferromagnetic electrode also flips its magnetization direction. As a result, magnetizations of those two ferromagnetic electrodes once again become parallel. Thus, the corresponding device resistance decreases again to R_P at $\mathrm{H} = -|H_2|$. Therefore, as shown in Fig. 2.19, during a single scan of magnetic field from H_{sat} to $-H_{sat}$ (blue line), a rectangular resistance peak appears in between the coercive fields of those two ferromagnetic electrodes (i.e., between $-|H_1|$ and $-|H_2|$). Likewise, with the variation of the magnetic field from $-H_{sat}$ to H_{sat}, as shown by red line in Fig. 2.19, an identical rectangular peak in resistance is observed between $|H_1|$ and $|H_2|$. We estimate the relative variation in device resistance, as the magnetization configurations of two ferromagnetic electrodes switch from parallel to antiparallel orientations, by the simple mathematical expression as $\Delta R/R = (R_{AP} - R_P)/R_P$. This is termed as the so-called spin valve peak.

2.10.4 *Physical Description of Spin Valve Effect*

In an attempt to make the discussion simple, let us assume that the detector ferromagnetic electrode allows the complete transmission of those spins, which are parallel to its own majority spins or the magnetization. Similarly, it *totally* blocks the spins, which are parallel to the minority spins of the ferromagnetic detector. The corresponding transmission probability (T) is then proportional to $\cos^2 \theta/2$, where the incident electron spin arriving at the detector interface makes an angle θ with the magnetization of the detector ferromagnet (Bandyopadhyay and Cahay 2008). This essentially implies that the conditions

(a) the magnetization (M) of the ferromagnetic electrodes being parallel;
(b) majority spins are injected by injector ferromagnetic electrodes;
(c) no spin flipping or spin relaxation process takes place in the spacer layer or at the interfaces;

lead to the transmission coefficient to be ***unity***, since $\theta = 0$. ***Quite expectedly, this would yield a small device resistance.*** Similarly, with the condition that the magnetization (M) of the ferromagnetic electrodes being antiparallel and the conditions stated at (***b***) and (***c***) remain as before, the transmission coefficient should be obtained as zero, since $\theta = \pi$. ***Thus, such conditions should yield large device resistance***. It can be understood that in a real device, values of θ are expected to have a wide range of distribution. This is because different electrons suffer different degrees of spin relaxation before they arrive at the interface of ferromagnetic detector and paramagnetic spacer layer. Consequently, we obtain finite resistance, instead of infinite one (as expected theoretically), even for the antiparallel orientation of magnetizations of the two ferromagnetic electrodes when we take average over the electron ensemble.

2.11 Spin Extraction

Spin extraction from a ferromagnet into a paramagnet is another important phenomenon that requires both experimental and theoretical thorough study. In order to describe spin extraction phenomenon, let us consider the analysis by Pershin and Di Ventra of electrical spin extraction from a ferromagnet into a semiconducting paramagnet in contact with it. Their analysis was based on the drift-diffusion model of spin transport (Saikin 2004; Yu et al. 2012). Following their analysis, let us suppose the spin extraction at the contact geometry, consisting of a half-metallic ferromagnet, has 100% spin polarization and a paramagnetic semiconductor has *non-degenerate* electron population that could be well described by Maxwell–Boltzmann statistics. Under the application of a suitable electrical bias across the interface, electrons flow takes place from the paramagnetic semiconductor into the half-metallic ferromagnetic contact. It can be well understood that electrons incoming from the bulk of the semiconductor are *spin unpolarized* and that the half-metallic ferromagnet accepts only majority spins electrons in the half-metal (say, up-spin electrons). This implies that minority spin electrons (here down-spin electrons) cannot pass through the ferromagnet, rather scatter into it. Therefore, a cloud of down-spin electrons must accumulate at the interface between the half-metallic ferromagnet and the paramagnetic semiconductor. This in turn results in the formation of a local spin-dipole configuration close to the interface. Now, at a critical current magnitude, the semiconductor region near the interface becomes completely depleted of a particular species of electrons, which are actually majority spins electrons in the half-metallic contact. Such phenomenon has been referred to as '***spin blockade***' by Pershin and Di Ventra. Furthermore, *spin flip scattering events* (discussed in detail in Chap. 3) at the interface are neglected in this case, and the interface is modelled as a planar junction. The one-dimensional drift-diffusion equations are used to study spin extraction from the paramagnetic semiconductor.

2.12 Field- and Heat-Driven Spintronics Effect

Spintronic devices depend on a systematic generation, transportation and detection of spin currents. These functionalities can be executed by different spintronic effects. Spin–orbit interaction has two significant consequences: (i) it can exert torques and thereby induce a precession of the electron spin and (ii) it can regulate the orbital motion of electrons leading to the transport of spin angular momentum. The latter includes the spin-dependent Hall effect and the spin-dependent Seebeck effect. Spintronic phenomena may be field-driven or heat-driven. Field-driven effects are spin-dependent Hall effect (e.g., the anomalous Hall effect, the spin Hall effect and inverse Hall effect), magnetoresistance effects (such as the giant, tunnelling and spin Hall magnetoresistance), spin pumping and its inverse. Significantly, their realization is not to be limited to DC driving currents only. It can also be extended to free-space terahertz electromagnetic radiation-driven ultrafast charge current.

Let us discuss two phenomena: (***i***) field-driven Hall effect and (***ii***) heat-driven Seebeck effect in detail.

2.12.1 *Field-Driven Hall Effect*

Background

As a prelude, we may mention the history of the ***Hall effect*** in 1879 after the discovery of a small transverse voltage appeared across a current carrying thin metal strip in the presence of an applied magnetic field by Edwin H. Hall. Significantly, the Hall effect made it feasible for a direct measurement of the carrier density. Moreover, it enables a relatively simple measurement of electrical resistivity and the mobility of carriers in semiconductor. Owing to simple measurement technique, cheap and fast reversal time, the Hall effect becomes an inevitable technique in the semiconductor industry.

The spin Hall effect (SHE) originates from the coupling of the charge and spin currents mediated by the spin–orbit coupling. In an original work, Russian physicists, Mikhail I. Dyakonov and Vladimir I. Perel, have predicted this effect in 1971. In order to predict the extrinsic SHE, they referred to the phenomena of Mott scattering (1929) and of the anomalous Hall effect (AHE) (1881). The main points, identified by them, are as follows:

1. Due to Mott scattering, spin-dependent asymmetric deflection of electron beams has been observed in vacuum.
2. Mott's skew scattering has been considered as one of the origins of the AHE, experienced by conduction electrons in ferromagnets.

They have proposed a technique to measure inverse spin Hall effect (ISHE) under optical spin orientations in semiconductors. The name 'spin Hall effect' has been first suggested by Hirsch in 1999.

What is Hall effect?

Hall effect is a very familiar effect, proposed by Edwin Hall. This is basically related to creation of potential difference across an electrical conductor under the application of a magnetic field. The magnetic field should be applied in a direction perpendicular to the flow of electrical current. In this case, accumulation of opposite charges take place due to the action of Lorentz force, mediated by the magnetic field, at the sample boundary. Hall effect plays a very significant role in physics. It provides us a means for effective measurement of the carrier density, magnetic field and the type of semiconductor.

What is spin Hall effect (SHE)?

An intriguing phenomenon of spin accumulation to opposite edges of a given sample, which may be magnetic material or not, is called **spin Hall effect (SHE)**. Such effect results in an entirely non-magnetic material to become magnetic, when an electrical current flows through it. However, this effect also plays an important role in a ferromagnetic material as well. Therefore, it comes out that SHE is the feature of both non-magnetic and magnetic metal. SHE does not accompany with Hall voltage. Instead, opposite signs of spin polarization may appear at the edges of the sample, even if no magnetic field is applied externally. Such feature is generic to both extrinsic and intrinsic SHE. At the onset, we may mention that the concept of SHE has been adopted from the extrinsic anomalous Hall effect (AHE). Noteworthy, AHE appears due to-spin dependent scattering process. Accordingly, we understand that in case of SHE, the up-and down-spin electrons are scattered by spin–orbit scattering itself or

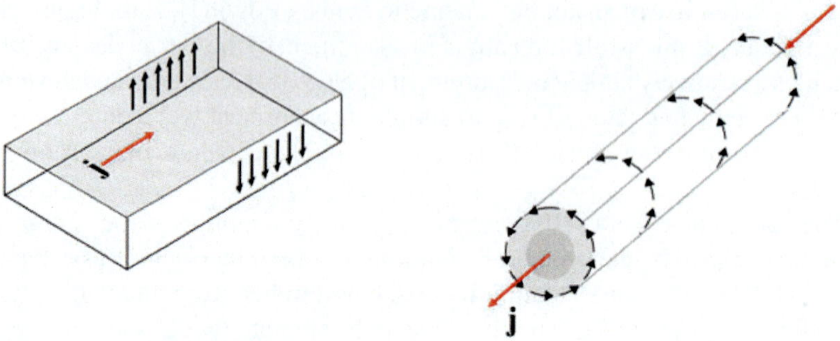

Fig. 2.20 Schematic diagram of spin Hall effect

by other scattering processes in the opposite direction. This gives rise to spin accumulation at the edges of the sample. Consequently, there is *spin-up* and *spin-down charge Hall current* in directions perpendicular to that of applied external electric field, as shown in Fig. 2.20. It can be understood that for non-magnetic metals, these two charge currents cancel out each other and as a result, no Hall voltage develops. However, spin-dependent scattering, arising out of non-vanishing spin–orbit interaction, indeed yields up- and down-spin currents, which flow in opposite directions. Now, the intrinsic imbalance in the population of *spin-up* and *spin-down* electrons in magnetic metal causes two branches of *spin-resolved* charge Hall current **asymmetric** and produces **Hall voltage**. This in turn results in a solely *spin-dependent Hall effect*, which is called ***spin Hall effect (SHE)***. A schematic diagram is given in Fig. 2.20.

Comparative discussion between Hall effect and spin Hall effect

It can be understood that the spin Hall effect is indeed somewhat similar to the normal Hall effect. However, there exist significant differences between them. Let us discuss the differences in the following sequence:

First, in case of SHE, for spin accumulation to occur no external magnetic field is required. On the contrary, in this case the application of a magnetic field, perpendicular to the current direction, might destroy the spin polarization.

Second, the value of the spin polarization across the boundary is limited by spin relaxation process. Spin polarization is supposed to exist in relatively thin layers of the boundary, which is determined by the spin diffusion length.

In case of normal Hall effect, the charge imbalance results in an offset between the Fermi levels of both sides of the sample, hence a voltage V_H appeared. Such voltage can be measured with a voltmeter. Similarly, in the case of SHE, Fermi levels corresponding to up- and down-spin electrons would be different on both sides of the sample. Interestingly, such difference should have opposite sign for both spins. However, the difficulty lies in the measurement and detection of this spin voltage V_{SH}, which also equivalently reflects the associated spin imbalance in the materials. In this direction, one possible way would be the measurement of magnetization difference between both edges of the slab. Perhaps, superconducting quantum interference device microscope, with high spatial resolution, might enable us for this kind of measurement by measuring local magnetic fields.

Physics of spin Hall effect (SHE)

The phenomenon ***spin Hall effect*** is related to the spin accumulation at opposite edges of a given slab. This seems to imply the generation of a current flowing perpendicular to electrical charge current. This effect may occur in both magnetic and non-magnetic materials. It is well known that when electrical current flows through a non-magnetic material it becomes magnetic. Some semi-classically defined mechanisms, such as intrinsic, skew scattering and side-jump scattering causes the generation of spin-polarized current, which is the origin of SHE. In order to understand these mechanisms, we have to discuss the types of perpendicular current, which are generated by SHE and *inverse* SHE (ISHE). Two types of perpendicular currents are as follows:

1. Band current (intrinsic current)
2. Scattering current (extrinsic current).

1. ***Band current***: This type of current occurs when the number of electrons moving in one direction is different from the number of electrons moving in the opposite direction. This happens when the probability of an electron scattering in one direction is different from the scattering probability in opposite direction. This is caused by the external electric field. For instance, under the application of electric field in x-direction, the scattering probability of electrons to a state ($k_x + \Delta k_x, k_y, k_z$) is larger than to a state ($k_x - \Delta k_x, k_y, k_z$). Such direction-dependent scattering probability is due to the conduction electrons gaining/losing energy when they are accelerating/decelerating in the electric field. Ordinary Hall effect and the current along electric field are the examples of band current.

2. Scattering current: In case of a conduction electron, after it suffers a scattering, in addition to the movement direction, its spatial position is also getting changed as well. Scattering current occurs when shift in position of the electron after it encounters scattering is different for two opposite directions. Consequently, the electron is found to be constantly shifted in one direction after consecutive scatterings. This is actually the case when because of the presence of an external field the scattering probability of an electron in one position is different from the scattering probability in opposite position. As an example, we may mention that under the application of an electric field in x-direction, scattering probability to a state ($x + \Delta x, y, z$) is larger than to a state ($x - \Delta x, y, z$). Consequently, the electrons are supposed to be continuously drifted along x-direction. The reason for spacial dependence of the scattering probability is that the energy of quantum states gradually becomes large along the electric field. Therefore, scattering of an electron to a lower energy state has a lower probability. Tunnelling current, hoping current and current in a low conductivity metal are the examples of such scattering current.

Coming back to the discussion of SHE, it essentially gives the description of the creation of spin-polarized current. As discussed in last paragraph, there are only two types of electron current in a conductor. Therefore, diversified physical mechanisms can yield only two kinds of spin-polarized currents. These mechanisms are:

1. **Orbital moment of a conduction electron**: This makes the magnetic field of spin orbit interaction to be dependent on the direction of electron movement.
2. **Skew scattering on defects**: This mechanism also makes the magnetic field of spin–orbit interaction to be dependent on the direction of electron movement.
3. **Side-jump scattering on defects**: This mechanism makes the magnetic field of spin–orbit interaction to be dependent on electron spatial position.
4. **Side-jump scattering across an interface**: This mechanism makes the magnetic field of spin–orbit interaction to be dependent on electron spatial position with respect to the interface.

Among these four mechanisms, the first one describes the creation of intrinsic spin Hall effect and the other three describe the extrinsic spin Hall effect.

Inverse spin Hall effect

Upon application of an electric field to a system consisting of non-magnetic metal or doped semiconductor with no external magnetic field, the electrons are directed towards the electric field. If the said system encounters spin–orbit coupling (SOC), the electrons can experience spin-dependent motion as shown in Fig. 2.21a. The up-spins and down-spins are deflected in the opposite direction, perpendicular to the electric field. As a result, there will be a transverse spin current in response to the electric field. If the system does not have SOC, there cannot be spin-dependent motions of electrons, and hence no spin Hall effect (SHE) will be observed. The reciprocal effect to the SHE is the inverse spin Hall effect (ISHE) and is shown in Fig. 2.21b. In this case, when the spin current is injected into the system, an electric field is induced. In the injected spin current, the up-spins and down-spins move in the opposite directions. The spin–orbit coupling causes deflections of the electrons, and the deflections have the same direction for up- and down-spins, causing a transverse charge current. The SHE and ISHE enable us to electrically manipulate or detect spin currents. Charge current can be converted to spin current through spin Hall effect, whereas spin current can be converted to charge current through inverse spin Hall effect.

2.12.2 *Heat-Driven Seebeck Effect*

Interestingly, temperature gradients can also lead to some encouraging effects, as observed in spin caloritronics (Seifert 2017; http://magnetism). Such heat-driven phenomenon, also observed in insulators, permits the transport of spin. Two types of spin caloritronic effects are important in spintronics. One is spin-dependent Seebeck effect (SDSE) which is mediated by conduction electron. Another one is spin Seebeck effect (SSE) which is mediated by magnon. These effects can be increased by breaking the inversion symmetry of magnetic heterostructures. As far as practical

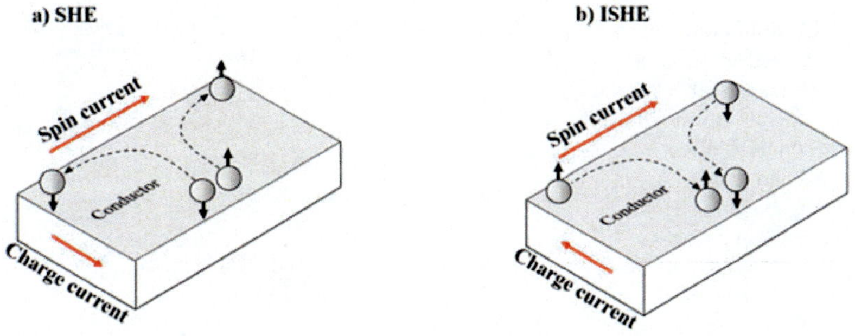

Fig. 2.21 Schematic representation of **a** SHE and **b** ISHE

implementation is concerned, it is done by ultrashort laser pulse which brings the system in highly non-equilibrium state in femtosecond time scale.

What is Seebeck effect?

In the Seebeck effect, two conductors (metals) with dissimilar transport properties are brought into electrical contact and heated to produce temperature difference ∇T. This temperature gradient drives a charge current in the conductors (as shown in Fig. 2.22a). If K_1 and K_2 are Seebeck coefficients of two different conductors then the charge current flows proportional to $\nabla T \cdot (K_1 - K_2)$. Seebeck effect thus converts temperature difference into an electrical voltage. The effect was first discovered by T. J. Seebeck in 1820s. It has already contributed a lot in the field of electronics. Major uses have been observed in infrared devices, thermoelectric generators etc.

In the field of spintronics, the spin version of the Seebeck effect, the spin Seebeck effect (SSE), has attracted much attention.

Spin-dependent Seebeck effect

In ferromagnetic materials, Seebeck current is spin dependent. The spin-dependent Seebeck effect (SDSE) can be acknowledged as the integration/combination of the two conductors exhibiting Seebeck effect within a single magnetic material. Conductors having net magnetic moment exhibit SSDE. Up-spin and down-spin electrons possess different transport properties leading to a spin-polarized current along the

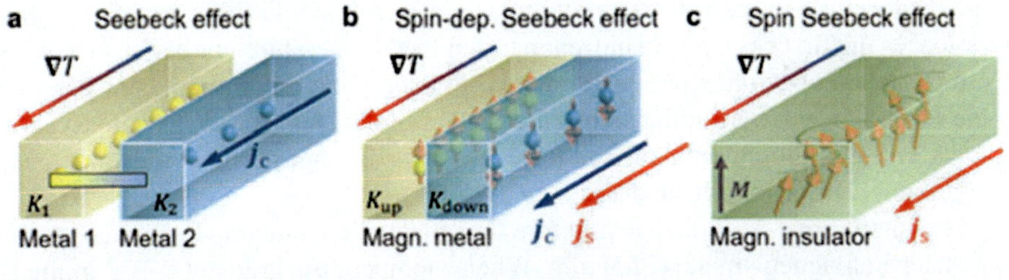

Fig. 2.22 Seebeck effects. In **a** ordinary metal, **b** magnetic metal **c** magnetic insulator. For clarity, the two spin channels are laterally displaced (Figure adapted from PhD dissertation of Seifert 2017)

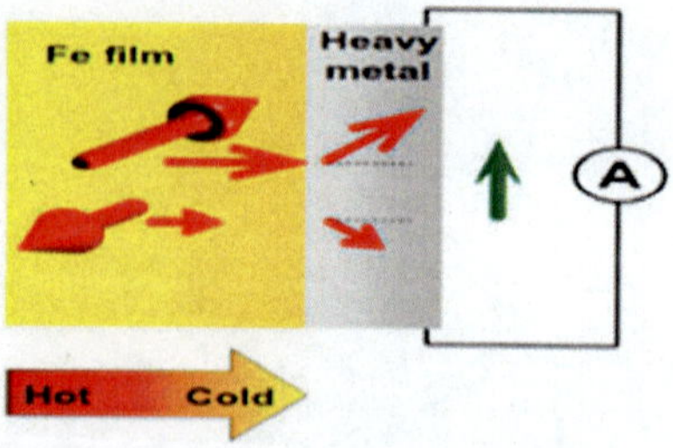

Fig. 2.23 Spin-polarized current is created through spin-dependent Seebeck effect, which is measured by inverse spin Hall effect (Adapted and redrawn from http://magnetism.eu/esm/2018/slides/kampfrath-slides.pdf; http://magnetism)

thermal gradient (as shown in Fig. 2.22b). If K_{up} and K_{down} are Seebeck coefficients of up-spin and down-spin electrons, respectively, then the spin-polarized current flows proportional to $\nabla T \cdot (K_{up} - K_{down})$. Spin-polarized current can be measured by ISHE, as shown in Fig. 2.23.

Spin Seebeck effect

The spin Seebeck effect (SSE) has been first observed in magnetic insulators. A temperature gradient imparts a pure spin current (j_s) carried by collective spin waves, known as magnons.

In static SSE, spin current density can be expressed as

$$j_s = K \cdot (T_N - T_F), \tag{2.55}$$

where K is the spin Seebeck coefficient and $(T_N - T_F)$ is the temperature difference between the metal electrons (T_N) and the magnetic insulator (T_F) magnons (see Fig. 2.22c). Detection of pure spin current is very difficult. The SSE is generally measured with the aid of an extra heavy-metal layer where the spin current is converted into a charge current through the ISHE.

2.13 Spin Current Measurement Mechanism

In present days, the SHE and its inverse (ISHE) are well-exploited tool for the measurement of spin currents. The observations of the SSE utilize the ISHE in two different device structures: (i) longitudinal configuration and (ii) transverse configuration. In the first case, a spin current parallel to a temperature gradient is measured, while a spin current flowing perpendicular to a temperature gradient is measured in the second case. Longitudinal configuration is simpler and used as insulator whereas transverse configuration is utilized for wide range of magnetic materials like metals and semiconductors to insulators.

In longitudinal SSE device structure ferromagnetic insulator (F) film is attached with a paramagnetic metal (PM) film. When a temperature gradient ∇T is applied to the F layer perpendicular to the PM/F interface, a spin voltage is thermally generated via magnetization (M) dynamics, which pumps a spin current J_s into an attached paramagnetic metal. In the paramagnetic metal, this spin current ($\mathbf{J_s}$) is converted into an electric field E_{ISHE} due to the inverse spin Hall effect (ISHE) and is given by

$$E_{ISHE} = [(\theta_{SH}\rho) \times (2e)\mathbf{J_s} \times \boldsymbol{\sigma}]/A\hbar, \tag{2.56}$$

where θ_{SH}, ρ and A are the spin Hall angle of PM, the electric resistivity of PM and the contact area between F and PM, respectively. By measuring E_{ISHE} in the PM film, the longitudinal SSE can be detected electrically.

2.14 Conclusions

In this chapter, we discussed some important notions, such as spin polarization, spin filter effect, spin injection, spin accumulation, spin relaxation, spin extraction etc. We have also discussed *passive* spintronic devices, such as spin valves in this chapter. Currently available commercial spintronic products, such as magnetic read heads for reading data in computer hard disks or entertainment systems such as Apple iPods, and magnetic random access memory utilize these passive devices. Hence, sufficient knowledge is essential for engineers to understand these devices. In this chapter, we have also presented different kinds of spin relaxation, in both time and space, of conduction electrons in metals and semiconductors. In this direction, four primary spin relaxation mechanisms such as the D'yakonov-Perel', the Elliott–Yafet, the Bir-Aronov-Pikus and hyperfine interactions with nuclear spins have been discussed. Since spin relaxation mechanism poses limitations to the performance of most of the spintronics devices, it is an important issue for thorough discussion. Ultimately, the objective of all device engineers and physicists is to minimize the spin relaxation rate in spintronics devices so that they become more robust and useful. We have also discussed *field and heat-driven spintronics effect*, i.e., spin Hall effect, Seebeck effect and spin current measurement mechanism. These effects are important in the context of terahertz spintronics devices and have been discussed in Chap. 8. Our objective is to introdduce important concepts about spintronics to the reader to make them well conversant with the topic.

2.15 Exercises

1. Define spin polarization.
2. What is spin filter effect in a ferromagnet? State the physical interpretation of *spin asymmetry 'A'*.
3. Define spin detection efficiency.
4. Why in case of ferromagnet the electronic transport is spin-dependent? Hence define spin filter effect in ferromagnet.
5. Define spin injection process.

6. Discuss spin injection process in case of ferromagnetic metal–paramagnetic semiconductor interface for both (i) Ohmic contact and (ii) tunnel/Schottky contact. Hence conclude which one is efficient for spin injection.
7. Derive the spin–orbit interaction Hamiltonian.
8. What is the role of spin–orbit interaction in the field of spintronics?
9. Does the electron in a rotating orbit around the nucleus accelerate constantly?
10. What is spin relaxation?
11. Define spin relaxation length.
12. Describe Elliott–Yafet spin relaxation mechanism?
13. What is the prerequisite condition for Elliott–Yafet mechanism?
14. Where does the Elliott–Yafet mechanism of spin scattering occur?
15. Describe the Rashba interaction?
16. What is the important role of Rashba interaction?
17. Describe the Dresselhaus interaction?
18. Describe briefly the basic concepts of D'yakonov-Perel' mechanism?
19. What is the prerequisite condition for D'yakonov-Perel' mechanism?
20. Considering D'yakonov-Perel' mechanism, is frequent momentum scattering actually be beneficial for spin longevity?
21. Where does D'yakonov-Perel' mechanism occur?
22. Describe briefly the basic concepts of Bir-Aronov-Pikus mechanism?
23. Where does the Bir-Aronov-Pikus mechanism of spin scattering occur?
24. Write down the Hamiltonian describing hyperfine interaction. How hyperfine interaction causes the dephasing of electron spins of a spin-polarized electron ensemble?
25. Can the magnetic field generated by a nuclear spin cause complete loss of electron spin polarization?
26. Consider two situations where an ensemble of electrons is drifting and diffusing in a solid. If the electrons velocity changes as delta function in one situation and Gaussian function with time in another situation, in which case will the spin relaxation process for D'yakonov-Perel' mechanism be faster? Explain why?
27. What is spin valve device?
28. Discuss the operation of spin valve effect with schematic diagram.
29. Describe spin valve device experiments.
30. Give a physical description of spin valve effect.
31. Define spin extraction process.
32. What is spin Hall effect?
33. What is heat-driven Seebeck effect?
34. Describe spin current measurement mechanism.
35. What is spin accumulation?
36. How will you estimate the spin accumulation length by a simple method?
37. How do we detect the injected spins into semiconductor?
38. What is called as conductance mismatch?
39. Explain how to overcome the conductance mismatch problem for efficient spin injection?

40. What are the main issues limiting the design of spintronic devices?
41. What is spin filtering?
42. How was the spin-polarized electron transport across the ferromagnetic and semiconductor interface verified?
43. What is the role of barrier layer on controlling the spin filtering?
44. What is spin Hall effect?
45. What is the origin of spin Hall effect?
46. What are the differences between the classical and spin Hall effect?
47. What are the factors determine the spin injection efficiency?
48. Why does the spin accumulate at the interface between the ferromagnetic and non-magnetic?
49. How can we control the spin polarization of the injected current?
50. How do we achieve large spin accumulation in the semiconductor?
51. Multiple Choice Questions

    (a) Density of states spin polarization is negative for Co and Ni. Which of the following statement is correct?
        (i) Minority spin contribution is dominant at Fermi level.
        (ii) Majority spin contribution is dominant at Fermi level.
        (iii) Majority and minority spin contribution is equal at Fermi level.

    (b) Current density carried by the majority and minority spin electrons in a ferromagnet is given by 70% and 30% of the total current density, respectively. Spin injection efficiency is given by
        (i) 0.6
        (ii) 0.4
        (iii) 0.7

    (c) Let us consider an Ohmic contact between a ferromagnetic (Co) and a paramagnetic (Cu) metal. The resistivity of Co and Cu is 10 and 1 $\mu\Omega$ cm, respectively. What will be the value of the ratio between spin injection efficiency and conductivity polarization of the Co?
        (i) 1.1
        (ii) 0.1
        (iii) 0.9

    (d) In case of a tunnelling barrier between ferromagnetic metal and paramagnetic semiconductor, the spin injection efficiency is 0.70. What is the value of the polarization of the tunnelling conductivity at the interface?
        (i) 0.3
        (ii) 0.5
        (iii) 0.7

    (e) In case of a Schottky barrier between a ferromagnetic metal and paramagnetic semiconductor interface, we obtain

(i) efficient spin injection and charge conduction into and out of the semiconductor.
(ii) poor spin injection but charge conduction into and out of a semiconductor.
(iii) efficient spin injection but charge conduction for one bias polarity, not the other.

(f) In case of Ni, spin polarization is 11%. The number of spin-up electrons is equal to
(i) 2.2 times the number of spin-down electrons.
(ii) 1.2 times the number of spin-down electrons.
(iii) 3.2 times the number of spin-down electrons.

(g) Let us consider an ensemble of electrons drifting and diffusing in a solid and the electrons velocity changes as *delta* function. Which of the following statement is correct?
(i) spin relaxation rate is high for D'yakonov-Perel' mechanism.
(ii) spin relaxation rate is low for D'yakonov-Perel' mechanism.
(iii) spin relaxation rate is high for Elliott–Yafet mechanism.

(h) In case of 3d transition metals,
(i) Fermi level intersects only 4s band.
(ii) Fermi level intersects only 3d band.
(iii) Fermi level intersects both 3d and 4s bands.

(i) Spin filtering effect in ferromagnet arises
(i) minority spin electrons have larger absorption coefficient than majority spin electrons in spin-dependent energy band.
(ii) minority spin electrons have smaller absorption coefficient than majority spin electrons in spin-dependent energy band.
(iii) minority and majority spin electrons have same absorption coefficient in spin-dependent energy band.

References

I. Appelbaum, B. Huang, D.J. Monsma, *Electronic measurement and control of spin transport in silicon*. Nature **447**, 295 (2007)

S. Bandyopadhyay, M. Cahay, *Introduction to Spintronics* (CRC Press Taylor & Francis 2008); *Electron spin for classical information processing: a brief survey of spin-based logic devices, gates and circuits*. Nanotechnology **20**, 412001 (2009)

Y. Bychkov, E.I. Rashba, Oscillatory effects and the magnetic susceptibility of carriers in inversion layers. J. Phys. C **17**, 6039 (1984)

G. Dresselhaus, Spin orbit coupling effects in zinc blende structures. Phys. Rev. **100**, 580 (1955)

A. Hirohataa, K. Yamadab et al., Review on spintronics: principles and device applications. J. Magn. Magn. Mater. **509**, 166711 (2020)

F.J. Jedema, A.T. Filip, B.J. van Wees, Electrical spin injection and accumulation at room temperature in an all-metal mesoscopic spin valve. Nature **410**, 345–348 (2001)

F.J. Jedema, H.B. Heersche, A.T. Filip, J.J.A. Baselmans, B.J. van Wees, Electrical detection of spin precession in a metallic mesoscopic spin valve. Nature **416**, 713–716 (2002)

M.W. Wu, J.H. Jiang, M.Q. Weng, Spin dynamics in semiconductors. Phys. Rep. **493**, 61 (2010)

T. Kampfrath, Probing and controlling spin dynamics with THz pulses. http://magnetism.eu/esm/2018/slides/kampfrath-slides.pdf

Y. Ohno et al., Electrical spin injection in a ferromagnetic semiconductor heterostructure. Nature **402**, 790 (1999)

S. Pramanik, C.G. Stefanita, S. Bandyopadhyay, Spin transport in self assembled all-metal nanowire spin valves: a study of the pure Elliott–Yafet mechanism. J. Nanosci. Nanotechnol. **6**, 1973–1978 (2006)

E.I. Rashba, Theory of electrical spin injection: tunnel contacts as a solution of the conductivity mismatch problem. Phys. Rev. B **62**, R16267 (2000)

S. Saikin, A drift diffusion model for spin polarized transport in a two-dimensional non-degenerate electron gas controlled by spin-orbit interaction. J. Phys: Condens. Matt. **16**, 5–71 (2004)

G. Schmidt, D. Ferrand, L.W. Molenkamp, A.T. Filip, B.J. van Wees, Fundamental obstacle for electrical spin injection from ferromagnetic metal into a diffusive semiconductor. Phys. Rev. B **62**, R4790 (2000)

Y. Suzuki, A.A. Tulapurkar, C. Chappert, Spin-injection phenomena and applications. Nanomagnet. Spintron. 978-0-444-53114-8 (2009)

H.X. Tang et al., *Spin Injection and Transport in Micro- and Nano-scale Devices* in *Semiconductor Spintronics and Quantum Computation*, ed. by D.D. Awschalom, N. Samarth, D. Loss (Springer, Berlin, 2002)

T.S. Seifert, Spintronics with Terahertz Radiation: Probing and driving spins at highest frequencies. PhD dissertation, Veritus Iustitia, Freie Universitat Berlin, 2017

Z.G. Yu, M.E. Flatt´e, Electric field dependent spin diffusion and spin injection into semiconductors. Phys. Rev. B **66**, 201202 (2002)

I. Zutic, J. Fabian, S. Das Sarma, Spintronics: fundamentals and applications. Rev. Modern Phys. **76**, 323–410 (2004)

Chapter 3
Giant Magnetoresistance (GMR)

3.1 Introduction to Magnetoresistance

Magnetoresistance (MR) property of a material/system refers to the change in its electrical resistance, when the magnetic field is turned on. MR $\left(\Delta\rho/_{\rho}\right)$ is usually defined by

$$\frac{\Delta\rho}{\rho} = \frac{R(H) - R(0)}{R(0)} \tag{3.1}$$

The effect was first discovered by William Thomson, more commonly known as Lord Kelvin, in 1856. He demonstrated a change in the electrical resistance of ferromagnetic materials, e.g., iron sample by applying an external magnetic field. He showed that the resistance increases by 0.2% when the current and the applied magnetic field are in the same direction, while it decreases by 0.4% when they are perpendicular to each other. In this case, MR occurs because the drift velocity (v) of all the charge carriers is not supposed to be identical. Now, application of magnetic field $\left(\vec{H}\right)$ to a crystal of thickness t gives rise to Hall voltage given by, $V = E_y t = \left|\vec{v} \times \vec{H}\right|$, where $\vec{v}$ denotes the drift velocity and $\vec{E}$ is the applied transport driving electric field. The details of Hall effect and the voltage created by this effect are not within the scope of this book. Such Hall voltage compensates exactly the Lorentz force for carriers with average velocity. This means that charge carriers, which are comparatively slow moving, will be compensated to a large extent and can be considered as overcompensated. On the other hand, charge carriers that are pretty fast-moving carriers will not be compensated to a reasonable extent, i.e., they can be considered as undercompensated. This in turn results in trajectories and those are not lying along the applied magnetic field. Consequently, there is a rise in resistivity caused by an effective fall in the mean free path. This leads to *positive MR*. Thus, in general, a *positive MR* is expected which is referred to as ordinary MR. Interestingly,

P. Dey and J. N. Roy, *Spintronics*,
https://doi.org/10.1007/978-981-16-0069-2_3

a free electron gas exhibits no MR. MR can be described by models, which involve multiple carriers and it is the topology of the Fermi surface that decides its high field behaviour.

In fact, the effect of MR could be realized in diversified ways. We may mention some varieties of MR take place in bulk of non-magnetic metals and semiconductors. For instance, in case of metals MR phenomena, which happen to take place, are geometrical magnetoresistance, Shubnikov de Haas oscillations, or as already mentioned common positive MR. Noteworthy, in case of magnetic metals, some other kinds of MR effects are observable, such as the very general negative magnetoresistance in ferromagnetic material and anisotropic magnetoresistance (AMR). Furthermore, in the last few decades (starting from 1970s) in multicomponent heterostructure or multilayer systems, giant magnetoresistance (GMR), tunnel magnetoresistance (TMR), colossal magnetoresistance (CMR) and extraordinary magnetoresistance (EMR) have been discovered. Thus MR can be mainly categorized into following varieties: (i) ordinary MR; (ii) anisotropic MR (AMR); (iii) giant MR (GMR); (iv) tunnel MR (TMR).

Magnetoresistance types	MR (%)
Ordinary magnetoresistance	~0.1
Anisotropic magnetoresistance (AMR)	~1–2
Giant magnetoresistance (GMR)	~10–50
Tunnel magnetoresistance (TMR)	*~100–600

It is quite evident from the above table that the magnitude of the MR effect is quite low, only about 1% at room temperature for ordinary MR and AMR, but goes to about 10% at room temperature and even 50% at low temperatures in giant magnetoresistive multilayer structures. For TMR structure or in some perovskite systems, MR effects of more than 95% have been observed. Thus besides direct effects of the magnetic field such as Lorentz magnetoresistance and the Hall effect, i.e., ordinary MR, of particular interest are the effects causing AMR, GMR, TMR and giant magnetoimpedance (GMI). Such MR or related phenomena arise from the interaction of conduction electrons with that of magnetization. In this context, we must mention that AMR arises in materials, whereas GMR and TMR occur at the nanostructured combination of magnetic and non-magnetic systems and the observation of GMI is subject to wires and tubes.

From application point of view, MR is a quite useful quantity because MR sensors are having wide technological implementation. For instance, MR sensors are employed to sense magnetic field at the magnetic strip on a credit card. Furthermore, MR may be used as a probe of material parameters. It may also reveal the physics of interaction of conduction electrons with magnetization and hence MR effect is a key in spintronic investigations. It may also play an instrumental role to elucidate the effect of current on magnetization by performing the spin-torque experiments.

3.2 Different Kinds of MR

As a prelude, let us first discuss briefly about different kinds of MR.

3.2.1 *Ordinary Magnetoresistance*

For a normal metal, when there is a flow of current, motion of electrons at various parts of the Fermi surface takes place in such a way so that the scattering becomes minimum. In such case, the electrons should follow the path that offers minimum scattering while traversing the sample. Now, the externally applied magnetic field forces the electrons to follow a *different* path instead of the least scattering one. As a result, electrons will suffer more scattering, which means that the change in resistivity is positive, hence leading to *positive MR* (Δρ). In both cases, when the magnetic field is parallel (where, $\Delta\rho_{\mathrm{II}}$ = corresponding positiveMR) and transverse (where, $\Delta\rho_T$ = correspondingpositiveMR) to the current direction, both $\Delta\rho_{\mathrm{II}}$ and $\Delta\rho_T$ are positive with $\rho_T > \rho_{II}$. Such ordinary MR can be classified into three distinct cases as discussed below, based on the structure of electron orbitals at Fermi surface:

1. In case of closed Fermi surfaces in metallic systems, the motion of electrons is confined to their orbit in *k*-space. In this case, the effect of magnetic field occurs as an enhancement in the cyclotron frequency of the electron in its closed orbit. Thus, in general, a *positive MR* is expected.
2. For metals with equal numbers of electrons and holes, *MR* increases with *H* up to the highest field measured and independent of crystallographic orientation. Bismuth falls in this class.
3. Metals containing Fermi surfaces having open orbits, oriented along some crystallographic directions, exhibit large MR when the magnetic fields are applied in those same directions. However, the resistance founds to saturate when the magnetic fields are applied in a direction, where those orbits are closed.

3.2.2 *Magnetoresistance of Ferromagnetic Transition Metals*

Observation of negative MR in ferromagnetic metal is quite interesting (Fert and Campbell 1976) and can be explained as follows: A very significant perception of this problem was introduced by Mott. He described the transport properties of Ni, which requires only a few electron volts to alter the configuration from ($3d^8 4s^2$) to ($3d^9 4s^1$) or ($3d^{10}$). In general, Ni is considered to be ($3d^{9.4} 4s^{0.6}$). It is well known that the *d*-band is very narrow, hence $m_d^* \gg m_e$, where m_d^*/m_e = mass of the electrons at *d*-band/free state. Since the *s*-band is nearly free, $m_s^* \sim m_e$, where m_s^* = mass of the electrons at *s*-band.

Expression of the conductivity $\boldsymbol{\sigma}$ is given by

$$\sigma = \frac{n_s e^2 \tau_s}{m_s^*} + \frac{n_d e^2 \tau_d}{m_d^*} \tag{3.2}$$

In Eq. 3.2, n_s/n_d = ratio of electrons' numbers in the s and d bands, τ_s/τ_d = scattering times of electrons in the *s*/*d* bands, where $\tau_S \sim \tau_d$. Thus, considering our discussion on the mass of electrons it comes out that the first term in Eq. 3.2 dominates σ, and s electrons mainly contribute to the conduction process. It is also understandable that it is the $s \rightarrow d$ transition that mainly contributes to the transition probability. Now at temperatures $T \ll T_C$, where T_C = para-ferromagnetic transition temperature, all the unoccupied d orbitals being antiparallel, resulting in only half of the s electrons that could possibly undergo the transitions. However, for $T > T_C$, since all the s electrons can undergo transitions, therefore, there is more scattering. Thus from Eq. 3.2, below T_C one may expect a decrease in the resistivity. Now, let us consider a magnetic field is applied to Ni. Such applied field might possibly rise the spin polarization and thereby allow less $s \rightarrow d$ transitions. This in turn leads to decrease in the resistivity; therefore, a negative MR is observed.

3.2.3 Anisotropic Magnetoresistance (AMR) of Ferromagnetic Transition Metals

The origin of AMR of a ferromagnetic transition metal lies on the fact that the electrons would suffer more scattering while travelling along the applied magnetic field than those travelling perpendicular to the applied field. This implies that in ferromagnetic materials electrical conductivity is decided by the relative alignment of the direction of the electrical current and the magnetization of the ferromagnetic materials. As is depicted in Fig. 3.1, resistivity depends on the direction of flow of current. This phenomenon, discovered by Lord Kelvin in the nineteenth century, is also known as *spontaneous magnetoresistance anisotropy* (SMA).

Physical origin of this MR phenomenon in a ferromagnetic material lies on the decrease of symmetry of the magnetized state compared to its non-magnetic state. This is due to the simultaneous presence of the magnetization and spin–orbit coupling. Resistivity of such ferromagnetic materials varies following cosine square function if the angle between the direction of magnetization and that of applied electrical bias is varied. The consequence of such MR effect is that maximum resistivity ($\rho_\parallel$) is achieved when the electric field and magnetic field are parallel to each other, whereas the resistivity ($\rho_\perp$) is minimum when they are perpendicular. As is commonly known, the ratio $(\rho_\parallel - \rho_\perp)/\rho_\perp$ is called the *MR ratio* (%). The most common AMR sensor material is permalloy ($Ni_{81}Fe_{19}$), which has very high permeability (>2000). Moreover, the AMR effect has been found to depend on the thickness of the film and its deposition method. For instance, in case of permalloy film deposited by ion beam

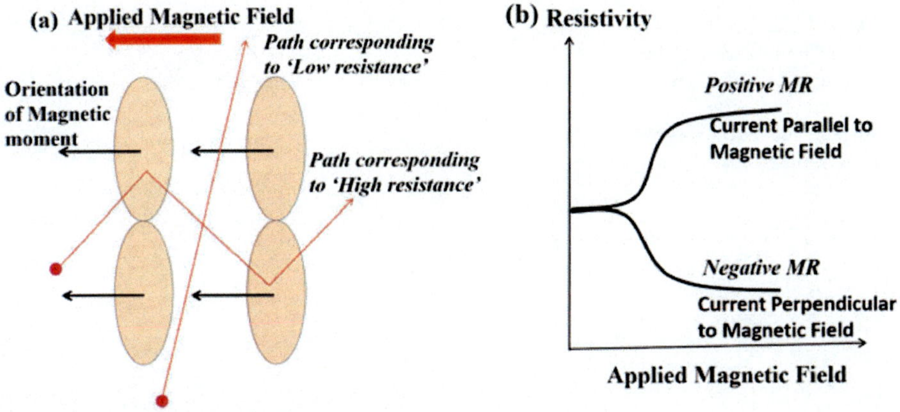

Fig. 3.1 Pictorial description of the origin of AMR: **a** Electrons are experiencing more scattering when current is flowing along the direction of applied magnetic field than those travelling perpendicular to the field. **b** Resistivity and hence MR depends on the current flow direction with respect to the applied magnetic field. Low resistivity and hence negative MR (from Eq. 3.1) is obtained when current is flowing perpendicular to the applied magnetic field. On the other hand, large resistivity and positive MR is obtained when current is flowing parallel to the magnetic field

sputtering on a tantalum seed layer, the MR ratios have been found to be 1.2, 2.0, 2.3 and 2.6% for 50, 100, 150 and 250 Å thick films, respectively. Technical applications of AMR have been realized in 1989 through the development of magnetic sensors and hard disks read heads, based on the AMR effect. Noteworthy, the signal output of MR read heads is directly proportional to the MR of AMR sensor material. The AMR effect has also been studied in nanowires and nanotubes; for instance, in Co; Co and Ni; single-crystal Ni nanowires. Similar studies have also been carried out for various tubes, such as Ni, CoFeB and permalloy. AMR has also been investigated in segmented wires, like alternating cobalt and nickel segments. Significantly, the presence of one or multiple domain walls between voltage terminals can be tracked employing AMR. Furthermore, AMR has also been proposed to be used to examine domain wall stochastic motion as well.

3.3 Introduction on Giant Magnetoresistance (GMR)

3.3.1 What Is GMR?

Similar to other magnetoresistive effect (described above) GMR is the change in electrical resistance in response to an applied magnetic field. However, the MR value in this case has been found to be much higher compared to both ordinary and anisotropic magnetoresistance. This is the reason such MR is referred to as 'giant magnetoresistance' or GMR. GMR refers to the giant decrease in electrical resistance (typically

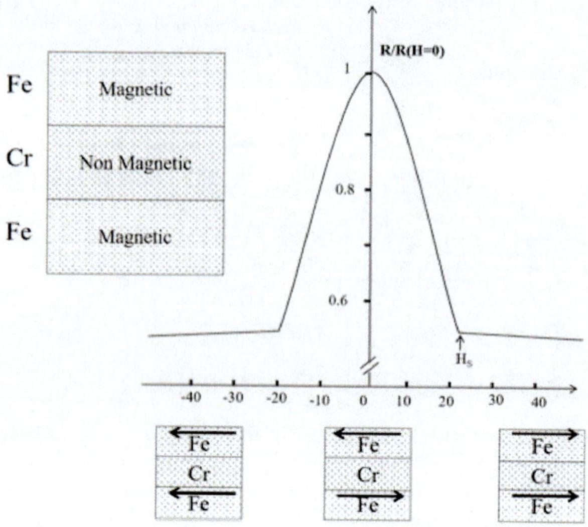

Fig. 3.2 Resistance of a Fe/Cr/Fe multilayer system as a function of the applied magnetic field (Figure adapted and redrawn from Ref. (Baibich et al. 1988))

10–80%) of a system under the application of magnetic field (Baibich et al. 1988; Grunberg et al. 1986; Tsymbal and Pettifor 2001; Hartmann 2000; Kristen Coyne 2015). This is basically a quantum mechanical MR effect observed in thin film structures composed of alternating ferromagnetic and non-magnetic layers. In general, GMR in multilayers is a result of interplay between resistance and magnetization. In this case, it is the relative directions of magnetizations of the adjacent ferromagnetic layers that decide the resistance of the multilayer structure. The phenomenon is pictorially demonstrated in Fig. 3.2.

This effect is generally formulated as

$$\frac{\Delta \boldsymbol{R}}{\boldsymbol{R}_P} = \left(\frac{\Delta \boldsymbol{R}}{\boldsymbol{R}_P}\right)_{\mathbf{max}} \frac{1 - \cos \boldsymbol{\theta}}{2} \tag{3.3}$$

where $\theta =$ angle between ferromagnetic layers magnetization.

The GMR effect was discovered in 1988 in Fe (ferromagnetic metal)/Cr (non-magnetic metal)/Fe multilayers, by two eminent scientists of Europe working independently: Peter Gruenberg from KFA Research Institute, Julich, Germany and Albert Fert from the University of Paris-Sud, France (Fig. 3.3a). An appreciably large change in resistance of the order of 6–50% was observed in artificially constructed nanostructure, consisting of alternating metallic ferromagnetic and non-magnetic very thin layers (Fig. 3.3b).

Those experiments were carried out at low temperatures and in the presence of sufficiently high magnetic fields on heterostructures having subtle fabrication and engineering details. Normally, such materials are not expected to be mass-produced. However, the large magnitude of GMR and its probable promising applications for hard disk drives invoked severe research interest among scientists around the world to investigate whether the power of the GMR effect could possibly be harnessed.

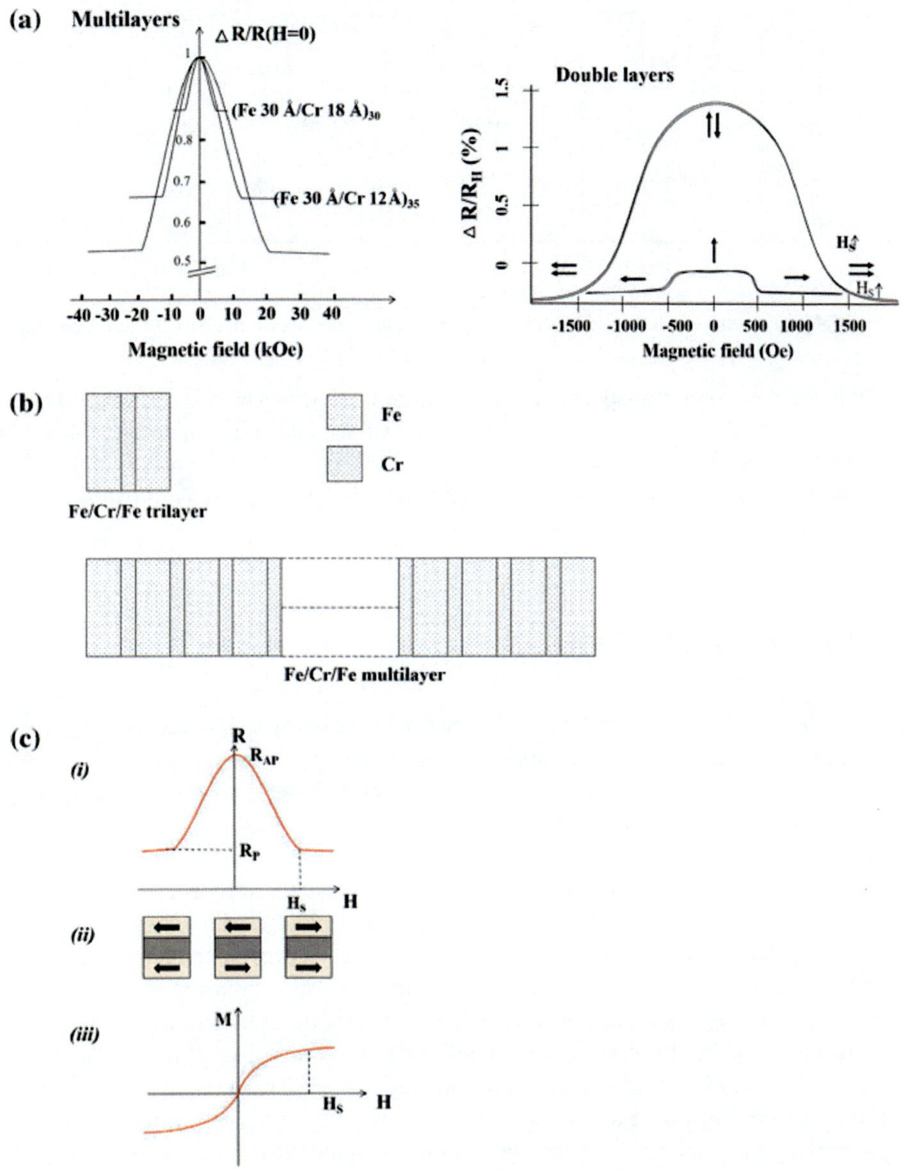

Fig. 3.3 **a** Left image shows magnetoresistance response $\left(\frac{\Delta R}{R}\right) = \left(\frac{R_{\uparrow\downarrow}-R_{\uparrow\uparrow}}{R_{\uparrow\uparrow}}\right)$ of three different Fe/Cr superlattices at 4.2 K (Figure adapted and redrawn from Ref. (Baibich et al. 1988)). Right image shows magnetoresistance response obtained from Fe double layers with Cr spacer layer (Figure adapted and redrawn from Ref. (Grunberg et al. 1986)). **b** Schematic diagram of Fe/Cr/Fe trilayer and multilayer structures. **c** Schematic representation of the relation between (i) GMR effect and (ii) & (iii) magnetization configurations of the GMR trilayer structures. Magnetization configurations are shown both by (ii) the arrows of the trilayer at various magnetic fields, and (iii) the magnetization curve. Magnetizations are aligned antiparallel at zero field and becomes parallel when the applied magnetic field H is larger than the saturation field (H_S) (Figure adapted and redrawn from Ref. (Tsymbal and Pettifor 2001))

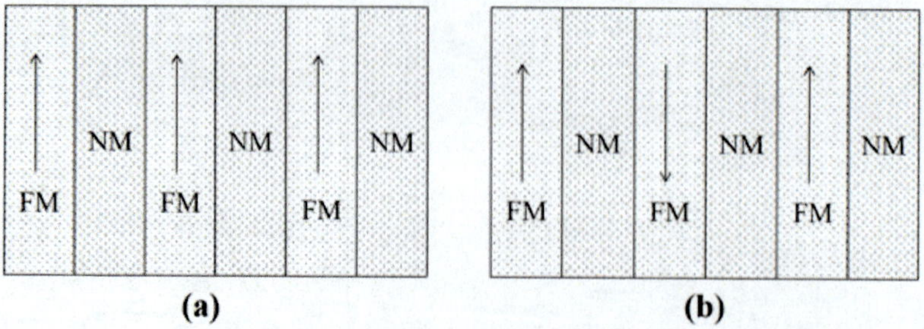

Fig. 3.4 **a** Ferromagnetic and **b** antiferromagnetic configurations of magnetic multilayers film

Such novel magnetotransport phenomenon, such as GMR (in magnetic multilayers), appears subject to the nanometric size scale of the magnetic materials. However, later it has been found that the effect is a very basic phenomenon which occurs in magnetic materials ranging from nanoparticles over multilayered thin films to permanent magnets.

3.3.2 *How Is GMR Effect?*

In Fe/Cr multilayers, the particular chosen thickness of the spacer layer (Cr) gave rise to indirect antiferromagnetic coupling between the Fe films when there is no applied magnetic field. That is in absence of any applied magnetic field, the orientation of magnetizations of the successive ferromagnetic layers is ordered antiparallel. This is obtained by the well-known quantum mechanical phenomenon, namely Ruderman-Kittel-Kasuya-Yosida (RKKY) indirect exchange interaction[1] between two ferromagnetic layers. Figure 3.3c presents a simplified picture of how application of a magnetic field aligns the magnetic moments of the successive ferromagnetic layers and saturates its magnetization, thereby yielding a corresponding change in resistance in multilayer structure as a function of magnetic field. Figure 3.4 shows the typical metallic multilayer structure, consisting of a series of very thin ferromagnetic films, separated by similarly thin non-magnetic films. The resistance of the multilayer has been observed to be changed as the multilayer passes from antiparallel to parallel magnetization configurations according to the illustration shown in Fig. 3.4. The resistance of the magnetic multilayer structure is low when all the ferromagnetic films attain parallel magnetization orientations (Fig. 3.4a), whereas the resistance is pretty high when the magnetizations of the neighbouring ferromagnetic layers are antiparallel (Fig. 3.4b).

In this case, the definition of MR ratio becomes

$$\left(\frac{\Delta R}{R}\right) = \left(\frac{R_{\uparrow\downarrow} - R_{\uparrow\uparrow}}{R_{\uparrow\uparrow}}\right) \tag{3.4}$$

[1]See References (Tsymbal and Pettifor 2001) and (Parkin et al. 1990)

where $R_{\uparrow\downarrow}$ and $R_{\uparrow\uparrow}$ are the resistances in case of antiferromagnetic (AFM) and ferromagnetic (FM) orientations of magnetization of successive ferromagnetic films, respectively.

The most commonly used combinations of ferromagnetic and non-magnetic layers are Fe/Cr. In case of Fe/Cr multilayer structures, as shown in Fig. 3.3b, the GMR ratio has been obtained to be approximately 50% (from Eq. 3.4). Similar effect has also been discovered simultaneously in Fe/Cr/Fe trilayer, as shown in Fig. 3.3b, as well. However, the GMR effect in this case has been found to be small. The GMR effect was also obtained in diversified multilayer systems containing various combinations of ferromagnetic and non-magnetic spacer layers, such as Co/Cu. Also, permalloy has been used widely as the ferromagnetic component of GMR multilayers.

In particular, the subject of spintronics received wide attention after the discovery of the fact that the electric current in a metallic multilayered nanostructure, composed of a series of alternating thin ferromagnetic and non-magnetic layer, is determined by the relative directions of the magnetizations of the successive ferromagnetic layers. This suggests that the spin magnetic moment of the electrons has significant role in the transport of electronic charge.

3.4 Types of GMR

3.4.1 Multilayer GMR

Multilayer GMR refers to the change in resistance in multilayer structure under the application of a magnetic field. A discussion on this topic has already been made in the 'Spin Valve' section in Chap. 2. A typical multilayer structure is shown in Fig. 3.5. As already mentioned in the last section, parallel magnetizations of all the ferromagnetic layers in a typical nanostructured *all-metallic* multilayer (Fig. 3.4a) correspond to the low resistance state, whereas antiparallel magnetizations of its neighbouring ferromagnetic layers correspond to the high resistance state (Fig. 3.4b).

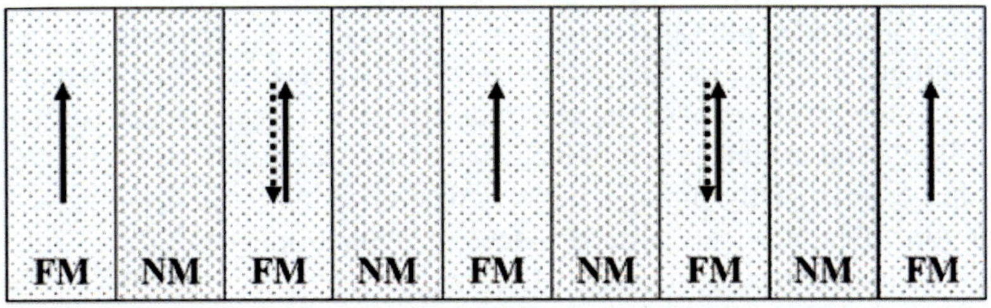

Fig. 3.5 Magnetic multilayer structure, consisting of alternating ferromagnetic (FM) and non-magnetic (NM) metallic layer

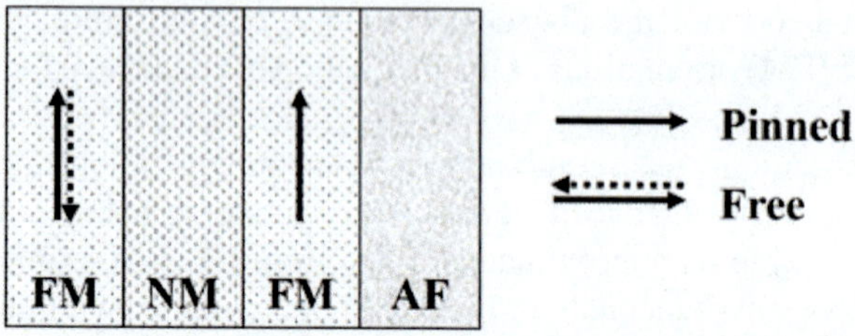

Fig. 3.6 Spin valve GMR structure, consisting of two ferromagnetic (FM) and one non-magnetic (NM) spacer layer. Additional antiferromagnetic (AF) layer has been introduced to pin magnetization of one of the FM layers. Magnetization of another FM layer is free

3.4.2 Spin Valve GMR

In spin valve GMR structure, a thin non-magnetic spacer layer (~3 nm) is sandwiched between two ferromagnetic layers, with no RKKY coupling between those two ferromagnetic layers (Fig. 3.6). In this case, the condition is that the coercive fields of the two ferromagnetic electrodes must be different so that it is possible to switch them independently. Therefore, parallel and antiparallel alignment of magnetization of successive ferromagnetic layers can be achieved. As discussed in case of multilayers, the resistance is higher in the antiparallel case and lower in parallel case. Such devices, as already mentioned in the last chapter, are referred to as spin valves.

Improvement of spin valve GMR can be obtained by increasing the spin relaxation length, or by enhancing the polarization effect on electrons by the ferromagnetic layers and the interface. In this context, the role of surface and interface effects are very critical due to high local ratio of atoms as compared to the bulk. Practical methods such as increasing the interfacial resistance, or by inserting half metallic layers into the spin valve stack can possibly lead to enhancement of spin valve GMR.

3.4.3 Pseudo-spin Valve GMR

Structure of pseudo-spin valve devices is very much like the conventional spin valve, with significant distinction lies in the magnetization reversal mechanism, i.e., coercivities of the two ferromagnetic electrodes (Fig. 3.7). Pseudo-spin valve structure consists of a soft magnet as one of its ferromagnetic layer, whereas a hard magnet as the other. Consequently, flipping of magnetization of those two ferromagnetic layers occur for different widely spaced values of applied magnetic field. This in turn enables us to obtain the requisite antiferromagnetic alignment of adjacent ferromagnetic layers that is required for GMR devices. Now, in order to minimize the exchange coupling between the ferromagnetic layers, the thickness of the non-magnetic spacer layer is generally kept thick enough. For the operation of pseudo-spin valve devices,

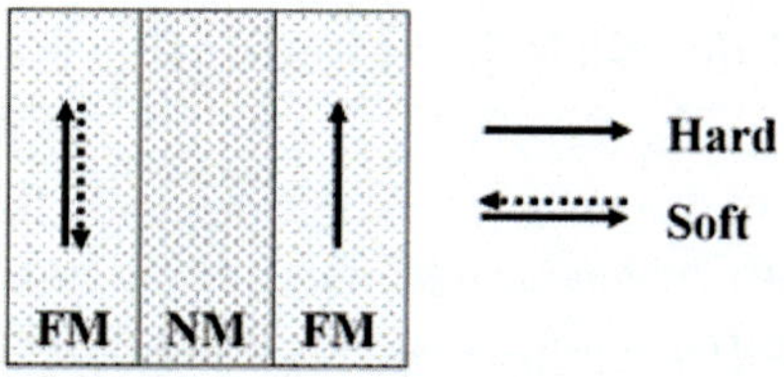

Fig. 3.7 Pseudo-spin valve structure, consisting of two ferromagnetic (FM) and one non-magnetic (NM) spacer layer. One ferromagnetic layer is hard, another is soft. In case of hard FM layer, the magnetization is pinned, whereas magnetization is free for the soft FM layer

it is essential to avoid such interaction between two ferromagnetic electrodes so that complete control over the device can be exercised.

3.4.4 Granular GMR

In a sample consisting of solid granules of magnetic material dispersed in a matrix of non-magnetic material (Fig. 3.8), granular GMR has been found to occur. For example, the granular GMR effect is observed in system, like copper matrices with cobalt granules dispersed on it. However, the GMR ratios, produced in granular materials, are not found to be high enough as obtained in the multilayer counterparts.

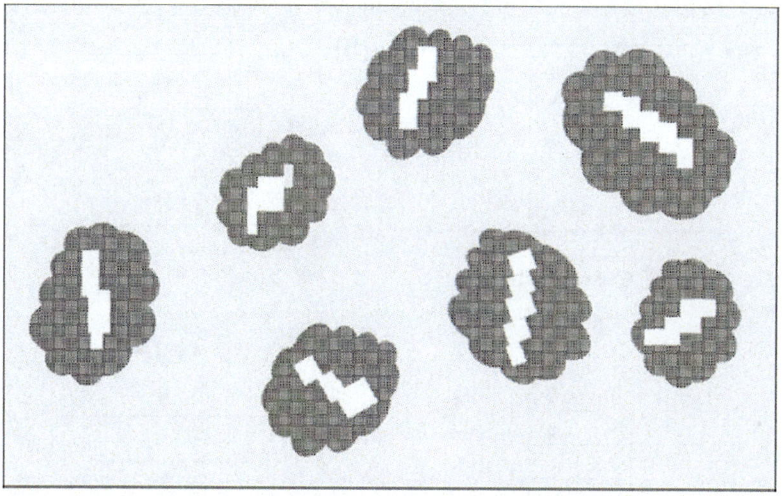

Fig. 3.8 Granular GMR structure with ferromagnetic material dispersed in a matrix of non-magnetic material

3.5 Physical Origin of GMR: Qualitative Explanation

It is clear from the above-made discussion that the GMR effect is a very basic phenomenon occurring in magnetic materials ranging from nanoparticles over multi-layered thin films to permanent magnets. In this contribution, we first focus on the links between the effect characteristic and underlying microstructure.

GMR effect can be qualitatively explained by adopting the well-known Mott model

The model was introduced in early 1936 to explain the sudden increase in resistivity of ferromagnetic metals when they are heated above their respective Curie temperature.

Electron conduction in two separate channels

At the onset, we should mention that the electron spin should remain conserved up to distances spanned over several tens of nanometres. It should be noted that such distances are much larger than the thickness of a typical multilayer, where the thickness of the layer is generally kept on the order of few Angstroms. Therefore, one can assume the electric current in a multilayer flowing in two separate channels: one corresponding to electrons with spin projection up (↑) and the other corresponding to electrons with spin projection down (↓). Electric current is passed in two different ways through magnetic superlattice structure (Fig. 3.9)—one is the current-perpendicular-to-plane (CPP) configuration, in which the electrodes are positioned on different sides of the device structure and the current is flowing perpendicular to the layers (Fig. 3.9a). Another is the current-in-plane (CIP) geometry in which

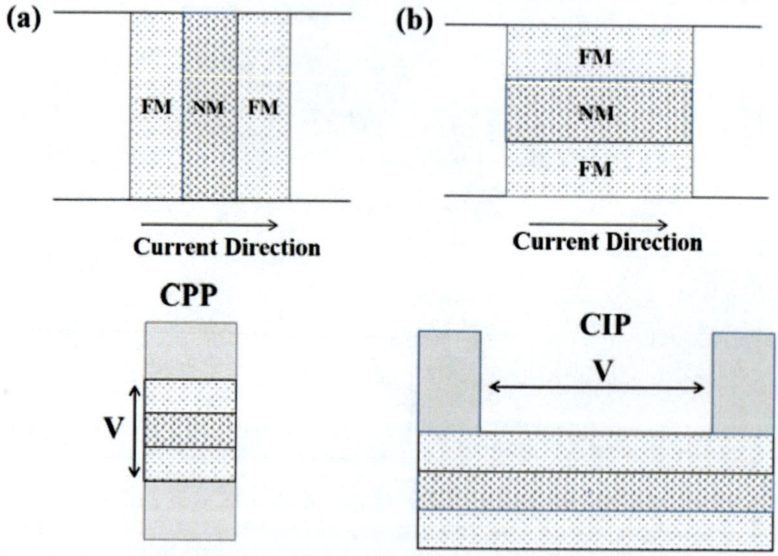

Fig. 3.9 Spin valves in the **a** current-perpendicular-to-plane (CPP) and **b** current-in-plane (CIP) GMR geometries

the electrodes are positioned on one side of the device structure and the current is supposed to flow along the layers (Fig. 3.9b). The GMR results from CPP geometry have been found to be more than double that of CIP configuration. However, CPP is more difficult to understand in practice.

To summarize, the following considerations can be made:

(1) Since the ↑ and ↓ spin channels are independent, they can be regarded as two parallelly connecting wires.

(2) Additionally, the majority and minority spin electrons (already discussed in Chap. 2), during their passage through a ferromagnetic layer, are scattered in different probabilities. However, the spin orientations of electrons would not change in this case. This phenomenon is referred to as *spin-dependent scattering*.

3.5.1 What Is Spin-Dependent and Spin-Flip Scattering of Electrons in Multilayers?

Before going into the details of origin of GMR, first we will discuss various categories of scatterings that electrons might suffer in magnetic multilayers. In our approach to understand the phenomenon considering Boltzmann equation, let us mainly suppose the elastic, i.e., energy conserved scattering, where in each scattering process electrons undergo change of its direction of propagation only. However, in real situation, the electrons may experience different types of scattering, which might flip the spin of electrons from up to down and vice versa. This type of scattering is generally referred to as *spin-flip scattering*. Now, distinction between spin-dependent scattering causing GMR and spin-flip scattering that plays detrimental role to GMR needs to be discussed to understand GMR properly. Such scattering processes are illustrated in Fig. 3.10.

Let us first give simple definition of two scattering processes:

Spin-dependent scattering: In ***spin-dependent scattering*** phenomenon, during each scattering event suffered by electrons, the scattering probabilities of ↑ and ↓ spin electrons are different. Noteworthy, in this scattering process, the orientation of the electrons' spin should be conserved.

Spin-flip scattering: In ***spin-flip scattering*** phenomenon, during each scattering event suffered by electrons, there is change of spin orientation of electrons from ↑ $\left(s_z = \frac{\hbar}{2}\right)$ to ↓ $\left(s_z = -\frac{\hbar}{2}\right)$ or vice versa.

Now, let us discuss the scattering processes one by one.

I. Spin-dependent Scattering

As already discussed, the key features of ***spin-dependent scattering*** process, encountered by conduction electrons, when they flow through the ferromagnetic layers are: (a) the scattering process is elastic; (b) the electron spin is conserved during scattering; (c) electrons having ↑ *or* ↓ spin orientations are scattered at different probabilities. First, we will discuss some general features of electron scattering, irrespective

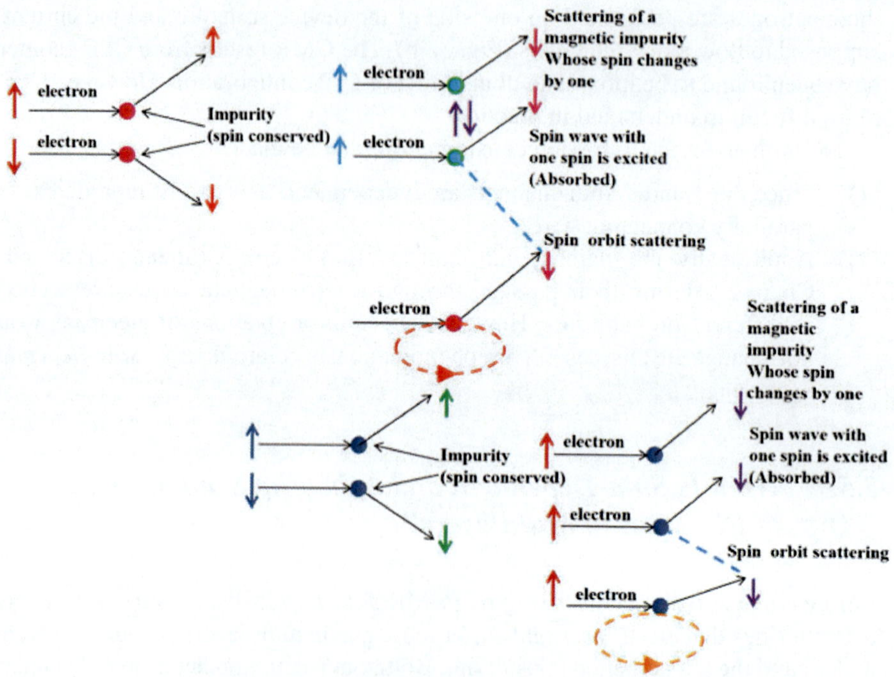

Fig. 3.10 Schematic representation of different types of scattering in magnetic multilayers

of spin, in a solid. It is well known that electrons obey Pauli exclusion principle. As a result, after experiencing scattering from impurity states, an electron can be transferred only to a *free* quantum states not occupied by similar other electrons. Now, at zero or low temperatures all the allowed energy levels (E) below Fermi energy (E_F) are occupied by electrons and the energy levels with $E > E_F$ are vacant. Because of elastic nature of impurity scattering, current carrying electrons at E_F after undergoing scattering process can be transferred only to the energy levels laying in the immediate vicinity to that of Fermi level.

In case of transition metals, the d band is occupied partially. Hence, the Fermi levels in these metals intersect not only at the conduction bands, i.e., 4 s bands, but also at the 3d bands. Furthermore, the atomic wavefunctions associated with d levels are known to be more localized than the wavefunctions associated with the outer s levels. As a result, energy levels of 3d bands overlap much less than that of 4 s bands. Consequently, d band should be narrow and accordingly its density of states $D(E_F)$ is supposed to be high. Hence, in the 3d band, an effective channel for scattering of conduction electrons can be supposed to open. In contrast, the Fermi level in noble metals does not intersect the 3d band, rather it only intersects the *4 s* conduction band having low $D(E_F)$, therefore results in less scattering of conduction electrons. This is the reason why noble metals are good conductors compared to 3d transition metals.

Now, we will bring 'spin' in this discussion. In case of ferromagnetic transitional metallic elements, one need to consider that ↑ and ↓ spin energy band, corresponding to 3*d* orbitals are almost rigidly shifted with respect to each other along the energy axis by the ferromagnetic exchange interaction (discussed in Chap. 1). Consequently, in case of a ferromagnetic metal, the potentials experienced by ↑ and ↓ spin electrons are different. This constitutes the basis of spin-dependent scattering in a ferromagnetic metal. In addition, in case of a multilayer structure, electrons flowing into the ferromagnetic layer from its underneath/adjacent non-magnetic spacer layer encounter a spin-dependent potential barrier. This in turn distinctly reflects ↑ and ↓ spin electrons. It has been found that electrons with spin magnetic moment antiparallel (minority spin electrons) to the magnetization (M) of the ferromagnetic layer are scattered more strongly than electrons with their spin magnetic moment parallel (majority spin electrons) to M. This means that in case of electronic conduction in a ferromagnetic material, majority spin electrons are more easily transmitted than minority spin electrons. Now this feature can be explained as follows: We note that there is more empty density of states available for minority spin electrons to scatter into than for majority spin electrons. As a result, the minority spin electrons encounter larger scattering than that of majority spin electrons. This *spin-dependent scattering*, in effect, causes GMR.

II. **Spin-flip scattering**

As already discussed, spin-flip scattering of an electron is characterized by changing or flipping of electron spin orientation. Several sources have been identified as responsible for spin-flip scattering:

1. ***Magnetic impurity at non-magnetic (NM) spacer layer***—During the fabrication process, due to top ferromagnetic layer deposition on non-magnetic spacer layer, diffusion of some magnetic atoms may take place into the non-magnetic spacer layer. Such ferromagnetic inclusions eventually form magnetic impurities in the non-magnetic spacer layer. If these magnetic impurities are located considerably far away from the ferromagnetic/non-magnetic interface, then the coupling of the spin magnetic moment associated with these impurities and that of the ferromagnetic layers is quite weak. As a result, spins associated with these impurities rotate freely. In this case, when there is scattering between an electron and magnetic impurity, interchanging of spins of the electron and that of the impurity may take place.
2. ***Spin waves in FM layer***—Spin waves can also be a source of scattering of electrons in ferromagnetic layers. Spin waves are quasi-particles with spin one. Therefore, creation or annihilation of a spin wave, as the case may be, due to scattering with an electron leads to a spin flipping of that electron. Since it involves the spin wave energy, this is an inelastic process which is important only at elevated temperatures.
3. ***Spin–orbit interaction due to gold (Au) impurity***—Let us consider the presence of impurities, offering strong spin–orbit coupling, such as gold (Au) in the

multilayer. In this case, the conduction electrons could be subjected to spin–orbit interaction due to such impurities. Consequently, spin orientations of these electrons, scattered by such impurity, may be reversed/changed.

3.5.2 *How Does Mott Model Describe GMR?*

In an attempt to explain GMR, there are two main points, mentioned below, as proposed by Mott:

(i) As per general consensus, in ferromagnetic metals the probability of *spin-conserved* spin-dependent scattering process is higher compared to that of spin-flip scattering process. Consequently, the identity of up-spin and down-spin electrons would be retained and would not get mixed over long distances. Hence, electrical conductivity in ferromagnetic metals takes place in parallel in two largely independent conduction channels, namely, up-spin and down-spin channels. It is well understood that the up- or down-spin orientations of electrons are defined as per projection of the spins along the quantization axis.

(ii) Ferromagnetic metals have exchange-splitting of its band structure, i.e., spin-dependent band structure, where at the Fermi energy the density of states are not the same for the two spin species of electrons. Quite intuitively, scattering rates are expected to be proportional to the density of states. Therefore, in ferromagnetic metals the up-spin and down-spin electrons scattering rates are quite different, irrespective of the nature of the scattering centres. Hence, the resistivities corresponding to the up-spin $(\rho^{\uparrow})$ and down-spin $(\rho^{\downarrow})$ electrons are different, i.e., $\rho^{\uparrow} \neq \rho^{\downarrow}$.

3.5.3 *Explanation of Negative MR of Ferromagnetic Transition Metal, Considering Spin Scattering Mechanisms*

As a prelude, let us explain the negative MR of any ferromagnetic transition metal, considering the above discussed spin scattering mechanisms. At temperatures very less than ferromagnetic Curie temperature, i.e., $T \ll T_C$, spin-flip scattering is not supposed to be a major spin scattering mechanism. As already mentioned, in case of a ferromagnet the conduction phenomenon can be nicely approximated by the two-current model, where currents set up by the $\uparrow$ and $\downarrow$ spin electrons can be considered to be an independent process. Now, in the presence of magnetic field, which decides the quantization axis of the ferromagnet, $\downarrow$ spin electrons are scattered more strongly than the $\uparrow$ spin electrons. This results in different resistivity for each spin state, which in turn leads to negative MR in ferromagnetic metal. At high temperatures, the spin-flip scattering of conduction electrons takes place due to collisions with spin waves. This in turn leads to spin mixing, which diminishes the distinction between the two spin channels. Hence, negative MR of ferromagnetic metal gets reduced with rise in temperature.

3.5.4 Explanation of GMR by Mott Model

As far as qualitative description is concerned, it is straightforward to explain GMR using Mott's arguments. In this direction, let us consider a simple trilayer magnetic film with two magnetic layers separated by a non-magnetic metallic spacer layer. Mott's second argument, which reflects asymmetric density of states at Fermi level, suggests for the spin-dependent scattering. We understand that for minority spin electrons with spin antiparallel to the magnetization, the scattering is expected to be pretty strong, whereas for majority spin electrons with spin parallel to the magnetization the scattering would be rather weak. Thus, it comes out that for parallel-aligned magnetic layers, majority spin (up-spin) electrons transit through the entire trilayer structure almost without suffering any scattering. On the contrary, minority spin (down-spin) electrons suffer strong scattering in both the ferromagnetic layers. As already mentioned, conduction process in ferromagnetic metals occurs in parallel in two spin channels, i.e., up-spin and down-spin channels, as shown in Fig. 3.11a. Therefore, the overall resistivity of the trilayer structure is decided primarily by the highly conductive majority spin/up-spin channel and hence found to be pretty low. In case of antiparallel-aligned trilayer, both the majority (up-spin) and minority (down-spin) spin electrons are getting scattered strongly within either one of the ferromagnetic layers, in which conduction electrons spins are antiparallel to its magnetization direction (Fig. 3.11b). This results in high total resistivity of the trilayer structure.

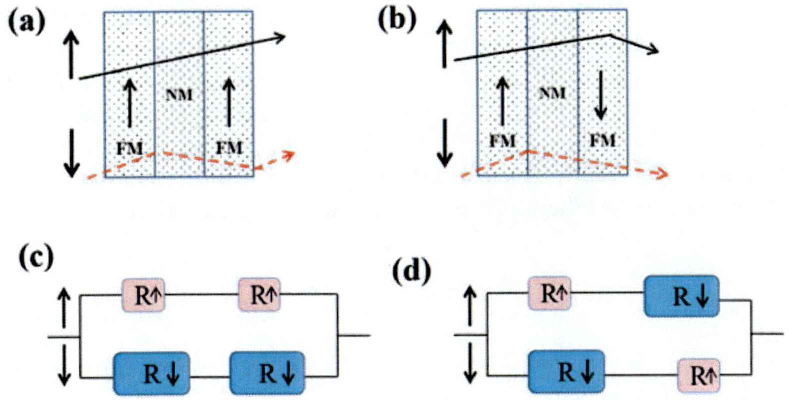

Fig. 3.11 Conduction process in ferromagnetic metals occurring in parallel in two spin channels, i.e., up-spin and down-spin channels in case of **a** parallel-aligned and **b** antiparallel-aligned trilayer GMR structure. Schematic representations of simple resistor model for describing GMR in case of **c** parallel-aligned and **d** antiparallel-aligned trilayer GMR structure

3.6 Quantitative Explanation of GMR

Such discussion on the origin of GMR in a trilayer structure, consisting of two ferromagnetic metallic layers separated by a non-magnetic metallic layer, can be efficiently put in the form of a simple resistor model as shown in Fig. 3.11c, d.

Let us first consider the ferromagnetic (FM) configuration (Fig. 3.11a) of the structure, i.e., when the magnetic moment of the ferromagnetic layers are parallel to each other. In this case, the electrons with ↑ spin suffer weak scattering and the electrons with ↓ spin undergo strong scattering both in the first and second ferromagnetic layers. This feature of scattering can be simulated in an equivalent resistor network, where two small resistors ($R_{\uparrow}$) represent ↑ spin channel and two large resistors ($R_{\downarrow}$) represent that of ↓ spin channel, as shown in Fig. 3.11c. ***Here, $R_{\uparrow}/R_{\downarrow}$ corresponds to the resistance when spin of conduction electron is parallel/antiparallel to the magnetization of the ferromagnetic layers. Obviously, $R_{\downarrow}$ should be much larger than $R_{\uparrow}$, since minority spin experiences large scattering compared to that of majority spin electrons.*** Thus, in parallel configuration, resistance of ↑ spin channel is much smaller compared to that of ↓ spin channel. For simplicity, consider the case where $R_{\uparrow} << R_{\downarrow}$, therefore the total resistance in the parallel configuration is

$$\boldsymbol{R}_{\uparrow\uparrow} = \frac{2\boldsymbol{R}_{\uparrow}\boldsymbol{R}_{\downarrow}}{\boldsymbol{R}_{\uparrow} + \boldsymbol{R}_{\downarrow}} \approx 2\boldsymbol{R}_{\uparrow} \tag{3.5}$$

Therefore, the total resistance in ferromagnetic configuration is determined by the small-resistance in the ↑ spin channel, which 'short-circuit' the high-resistance ↓ spin channel.

Let us now consider the antiferromagnetic (AFM) configuration of the structure (Fig. 3.11b), i.e., when the magnetic moment of the ferromagnetic layers is antiparallel to each other. In this case, ↓ spin electrons suffer strong scattering in the first ferromagnetic layer and weak scattering in the second ferromagnetic layer. Similarly, ↑ spin electrons suffer weak scattering in the first ferromagnetic layer and strong scattering in the second. In this case, the equivalent resistor network can be modelled by associating one large ($R_{\downarrow}$) and one small ($R_{\uparrow}$) resistor corresponding to strong and weak scattering, respectively, as shown in Fig. 3.11d. No shorting feasibility can be sensed in this case and therefore, total resistance in this case becomes,

$$\boldsymbol{R}_{\downarrow\uparrow} = \frac{\boldsymbol{R}_{\uparrow} + \boldsymbol{R}_{\downarrow}}{2} \approx \frac{\boldsymbol{R}_{\downarrow}}{2} \tag{3.6}$$

Thus, $R_{\uparrow\downarrow} > R_{\uparrow\uparrow}$. We can express GMR by defining a new term, spin asymmetry, i.e., $\propto = \frac{R_{\uparrow}}{R_{\downarrow}}$. Thus we get,

$$GMR\,(\%) = \frac{\Delta \boldsymbol{R}}{\boldsymbol{R}} = \frac{\boldsymbol{R}_{\downarrow\uparrow} - \boldsymbol{R}_{\uparrow\uparrow}}{\boldsymbol{R}_{\uparrow\uparrow}} = \frac{(\boldsymbol{R}_{\uparrow} - \boldsymbol{R}_{\downarrow})^2}{4\boldsymbol{R}_{\uparrow}\boldsymbol{R}_{\downarrow}} = \frac{(\propto - 1)^2}{4\propto} \tag{3.7}$$

To summarize, GMR characterizes the gradual transit of magnetization of adjacent ferromagnetic layers from the antiparallel alignment in zero magnetic field to the parallel alignment under the application of a strong magnetic field. In order to obtain a large GMR, one must seek a good alignment of energy bands, as much as possible, between the ferromagnetic layers and the spacer layer in one spin channel; on the other hand, a large mismatch between them to the best of the possibility in another spin channel. This means that $R_\uparrow$ should be as small as possible and $R_\downarrow$ should be as large as possible, which in turn means large spin asymmetry. Also, it needs to be remembered that spin-flip scattering processes are detrimental to obtain the GMR effect as it causes mixing of the $\uparrow$ and $\downarrow$ spin channels.

3.7 Magnetoresistance Theory

3.7.1 Resistor Network Theory of GMR

Although conceptually correct, simple resistor model of GMR in trilayer structure should be transformed into a quantitative theory so that one could possibly explain the differences between the observed GMR effect on CIP and CPP geometries, the observed dependence of GMR on the layer thicknesses of the device and also on the material.

In order to determine the effect of spin-dependent scattering on electrical conduction in the multilayer films, spin-dependent scattering process needs to be incorporated into the ***Boltzmann Equation***. In general, electrons in a metal experience a constant force $\mathbf{F} = -e\,\mathbf{E}$, where e is the electron charge, under the application of an electric field $\mathbf{E}$. The scattering, suffered by electrons in a metal from its imperfections, is supposed to modify the simple accelerated motion. Now, during *mean free time*, i.e., the time interval when each electron with mass m moves without scattering, electrons continue to accelerate and finally acquire a velocity $\boldsymbol{v}$ in the direction of $\mathbf{E}$. It is well known that the distance covered during this time is called as *mean free path*. After experiencing each scattering events, electrons undergo acceleration process to again attend its velocity $\boldsymbol{v}$. This in turn leads to a steady-state condition where all the electrons flow in the direction of $\mathbf{E}$ with a velocity $\boldsymbol{v}$ given by

$$\boldsymbol{v} = -\frac{\boldsymbol{eE\tau}}{\boldsymbol{m}} \tag{3.8}$$

Considering n electrons per unit volume, the current density becomes

$$\boldsymbol{j} = -\boldsymbol{nev} = -\frac{\boldsymbol{ne^2E\tau}}{\boldsymbol{m}} \tag{3.9}$$

This equation is like Ohm's law. Again, electrical conductivity (σ) is defined as $j = \sigma E$. Hence, resistivity ρ can be clearly given by

$$\rho = \frac{m}{ne^2\tau} \tag{3.10}$$

ρ is clearly found to depend on the electron density, n and on the mean free time, τ. It is straightforward to understand that τ should be inversely proportional to the scattering probability. In case of magnetic multilayers films, probability of scattering is governed by the following two factors: (i) strength of the potential related to scattering and (ii) density of states available for scattering at the vicinity of E_F. The first factor is governed by the *interfacial spin-dependent scattering* in multilayers structure. The second factor results in bulk spin-dependent scattering, which is basically the well-known Mott mechanism.

Bulk spin-dependent scattering

Let us first focus on the bulk spin-dependent scattering process. If we consider, $\rho_{FM} =$ total resistivity of the bulk ferromagnetic materials, thus considering two independent spin channels, we can write

$$\frac{1}{\rho_{FM}} = \frac{1}{\rho_{FM}^{\uparrow}} + \frac{1}{\rho_{FM}^{\downarrow}} \tag{3.11}$$

where $\rho_{FM}^{\uparrow} = \frac{2\rho_{FM}}{1+\beta}$, $\rho_{FM}^{\downarrow} = \frac{2\rho_{FM}}{1-\beta}$ and $\beta = \frac{\rho_{FM}^{\downarrow}-\rho_{FM}^{\uparrow}}{\rho_{FM}^{\downarrow}+\rho_{FM}^{\uparrow}} =$ bulk scattering asymmetry.

Interfacial spin-dependent scattering

Magnetic multilayers consist of thin interfacial layers, which constitute spin-dependent potential barriers at the interface of a ferromagnet and a non-magnet. As a result, the corresponding resistivity of the concerned interface is spin-dependent. Hence, similar to the bulk spin-dependent scattering, an interfacial scattering asymmetry, γ can also be introduced, which is defined by

$$\rho_{FM-NM}^{\uparrow} = \frac{2\rho_{FN-NM}}{1+\gamma},\ \rho_{FM-NM}^{\downarrow} = \frac{2\rho_{FN-NM}}{1-\gamma}$$

where $\rho_{FM\text{-}NM}$ is the total resistivity of the interfacial layer, $\rho_{FM-NM}^{\uparrow/\downarrow} =$ interfacial resistivity corresponding to $\uparrow/\downarrow$ spin channel.

3.7.1.1 Periodic Superlattice Structure

Before going into the details of the calculation of GMR, let us introduce the concept of ***Periodic Superlattice***. Actually, GMR devices are nothing but an extension of spin valves. The idea behind the forming of a 'spin-valve-superlattice structure', as depicted in Fig. 3.12, is to accrue any advantage by increasing the number of layers in the basic unit cell of a spin valve, while keeping the width of each layer a few nanometres thick. A superlattice structure, which is periodic in nature, composed of

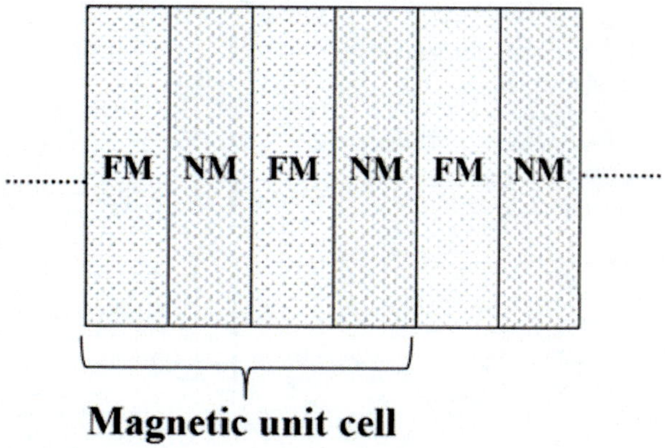

Fig. 3.12 Schematic illustration of magnetic superlattice

an arrangement of successive ferromagnetic and non-magnetic layers (Fig. 3.12) is subject to spin-dependent scattering of conduction electrons in the bulk portion of the ferromagnetic layers. Such entire periodic structure consisting of identical structural units is called as superlattice unit cell. Electrons of given spin orientations, while traversing through different superlattice regions, encounter various local resistivities. For instance, regions having high density of states at E_F offer high resistivity since there are large available energy states for scattering. Regions having three different local resistivities can be identified in the superlattice unit cell:

(a) ρ_{NM} = resistivity corresponding to the non-magnetic spacer layer,
(b) $\rho_{\mathrm{FM}}^{\mathrm{L}}$ = low resistivity for the parallel spin orientation configuration. This is almost the same as resistivity of non-magnetic space layer.
(c) ρ_{FM}^{H} = high resistivity for the antiparallel spin orientation configuration.

Therefore, distribution of such regions with different resistivities in a periodic superlattice structure with ferromagnetic and antiferromagnetic configurations can be schematically drawn as follows:

The schematic diagram (Fig. 3.13) clearly shows that a system of eight resistors, with four resistors in each spin channel, constitutes the unit cell of a magnetic superlattice. Therefore, in order to calculate MR, we have to set up a procedure for summing up four resistors in each spin channel and extend it for both spin channels. Then, resistors of both spin channels can simply be added following the addition rule for resistors connected in parallel in order to obtain an overall resistance of the magnetic unit cell. Same process has to be repeated for both (a) ferromagnetic (↑↑) and (b) antiferromagnetic (↑↓) configurations. Thus, the resistance for ferromagnetic (↑↑) and antiferromagnetic (↑↓) configurations as denoted by $R_{\uparrow\uparrow}$ and $R_{\uparrow\downarrow}$, respectively, can be defined as follows:

$$\frac{1}{R_{\uparrow\uparrow}} = \left(\frac{1}{R_{\uparrow}} + \frac{1}{R_{\downarrow}}\right)_{\uparrow\uparrow} \text{ and } \frac{1}{R_{\uparrow\downarrow}} = \left(\frac{1}{R_{\uparrow}} + \frac{1}{R_{\downarrow}}\right)_{\uparrow\downarrow} \tag{3.12}$$

where $R_{\uparrow}/R_{\downarrow}$ is the resistance of the unit cell for ↑ and ↓ spin channels, respectively.

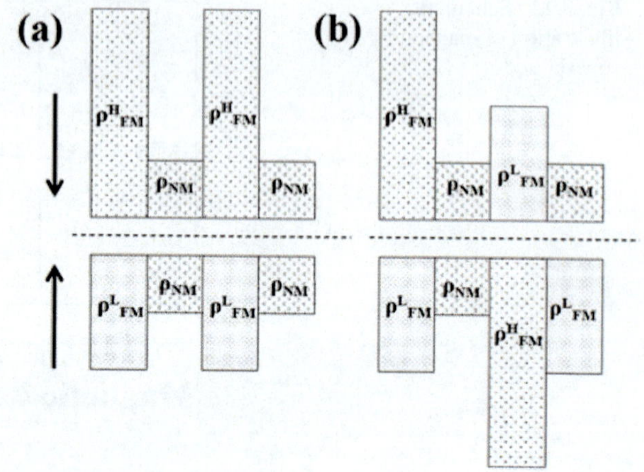

Fig. 3.13 Schematic illustration of distribution of local resistivities in a magnetic unit cell. **a** Ferromagnetic (FM) and **b** antiferromagnetic (AFM) configurations

More precisely,

- for the ferromagnetic configuration, the process could be reduced to the estimation of the resistance only for a two-component superlattice.
- for the antiferromagnetic configuration, one has to consider the whole four-component superlattice.

3.7.2 Calculation for Ferromagnetic Configuration

For simplicity, let us suppose that in a two-component superlattice structure, ***a*** and ***b*** are the thicknesses of the alternating regions having high (ρ^a) and low (ρ^b) resistivity, respectively. Thus, in a two-component superlattice structure, average resistivity experienced by conduction electrons is given by

$$\bar{\rho} = \frac{a\rho^a + b\rho^b}{a + b} \tag{3.13}$$

Now, using the above generalization, we can go for evaluating the MR:

$$(R_\uparrow)_{\uparrow\uparrow} = \frac{M\rho_{FM}^L + N\rho_{NM}}{M + N}, (R_\downarrow)_{\uparrow\uparrow} = \frac{M\rho_{FM}^H + N\rho_{NM}}{M + N} \tag{3.14}$$

where M and N are the number of atomic planes of each ferromagnetic and non-magnetic layers, respectively. Adding both parts in Eq. (3.14) results the two-component superlattice resistance for ferromagnetic configuration given by

$$\left(\frac{1}{R}\right)_{\uparrow\uparrow} = (M + N)\left[\frac{1}{M\rho_{FM}^L + N\rho_{NM}} + \frac{1}{M\rho_{FM}^H + N\rho_{NM}}\right] \tag{3.15}$$

Thus for four-component superlattice, Eq. (3.15) turns out to be

$$\left(\frac{1}{R}\right)_{\uparrow\uparrow} = 2(M+N)\left[\frac{1}{M\rho^L_{FM} + N\rho_{NM}} + \frac{1}{M\rho^H_{FM} + N\rho_{NM}}\right] \tag{3.16}$$

3.7.3 Calculation for Antiferromagnetic Configuration

Adopting similar procedure as that followed in case of ferromagnetic configuration, the calculation of four-component superlattice resistance for antiferromagnetic configuration using Eq. (3.12) results in

$$\left(\frac{1}{R}\right)_{\uparrow\downarrow} = \left[\frac{8(M+N)}{M\rho^L_{FM} + M\rho^H_{FM} + N\rho_{NM}}\right] \tag{3.17}$$

Recalling **GMR (%)** $= \frac{\Delta R}{R} = \frac{R_{\downarrow\uparrow} - R_{\uparrow\uparrow}}{R_{\uparrow\uparrow}}$, substituting Eqs. (3.16) and (3.17), we get

$$\left(\frac{\Delta R}{R}\right) = \left[\frac{M^2(\rho^L_{FM} - \rho^H_{FM})^2}{4(M\rho^L_{FM} + N\rho_{NM})(M\rho^H_{FM} + N\rho_{NM})}\right] \tag{3.18}$$

An attempt to simplify the expression in R.H.S. of Eq. 3.18 results in

$$\left(\frac{\Delta R}{R}\right) = \left[\frac{\left(1 - \frac{\rho^H_{FM}}{\rho^L_{FM}}\right)^2}{4\left(1 + \frac{N}{M}\frac{\rho_{NM}}{\rho^L_{FM}}\right)\left(\frac{\rho^H_{FM}}{\rho^L_{FM}} + \frac{N}{M}\frac{\rho_{NM}}{\rho^L_{FM}}\right)}\right]$$

$$\text{or, } \left(\frac{\Delta R}{R}\right) = \left[\frac{(1-\beta)^2}{4\left(1 + \frac{N}{M\mu}\right)\left(\beta + \frac{N}{M\mu}\right)}\right] \tag{3.19}$$

where $\beta = \frac{\rho^H_{FM}}{\rho^L_{FM}}$ and $\mu = \frac{\rho^L_{FM}}{\rho_{NM}}$.

From the above equation, it is now straightforward to identify the prime factors that decide GMR. Equation (3.19) evidently suggests the following:

1. Two variables, β and ($M\mu/N$), govern GMR (ΔR/R).
2. Spin asymmetry ratio, β should be pretty large in order to yield large GMR.
3. For a given value of β, GMR is supposed to increase with increasing ($M\mu/N$) values and saturates eventually.

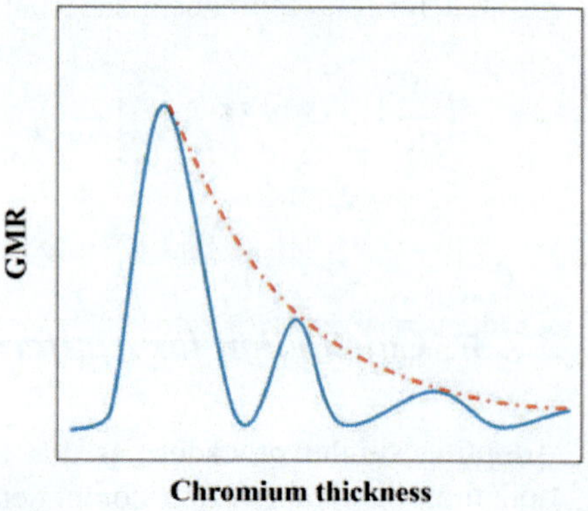

Fig. 3.14 Schematic variation of GMR ratio with the non-magnetic Cr layer thickness in Fe/Cr multilayer films. The dotted line indicates the variation as ($1/N^2$) (Figure adapted and redrawn from Ref. (Parkin et al. 1990))

4. On the other hand, GMR value decreases with increasing non-magnetic spacer layer thickness, as shown in red line in Fig. 3.14. However, as demonstrated by Stuart Parkin in his original experiment for Fe/Cr multilayers, GMR decreases in an oscillatory way with the thickness of the non-magnetic chromium spacer layer, as shown by green curve in Fig. 3.14.

Such oscillations of GMR as a function of spacer layer thickness are quite intriguing (Parkin et al. 1990). This seems to occur since MR effect is detectable only for some certain thicknesses of the chromium spacer layer. For those particular thicknesses of the spacer layer, the interlayer RKKY exchange interaction yields antiferromagnetic alignment of the magnetic moments of the adjacent ferromagnetic iron layers, which is prerequisite condition for GMR effect to be observed. For other thickness values such alignment becomes parallel and thereby does not cause GMR.

3.8 Conclusions

In this chapter, we have introduced and reviewed the concept of giant magnetoresistance. As a prelude, ordinary magnetoresistance, i.e., magnetoresistance of ferromagnetic transition metals and anisotropic magnetoresistance of ferromagnetic transition metals have been discussed. We have addressed the definition of GMR and its effect. Different kinds of magnetoresistance, such as multilayer GMR, spin valve GMR, pseudo-spin valve GMR, granular GMR, have also been reviewed briefly. We have presented an elaborate explanation on physical origin of GMR. In this direction, spin-dependent and spin-flip scattering of electrons in multilayers have been considered. We have also presented explanation on negative magnetoresistance of ferromagnetic transition metal, considering spin scattering mechanisms. We have explained *qualitatively* how Mott model could describe GMR. Quantitative explanation has also been given here. Magnetoresistance theory, based on *Resistor Network Theory of GMR*, has been discussed. Furthermore, *periodic superlattice structure* has also been highlighted. Most currently available commercial spintronic products, such as magnetic read heads for reading data in computer hard disks, magnetic random access memory etc. are based on giant magnetoresistance effect.

3.9 Exercises

1. What is magnetoresistance? How does the external applied magnetic field affect the resistivity in a metal?
2. What is ordinary or normal magnetoresistance?
3. Describe magnetoresistance of ferromagnetic transition metals.
4. What is anisotropic magnetoresistance (AMR) of ferromagnetic transition metals?
5. Calculate the normal magnetoresistance in Cu and in Co at room temperature for an applied field of 1 T? The resistivity of Co (Cu) is 10 (1) μΩcm and R_H (Cu) $= -5 \times 10^{-10}$ C^{-1} m^{-3} and R_H (Co) $= 1 \times 10^{-9}$ $C^{-1}m^{-3}$ [Ans: MR for Cu and Co is of the order of 10^{-3} and 10^{-4} at room temperature, respectively].
6. How do we determine the magnetoresistance in multilayer films?
7. What are the different types of scattering that destroy the magnetoresistance in multilayer films?
8. Are 3*d* transition metal good conductors as compared to copper?
9. What are the factors that determine the GMR in multilayer samples and how to enhance the GMR in multilayer films?
10. Define magnetoresistance (MR).
11. What are spin-dependent and spin-flip scattering? What are different types of scattering that destroy magnetoresistance in multilayer films?
12. What is GMR?
13. How is GMR effect?
14. Define different types of GMR.
15. Differentiate between multilayer GMR and spin valve GMR.
16. What is the difference between spin valve GMR and pseudo-spin valve GMR?
17. What is granular GMR?
18. What is spin-dependent and spin-flip scattering of electrons in multilayers?
19. How does Mott model describe GMR?
20. Explain negative magnetoresistance of ferromagnetic transition metal, considering spin scattering mechanisms.
21. What is periodic superlattice structure?
22. Derive the equation for determining GMR using resistor network theory.
23. Distinguish between *spin-dependent scattering* and *spin-flip scattering* of electrons in multilayers.
24. Define GMR effect. Discuss the physical origin of GMR in a trilayer structure using a simple resistor model.

25. Show the experimental variation of GMR ratio with the spacer layer thickness of a GMR structure. Explain the feature.
26. Using the resistor network model, derive an expression of the magnetoresistance ratio of GMR in terms of the spin asymmetry.
27. Multiple Choice Questions:
(a) The normal magnetoresistance is defined as

$$\frac{\Delta\rho}{\rho} = \left(\frac{ne^2\tau}{m}\frac{1}{ne}B\right)^2$$

The resistivity of Co (Cu) is 10 (1) μΩcm and Hall coefficient is R_H (Cu) = −5 × 10^{-10} C^{-1} m^{-3} and R_H (Co) = 1 × 10^{-9} $C^{-1}m^{-3}$. What is the order of normal magnetoresistance in Cu and Co at room temperature for an applied field of 1 T?

(i) MR for Cu and Co is of the order of 10^{-3} and 10^{-4} at room temperature, respectively.
(ii) MR for Cu and Co is of the order of 10^{-4} and 10^{-5} at room temperature, respectively.
(iii) MR for Cu and Co is of the order of 10^{-5} and 10^{-6} at room temperature, respectively.

(a) GMR is caused by

(i) spin-dependent scattering phenomenon.
(ii) spin-dependent scattering phenomenon.
(iii) combination of both.

References

M.N. Baibich, J.M. Broto, A. Fert, F.N. van Dau, F. Petro, P. Eitenne, G. Creuzet, A. Friederich, J. Chazelas, Giant Magnetoresistance of (001)Fe/(001)Cr Magnetic Superlattices. Phys. Rev. Lett. **61**, 2472 (1988)

A. Fert, I.A. Campbell, Electrical resistivity of ferromagnetic nickel and iron based alloys. J. Phys. F: Metal Phys. **6**, 849 (1976)

U. Hartmann, *Magnetic Multilayers and Giant Magnetoresistance—Fundamentals and Industrial Applications* (Springer Verlag/Berlin, Germany, 2000)

P. Grunberg et al., Layered magnetic structures: evidence for antiferromagnetic coupling of Fe layers across Cr interlayers. Phys. Rev. Lett. **57**, 2442 (1986); G. Binasch et al., Enhanced magnetoresistance in layered magnetic structures with antiferromagnetic interlayer exchange. Phys. Rev. B **39**, 4828 (1989)

S.S.P. Parkin, N. More, K.P. Roche, Oscillations in exchange coupling and magnetoresistance in metallic superlattice structures: Co/Ru, Co/Cr, and Fe/Cr. Phys. Rev. Lett. **64**, 2304 (1990)

E.Y. Tsymbal, D. Pettifor, Perspectives of giant magnetoresistance. Solid State Phys. **56**, 113–237 (2001)

E.S. Kristen Coyne, Giant magnetoresistance: the really big idea behind a very tiny tool (2015). https://nationalmaglab.org/education/magnet-academy/learn-the-basics/stories/giantmagnetoresistance

Chapter 4
Tunnelling Magnetoresistance (TMR)

4.1 Introduction on Tunnelling Magnetoresistance

4.1.1 Introductory Note

In the previous chapter, we have discussed ***the effect of spin-dependent scattering on the electric current*** passing through the magnetic multilayer films to produce GMR. Similarly, the magnetic field dependence of the electrical current passing through the magnetic tunnel junctions (MTJs) is another subject of great interest.

4.1.2 What Is Magnetic Tunnel Junction (MTJ)?

Magnetic tunnel junction (MTJ) is basically two ferromagnetic layers, separated by a non-magnetic insulating layer.

4.1.3 What Is Tunnelling Magnetoresistance (TMR)?

In order to obtain the GMR effect, the non-magnetic spacer layer between two ferromagnetic layers is supposed to be a conducting material, i.e., metallic. This means GMR is an *all-metallic* device structure. However, if the metallic spacer layer is replaced by an insulating layer, as in the case of MTJ, the conduction electrons originating from one ferromagnetic electrode would undergo quantum tunnelling across the insulating barrier to reach another ferromagnetic electrode. This means that quantum mechanical tunnelling would be the conduction mechanism in this case. More subtle feature is that such tunnelling process has been found to be ***spin-dependent*** and ***depends on the relative angle between the magnetization of those two ferromagnetic layers***. This phenomenon was first observed by Julliere in a Fe/Ge/Co

P. Dey and J. N. Roy, *Spintronics*,
https://doi.org/10.1007/978-981-16-0069-2_4

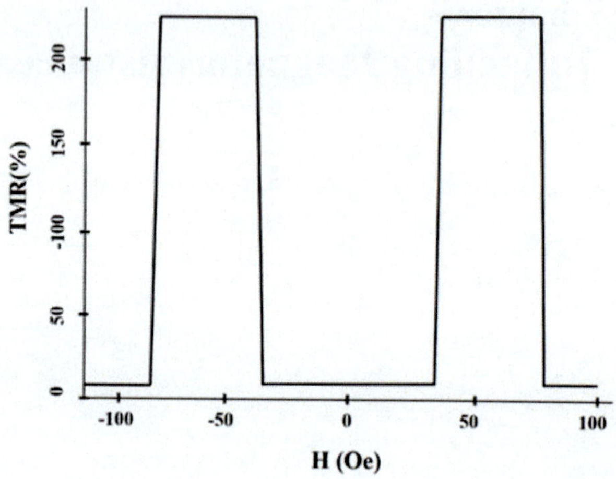

Fig. 4.1 TMR graph as a function of applied magnetic field (Figure adapted and redrawn from Ref. (Ming Loong et al. 2016))

junction in 1975 and is called the ***tunnelling magneto resistance*** (**TMR**). Spin-dependent tunnelling phenomenon possesses several fascinating scientific queries. Also, the number of applications for MTJ is continuously growing (Meservey and Tedrow 1994). Hence, it is indeed essential to understand the effect of magnetic field on the tunnelling current passing through magnetic junctions.

TMR may be regarded as the new candidate for mesoscopic-scale magnetic sensors and magnetic random access memory (MRAM) elements. TMR-based spintronic devices have many advantages over GMR devices for two reasons: ***First***, they are easier to fabricate. ***Second***, they can provide much larger signal (Fig. 4.1). Indeed, recently significant TMR values have been observed, up to 600% at room temperature and more than 1100% at 4.2 K in junctions of CoFeB/MgO/CoFeB. The tunnel barriers of MgO attracted attentions since 2001, when it has been predicted theoretically that the TMR can reach several thousand percent in Fe/MgO/Fe (Ming Loong et al. 2016). However, experimentally the highest value observed in such system is around 200% at room temperature. This disagreement between theory and experiment stimulated the researcher to investigate the reason behind it. Figure 4.1 shows schematic illustration of TMR (%) as a function of applied magnetic field of a $\mathbf{Co_{40}Fe_{40}B_{20}/MgO/Co_{40}Fe_{40}B_{20}}$ magnetic tunnel junction grown on PET substrate.

4.2 Magnetic Junctions

At first, we shall focus on the simple classification of magnetic junctions. In general, magnetic junctions consisting of two bulk ferromagnetic metallic electrodes separated by a non-magnetic insulating spacer layer can be classified into two types: (a) Tunnel-type junction and (b) contact-type junction.

4.2.1 Tunnel-Type Junction

When ferromagnetic electrodes are separated by distance exceeding a few angstroms (Fig. 4.2), transfer of conduction electrons between those ferromagnetic electrodes takes place by quantum tunnelling process.

From quantum mechanics, the probability that any one electron tunnels through a barrier of height V and length l is given by:

$$\boldsymbol{J} = \mathbf{exp}\left[-\boldsymbol{cl}\sqrt{\frac{2\boldsymbol{mV}}{\hbar^2}}\right], \tag{4.1}$$

where c is a constant of the order unity, depending on the detailed shape of the barrier and on the electronic wavefunctions.

This barrier between two ferromagnetic electrodes can be formed by two ways:

(i) created by vacuum between the electrodes;
(ii) insertion of an insulating layer between the two electrodes.

In the first case, the work function of electrodes plays the role of the height of the vacuum-created barrier. In the latter case, the barrier height is decided by the position of the edges of the band gap of the insulating material with respect to the *Fermi level* of the electrodes. It is evident from Eq. 4.1 that tunnelling depends exponentially on the spacing between the electrodes. Furthermore, in case of a junction of macroscopic size, it is expected that tunnelling events take place at the bulging of the interface. Hence, a slight variation of the spacing between the electrodes by few angstroms can even greatly modify the tunnelling probability. Accordingly, the conductance at any of these points is given by

$$\boldsymbol{g} \sim \frac{\boldsymbol{e}^2}{\boldsymbol{h}}\boldsymbol{J} \sim \frac{\boldsymbol{e}^2}{\boldsymbol{h}}\mathbf{exp}\left[-\boldsymbol{cl}\sqrt{\frac{2\boldsymbol{mV}}{\hbar^2}}\right]. \tag{4.2}$$

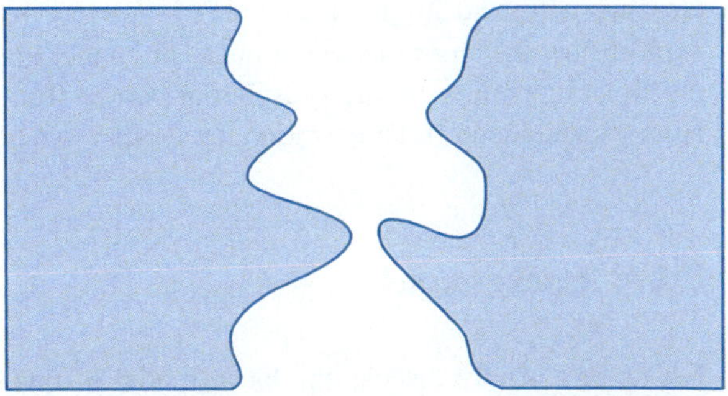

Fig. 4.2 Schematic diagram of a tunnel-type junction

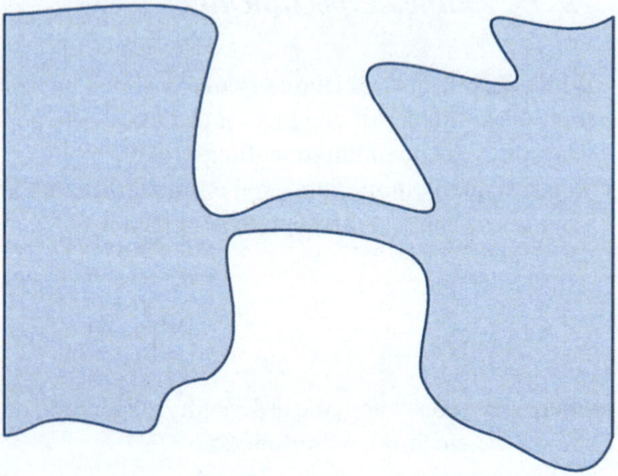

Fig. 4.3 Schematic diagram of a contact-type junction

4.2.2 Contact Type Junction

In this type of junction, two electrodes can be in contact at some points, as shown in Fig. 4.3.

Here, conductance of each contact is given by e^2/h times the number of electrons channelling through the contact. It is roughly given by the cross section of the contact expressed in units of inverse square Fermi wavevector (k_F^{-2}). Therefore,

$$g \sim \frac{e^2}{h} k_F^2 A, \tag{4.3}$$

where A is the area of the junction.

4.3 Physical Explanation

Here we will concentrate on the tunnel-type junction. Subsequently, we will try to explore how the basic models of quantum tunnelling need to be modified in case of electrons tunnelling through insulating barrier from one ferromagnetic electrodes, having spontaneous magnetization, to another in a MTJ.

4.3.1 Background

Let us first briefly discuss the basic model of quantum mechanical tunnelling of particles over a barrier having potential (V_0) greater than that of the energy € of the particle. In classical picture, the particle would be reflected from the boundary of such barrier. However, from the point of view of quantum mechanics, the particle essentially would behave like a matter wave. It can be shown that such matter wave has finite probability for penetrating the barrier followed by continuation of its travel as a wave on the other side. The probability of such journey of the particle through the barrier is characterized by the transmission coefficient. This phenomenon for $E < V_0$, i.e., even when the energy of the particle is less than the barrier height, there is still a *finite* probability for the particle to be transmitted through the barrier and appear at the other side of the barrier. This intriguing result, which differs from the classical case, is called *quantum tunnelling*.

4.3.2 Spin-Dependent Conductance of Charge Carriers

Ferromagnetic electrodes possessing spontaneous magnetization suggest the following:

First, the number of up-spin and down-spin conduction electrons is not equal because of spin-dependent band structure of the ferromagnet;

Second, the conductance **(*G*)** must be spin-dependent. Therefore, the expressions for conductance, ***G*** should be modulated by the density of states (DOS) for each type of electrons.

As per the general consensus, for a given bias voltage V, the electrons which participate in the conduction process come from the allowed energy levels located at a distance eV from the *Fermi energy* (E_F) (Fig. 4.4). Therefore, to understand the transport mechanism under the application of small bias voltages, we need to know DOS at E_F, i.e., $D_\uparrow(E_\mathrm{F})$ and $D_\downarrow(E_\mathrm{F})$. Hence, more specifically we may say that the expressions for conductance (G) should be modulated by DOS at E_F, i.e., $D_\uparrow(E_\mathrm{F})$ and $D_\downarrow(E_\mathrm{F})$ for the two spin species of electrons. Thus, it is quite expected that the tunnelling magnitude would be spin-dependent.

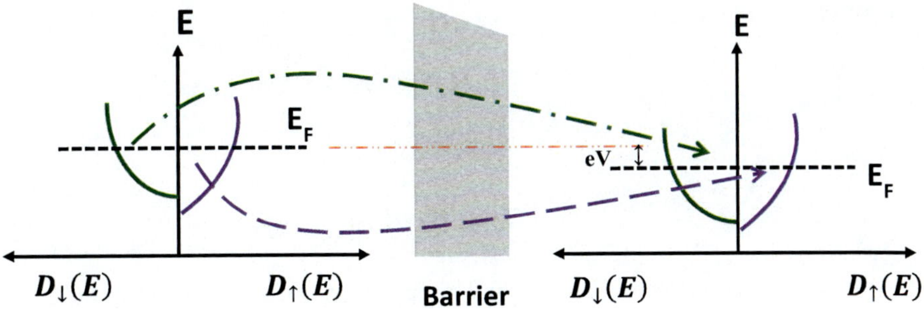

Fig. 4.4 Schematic illustration of DOS at the two sides of a ferromagnetic junction

4.3.3 Tunnelling Process

The TMR effect can be understood using the same two-channel electron spin model as that for GMR. While GMR occurs due to the spin-dependent scattering asymmetry through the entire structure, "***TMR is due to the difference between the tunnelling conductances in the two spin channels, i.e., up-spin and down-spin channel.***"

A Transfer Matrix Model

In general, interfaces between ferromagnetic contacts and the non-magnetic spacer layers have a profound effect in generating spin valve MR of any spin valve device. In this direction, we may adopt a simple one-dimensional model for clear understanding of the effect of interfacial layer on spin valve or TMR effect (Bandyopadhyay and Cahay 2008). In case of TMR, the intermediate non-magnetic semiconductor or insulator spacer layer is kept very thin so that tunnelling of conduction electrons must take place.

Therefore, such spacer layer is approximately referred to as a *delta scatterer*, as shown in Fig. 4.5a. The scatterer and accordingly the scattering potential are expected to have a *spin-independent* and a *spin-dependent* part of strength Γ and Γ', respectively. The spin valve MR is calculated employing *transfer matrix formalism,* considering spin transport to take place in purely one-dimension. Let us suppose that only one energy sub-band is occupied in case of one-dimensional semiconductor quantum wire. Noteworthy, in order to evaluate TMR spin valve, we may assume absolute zero temperature. Furthermore, for the calculation of TMR magnetoresistance, the conductance of the majority and minority spin sub-bands are estimated using Landauer conductance formula.

Let us consider two separate cases:

Case 1: No spin-scattering at the interface

In this case, let us set both Γ and Γ' equal to zero. Thus, for parallel configuration of magnetization of the two ferromagnetic electrodes, the conductance for both the majority and minority spins is given by

$$\boldsymbol{G_\uparrow = G_\downarrow = G_P}, \tag{4.4}$$

and

$$\boxed{\boldsymbol{G_P = 2e^2/h}.} \tag{4.5}$$

Now, let us consider the antiparallel configuration of magnetization of the two ferromagnetic electrodes. In this case, the majority spin electrons originating from the left ferromagnetic electrode, while transmitting across the interface, must experience an energy barrier of height Δ, which is actually the exchange splitting energy. However, for the electrons deriving from the minority spin band from the same ferromagnetic electrode, such energy barrier appears as a step down of size Δ (Fig. 4.5b).

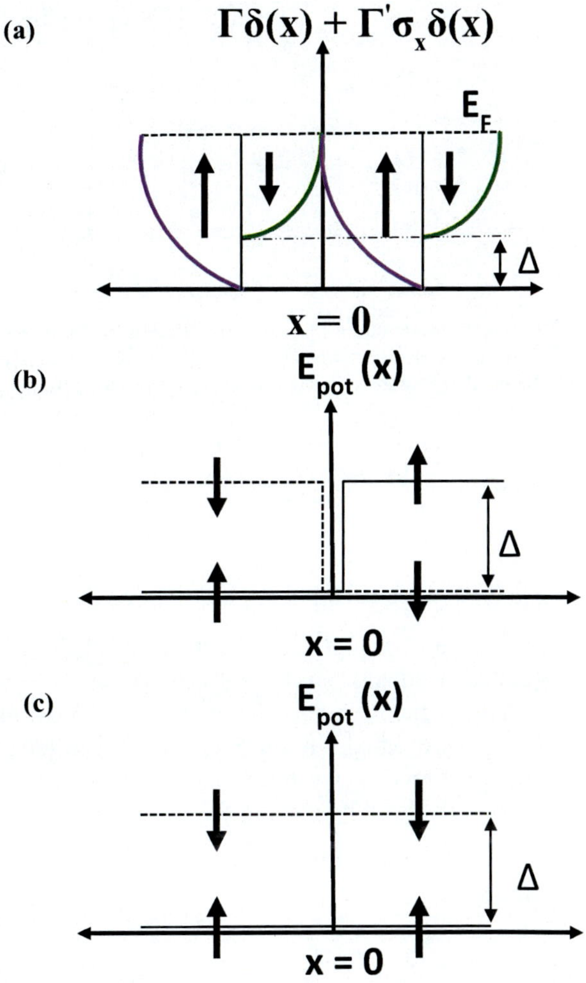

Fig. 4.5 **a** Illustration of a simple one-dimensional model of a barrier between two identical ferromagnetic electrodes. In the absence of any scattering potential at the interface: **b** in case of antiparallel alignment of magnetizations of the ferromagnetic electrodes, the majority spins impinging from the left electrode experience a step barrier of height $\boldsymbol{\Delta}$ (continuous line). The dashed line is the step-down barrier for the minority spins incident from the left electrode. **c** In case of parallel alignment of magnetizations of the ferromagnetic electrodes, either spins do not encounter any potential step in the path when flowing from one electrode to another (Figure adapted and redrawn from Ref. (Bandyopadhyay and Cahay 2008))

As it is well known, the transmission coefficients for the majority and minority spin sub-bands are identical and are given by

$$\boldsymbol{T}_{\uparrow} = \boldsymbol{T}_{\downarrow} = \frac{4\boldsymbol{k}_{\uparrow}^{F}\boldsymbol{k}_{\downarrow}^{F}}{\left[\boldsymbol{k}_{\uparrow}^{F} + \boldsymbol{k}_{\downarrow}^{F}\right]^{2}}. \tag{4.6}$$

Here, $k_{\uparrow}^{F}$ and $k_{\downarrow}^{F}$ are the Fermi wavevectors at the Fermi energy for the majority (up-spin) and minority (down-spin) electrons, respectively. Consequently, the conductance in the antiparallel configuration takes the form

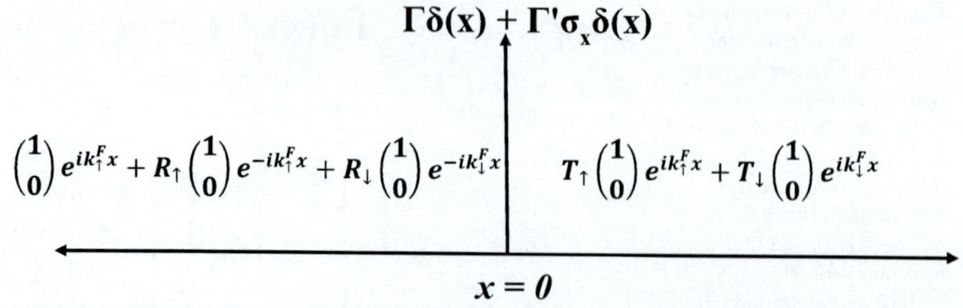

Fig. 4.6 Representation of scattering problem for majority spin electron incident from the left ferromagnetic electrode when the magnetizations alignment between the two electrodes are assumed to be parallel (Figure adapted and redrawn from Ref. (Bandyopadhyay and Cahay 2008))

$$G_{AP} = \frac{8k_{\uparrow}^{F}k_{\downarrow}^{F}}{\left[k_{\uparrow}^{F} + k_{\downarrow}^{F}\right]^{2}}\frac{e^{2}}{h}. \tag{4.7}$$

Case 2: Spin-scattering takes place at the interface

In this case, Γ and $\Gamma' \neq 0$. Let us calculate the conductance for majority spin electrons, incident from the left ferromagnetic electrode with magnetizations in the two ferromagnetic contacts are assumed to be parallel. The incoming and outgoing components of a majority spin incident from the left electrode are depicted in Fig. 4.6.

In this direction, the reflection ($R_{\uparrow}$, $R_{\downarrow}$) and transmission ($T_{\uparrow}$, $T_{\downarrow}$) amplitudes must be calculated by solving the *Pauli equation*, which is subject to the boundary equations for the spinor at the interface as given below:

$$\psi(0_{-}) = \psi(0_{+})$$

$$\frac{d\psi}{dx}(0_{+}) = \frac{d\psi}{dx}(0_{-}) + \frac{2m_0\Gamma}{\hbar^2}\left[1 + \left(\frac{\Gamma'}{\Gamma}\right)\sigma_x\right]\psi(0_{-}). \tag{4.8}$$

Now, substituting the expressions for the spinors, as shown in Fig. 4.6, we obtain a set of four equations for $R_{\uparrow}$, $R_{\downarrow}$, $T_{\uparrow}$ and $T_{\downarrow}$. By easily solving such set of equations one may obtain

$$R_{\uparrow} = \frac{-i\hbar^2 k_{\uparrow}^{F} m_0 \Gamma + m_0^2\left(\Gamma^2 - \Gamma'^2\right)}{A},$$

$$R_{\downarrow} = \frac{-i\hbar^2 k_{\uparrow}^{F} m_0 \Gamma'}{A}$$

$$T_{\uparrow} = \frac{\hbar^2 k_{\uparrow}^{F}(\hbar^2 k_{\uparrow}^{F} + i m_0 \Gamma)}{A}$$

$$T_{\downarrow} = \frac{-i\hbar^2 k_{\uparrow}^F m_0 \Gamma'}{A}. \tag{4.9}$$

where

$$A = \hbar^4 k_{\uparrow}^F k_{\downarrow}^F + i\hbar^2 m_0 \Gamma \left(k_{\uparrow}^F + k_{\downarrow}^F\right) + m_0^2 \left(\Gamma'^2 - \Gamma^2\right). \tag{4.10}$$

Thus, the conductance of the majority spin band in units of e^2/h is given by

$$G_{\uparrow} = \left|T_{\uparrow}\right|^2 + \frac{k_{\downarrow}^F}{k_{\uparrow}^F} \left|T_{\downarrow}\right|^2. \tag{4.11}$$

Following exactly the same procedure, involving scattering problem as presented above, the conductance of electrons incident from the minority spin sub-band can also be calculated. In this case, the expressions for reflection and transmission amplitudes are as follows:

$$\begin{aligned} R'_{\uparrow} &= \frac{-i\hbar^2 k_{\downarrow}^F m_0 \Gamma'}{A}, \\ R'_{\downarrow} &= -1 + \frac{\hbar^4 k_{\uparrow}^F k_{\downarrow}^F + i\hbar^2 k_{\uparrow}^F m_0 \Gamma}{A} \\ T'_{\uparrow} &= \frac{-i\hbar^2 k_{\downarrow}^F m_0 \Gamma'}{A} \\ T'_{\downarrow} &= \frac{\hbar^4 k_{\uparrow}^F k_{\uparrow}^F + i\hbar^2 m_0 \Gamma k_{\downarrow}^F}{A}. \end{aligned} \tag{4.12}$$

Therefore, the conductance of electrons originating from the minority spin sub-band in units of e^2/h is found to be

$$G_{\downarrow} = \left|T_{\downarrow}\right|^2 + \frac{k_{\uparrow}^F}{k_{\downarrow}^F} \left|T_{\uparrow}\right|^2. \tag{4.13}$$

Thus, the total conductance of electrons in the parallel configuration of magnetization of the ferromagnetic electrodes, obtained by adding the conductance $G_{\uparrow}$ and $G_{\downarrow}$ in units of e^2/h becomes

$$\boxed{G_P = G_{\uparrow} + G_{\downarrow} = \left(\frac{k_{\uparrow}^F}{k_{\uparrow}^F} + 1\right) \left|T_{\uparrow}\right|^2 + \left(\frac{k_{\downarrow}^F}{k_{\downarrow}^F} + 1\right) \left|T_{\downarrow}\right|^2.} \tag{4.14}$$

Likewise, total conductance of electrons in the antiparallel configuration of magnetization of the ferromagnetic electrodes, i.e., G_{AP} can be calculated by repeating the same procedure. Finally, we may calculate the magnetoresistance ratio by using G_P and G_{AP}.

4.3.4 *The Jullière Formula*

The TMR effect was first explained by Jullière in his famous 1975 paper (Julliere 1975). In fact, a systematic study of spin-valve magnetoresistance was carried out by Jullière in his innovative effort made in the early years. In the context of TMR, we may discuss one of his pioneering work where Jullière proposed a formula for the junction magnetoresistance (JMR) ratio. Such derivation of JMR ratio takes place from the transfer matrix model as discussed above. The intriguing point is that Jullière formula establishes a relation between JMR and spin polarizations of charge carriers at the Fermi level of the ferromagnetic electrodes. The paramount importance of this formula lies on the fact that this is widely employed to analyse data obtained from spin valve experiments. Moreover, such formula can also efficiently take account of spin relaxation in the intermediate paramagnetic spacer layer. Considering its historical significance, we present here the derivation of Jullière formula.

Before going into the details of Jullière formula, let us mention different kinds of spin valve effect investigated for a wide variety of ferromagnetic electrodes with various types of intermediate paramagnetic layers. Corresponding magnetoresistances have been defined in terms conductances of the device in parallel and antiparallel configuration of magnetizations of its ferromagnetic electrodes. Three commonly used magnetoresistance ratios are as follows:

(a) **Tunnelling magnetoresistance (TMR) ratio,**

$$\boxed{\text{TMR} = \frac{G_P - G_{AP}}{G_{AP}}}$$

(b) **Junction magnetoresistance (JMR) ratio,**

$$\boxed{\text{JMR} = \frac{G_P - G_{AP}}{G_P}}$$

(c) **The spin conductance ratio,**

$$\boxed{\textbf{Spin Conductance Ratio} = \frac{G_P - G_{AP}}{G_P + G_{AP}}}$$

These three magnetoresistance ratios are related to each other.

In 1975, Jullière carried out that revolutionary magnetoresistance experiments on Fe-Ge-Co junctions. The experiments have been carried out at low temperature region ($T \leq 4.2\ K$) and with the variation of the electrical bias applied between Fe and Co ferromagnetic contacts. In an attempt to explain the observed experimental results, he carried out the required modification of the derivation of current, tunnelling between two paramagnetic metallic electrodes separated by a thin insulating barrier, as done by Bardeen.[1] Such formula due to Bardeen is quite analogous to that of TsuEsaki formula. It should be mentioned in this context that in order to study the resonant tunnelling diodes and related devices, TsuEsaki formula has been consistently used by device engineers and physicists.

Let us discuss the required considerations for formulation of the theory as follows:

When no bias voltage is applied to the ferromagnetic electrodes, there will be no current flowing through the tunnel barrier across the device. With the application of voltage V across the ferromagnetic electrodes, there is a shift of the energy level of one ferromagnetic electrode relative to the second one by an amount of eV. Such energy shift eV between the Fermi levels gives rise to the tunnelling current. According to Bardeen approach, one must estimate the current $I(V,E)$ flowing under the application of a particular bias voltage V and energy E between the two ferromagnetic left (L) and right (R) electrodes using Fermi's golden rule, i.e.,

$$I(V, E) \propto |T(E)|^2 D_L(E) D_R(E + eV_{\text{bias}})\left[f(E) - f(E + eV_{\text{bias}})\right], \quad (4.15)$$

where $D_L(E)$ and $D_R(E)$ are the densities of states of the left and right ferromagnetic electrodes, respectively. $|T(E)|^2$ is the square of the tunnelling matrix element and $f(E)$ is the Fermi function. Let us simplify the thing by considering that the tunnelling matrix element does not depend on energy over the relevant range $\approx eV_{\text{bias}}$. Thus, the total tunnelling current can be found by taking integration of Eq. 4.15 with respect to energy

$$I(V) \propto |T|^2 \int_{-\infty}^{\infty} D_L(E) D_R(E + eV_{\text{bias}})\left[f(E) - f(E + eV_{\text{bias}})\right] dE. \quad (4.16)$$

Consideration of both low-bias regime, i.e., $V \to 0$ and low temperatures, i.e., $T \to 0$ we get

$$\lim_{V_{\text{bias}}, T \to 0} \frac{\left[f(E) - f(E + eV_{\text{bias}})\right]}{eV_{\text{bias}}} = \delta(E - E_F), \quad (4.17)$$

where E_F is the Fermi energy. Therefore, we obtain

$$G = \frac{dI}{dV} \propto |T|^2 D_L(E_F) D_R(E_F), \quad (4.18)$$

[1]This problem can be solved using software tools such as MATLAB or MATHEMATICA.

where $dI/dV = G$ is the zero-bias conductance of the tunnelling junction. It is quite evident from Eq. 4.18 that in the low-bias regime current should be proportional to the voltage, i.e., $I \propto V$, which implies the ohmic behaviour of the junction.

Jullière adopted Eq. 4.18 to explain the magnitude of tunnelling magnetoresistance in case of a MTJ, consisting of ferromagnetic (FM)–insulator (I)–ferromagnetic (FM) layer structures. As it has already been discussed in case of ferromagnetic electrodes, densities of states related to the majority (spin parallel to the magnetization) and minority (spin antiparallel to the magnetization) spin electrons, denoted by D↑(E) and D↓(E), respectively, are different. This is because the up-spin and down-spin energy bands are shifted with respect to one another along the energy axis by exchange interaction. In an attempt to account a crucial point of concern that tunnelling current depends on the relative orientation of magnetization of the ferromagnetic electrodes, an additional assumption that ***electron spin is conserved in tunnelling*** has been proposed by Jullière. Thus, tunnelling of up-spin and down-spin electrons can be considered as two independent channels of current resembling two wires connected parallelly. This implies that the tunnelling current flows in two separate up-spin and down-spin channels, which is the well-known ***two-current model***. Such model is also successfully employed for the interpretation of the giant magnetoresistance effect, which is closely related to TMR effect, in magnetic multilayers. As generally considered in any kind of spin valve effect, the magnetization of the ferromagnetic electrodes is ***antiparallel*** without application of any magnetic field, i.e., when the magnetic field is zero and ***parallel*** under the application of a saturating magnetic field H_s. Thus, employing Eq. (4.18) the conductance of the junction in zero field is

$$G_{AP} \propto D_L^{\uparrow}(E_F)D_R^{\downarrow}(E_F) + D_L^{\downarrow}(E_F)D_R^{\uparrow}(E_F), \tag{4.19}$$

and its conductance at the saturating field is

$$G_P \propto D_L^{\uparrow}(E_F)D_R^{\uparrow}(E_F) + D_L^{\downarrow}(E_F)D_R^{\downarrow}(E_F), \tag{4.20}$$

Furthermore, two new parameters P_1 and P_2 have been introduced in order to characterize the spin polarization of left and right ferromagnetic electrodes

$$P_1 = \frac{\left[D_L^{\uparrow}(E_F) - D_L^{\downarrow}(E_F)\right]}{\left[D_L^{\uparrow}(E_F) + D_L^{\downarrow}(E_F)\right]}, \tag{4.21}$$

and

$$P_2 = \frac{\left[D_L^{\uparrow}(E_F) - D_L^{\downarrow}(E_F)\right]}{\left[D_L^{\uparrow}(E_F) + D_L^{\downarrow}(E_F)\right]}, \tag{4.22}$$

Substituting the values of G_P and G_{AP} and using the definitions of the spin polarizations (P_1, P_2) of the ferromagnetic electrodes from Eqs. 4.21 and 4.22, the junction magnetoresistance ratio (*JMR*) is found to be

$$\mathbf{JMR} = \frac{2\boldsymbol{P}_1\boldsymbol{P}_2}{1 + \boldsymbol{P}_1\boldsymbol{P}_2}. \tag{4.23}$$

Equation 4.23 gives *JMR ratio* as derived by Jullière. The parameters P_1 and P_2 can be determined separately from the measurements of the tunnelling current in ferromagnet (FM)–insulator (I)–superconductor (S) junctions. This compact result, as shown in Eq. 4.23, is the famous Jullière formula.

Similarly, the TMR ratio obtained using Jullière formula is as follows:

$$\mathbf{TMR} = \frac{2\boldsymbol{P}_1\boldsymbol{P}_2}{1 - \boldsymbol{P}_1\boldsymbol{P}_2}. \tag{4.24}$$

4.3.5 Simple Description of Tunnelling Phenomenon

Let us first recall and summarize from the above-made discussions, ***the two important assumptions***, based on which tunnelling effect has been explained:

First, spins of electrons remain conserved in the tunnelling process. Given the fact that in ferromagnetic electrodes, densities of states of majority $D_{\uparrow}(E)$ and minority $D_{\downarrow}(E)$ spin electrons are different; tunnelling of up- and down-spin electrons should be two independent processes. Therefore, conductance is supposed to take place in two independent spin channels, constituted by up-spin and down-spin electrons (Fig. 4.4). It is obvious that electrons deriving from one kind of spin orientation (either up or down spin) of the first ferromagnetic electrode are received by unoccupied energy levels of the same spin orientation of the second ferromagnetic electrode. This means that when the magnetizations of the two ferromagnetic electrodes are parallel, the minority spins (spin antiparallel to the magnetization) should tunnel to the minority spin states, whereas the majority spins (spin parallel to the magnetization) tunnel to the majority spin states. However, in the opposite case when magnetizations of the two ferromagnetic electrodes are antiparallel, the identity of the majority- and minority-spin electrons is reversed. This means that if up-spin electron is the majority spin electrons (spin parallel to the magnetization direction) for the first ferromagnetic electrode, this will be the minority spin electrons (spin antiparallel to the magnetization) for the second ferromagnetic electrode. In the same picture, down-spin electrons acting as the minority spin electrons for the first ferromagnetic electrode would be the majority spin electrons for the second ferromagnetic electrode. Therefore, it comes out that for the antiparallel alignment of magnetizations of the ferromagnetic electrodes, majority spin electrons of the first ferromagnetic electrode tunnel to the available minority spin states in the second

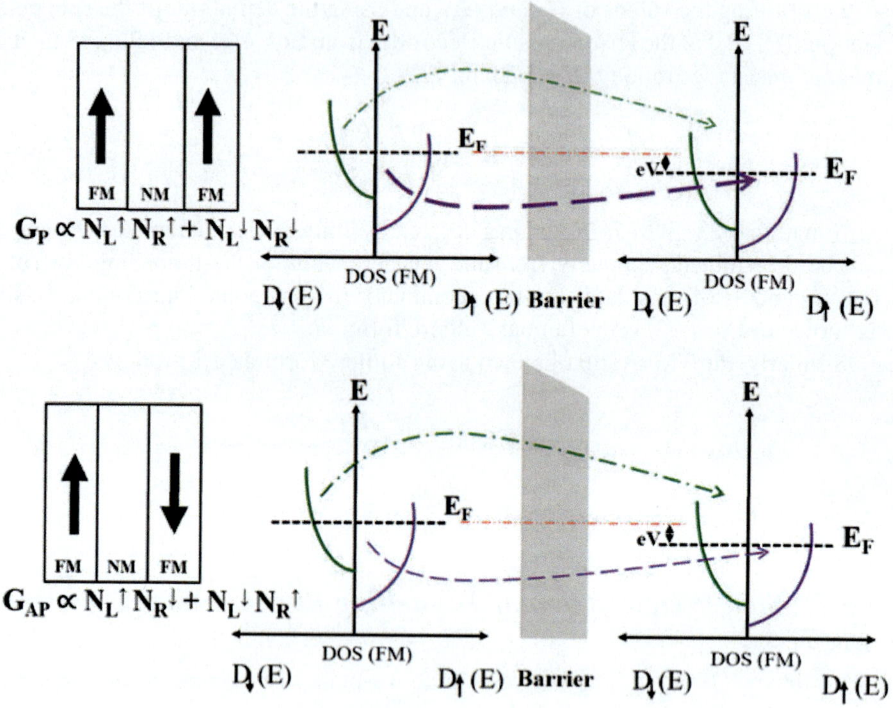

Fig. 4.7 Schematic description of the TMR as explained by Jullière model

ferromagnetic electrode and vice versa. This model is clearly depicted in Fig. 4.7. In order to simplify the calculation, let us suppose negligible direct magnetic coupling between the ferromagnetic electrodes and that the bulk effects are determining the alignment of magnetization in each ferromagnetic electrode.

Furthermore, it is understandable that

$$\begin{aligned} D_{\downarrow}(E_F) &\propto N_{\downarrow} \\ D_{\uparrow}(E_F) &\propto N_{\uparrow}. \end{aligned} \tag{4.25}$$

where $N_{\downarrow}$ and $N_{\uparrow}$ are the number of electrons with down- and up-spin, respectively.

Second, conductance (G) of the electron for a particular spin orientation tunnelling from one ferromagnetic electrode to another is proportional to the product of the effective density of states at Fermi level (E_F) of the two ferromagnetic electrodes.

Thus, considering Eq. (4.25) one may expect that

$$\begin{aligned} G_P^{\uparrow} &\infty N_L^{\uparrow} N_R^{\uparrow}, \\ G_P^{\downarrow} &\infty N_L^{\downarrow} N_R^{\downarrow}, \\ G_{AP}^{\uparrow} &\infty N_L^{\uparrow} N_R^{\downarrow}, \end{aligned}$$

$$G_{AP}^{\downarrow} \infty N_L^{\downarrow} N_R^{\uparrow}, \tag{4.26}$$

where superscripts L and R stand for the left and right ferromagnetic electrodes. N_L and N_R are the total number of electrons of the left and right ferromagnetic electrodes, respectively.

Therefore, conductance of the junction for parallel magnetization alignment of ferromagnetic electrodes (Fig. 4.7) can be written as

$$G_P = G_P^{\uparrow} + G_P^{\downarrow}, \tag{4.27}$$

and that for antiparallel magnetization configuration of ferromagnetic electrodes is

$$G_{AP} = G_{AP}^{\uparrow} + G_{AP}^{\downarrow}. \tag{4.28}$$

Evidently, G_p should be pretty large because $G_{AP}^{\uparrow}$ is supposed to be very large, whereas G_{AP} should be very much small since both $G_{AP}^{\uparrow}$ and $G_{AP}^{\downarrow}$ are having smaller values. Large and small values of conductance, corresponding to the relevant spin channel, are clearly depicted in Fig. 4.7 by thick and thin curved arrows, respectively, denoting tunnelling of electrons.

i.e., $\mathbf{G_p >> G_{AP}}$.

As already mentioned, TMR can be defined as the ratio of the change in conductance to the minimum conductance as follows:

$$\text{TMR} = \frac{G_P - G_{AP}}{G_{AP}}. \tag{4.29}$$

Again one can define the polarization of the left and right electrodes by

$$P_{L,R} = \frac{N_{L,R}^{\uparrow} - N_{L,R}^{\downarrow}}{N_{L,R}^{\uparrow} + N_{L,R}^{\downarrow}} = \frac{\Delta N_{L,R}}{N_{L,R}}. \tag{4.30}$$

Using Eqs. (4.26) and (4.30) in Eq. (4.29), we obtain the expression for TMR as follows:

$$\text{TMR} = \frac{2P_L P_R}{(1 - P_L P_R)}. \tag{4.31}$$

Equation (4.31) clearly shows that TMR is directly proportional to the polarization of the electrodes. Such simplified analysis roughly provides explanations on the experimental observations, obtained in pioneering experiments of spin tunnelling (Julliere 1975; Stearns 1977). It comes out that following Jullière model, one can estimate the magnitude of TMR in MTJs from the known values of the spin polarization of ferromagnetic electrodes. In this context, we must mention that spin polarizations of ferromagnetic electrodes can be obtained from the experiments on superconductors. Jullière formula is the most suitable for comparing TMR values for MTJs, consisting of different ferromagnetic electrodes but the same spacer layers because TMR in this case is evaluated in terms of the spin polarization. To summarize, in the light of Jullière's model, (a) spin orientations remain conserved, i.e., spin reorientation/flipping does not take place during electron tunnelling. As a result, tunnelling phenomenon of ↑and ↓ spin electrons can be supposed to take place independently with respect to each other; (b) process of tunnelling is spin independent. Hence, *conductance process corresponding to a particular spin species (↑/↓) is only decided by the appropriate density of states of the two ferromagnetic electrodes.* In order to obtain a more credible theoretical formulation, two additional effects should be included: (1) At the Fermi level, DOS is not necessarily proportional to the total polarization; (2) At the vicinity of the barrier, the wavefunctions of the majority and minority electrons need not be the same. In that case, transmission coefficient is supposed to acquire a spin dependence which in turn influences MR.

4.4 Effect of Various Parameters on Tunnel Magnetoresistance

Here we will discuss, (i) the changes in the basic elastic processes induced by the interface, (ii) inelastic tunnelling phenomenon produced by magnetic excitations both at the ferromagnetic electrodes or at the interfaces, (iii) the changes in the magnetic structure of the surface, and (iv) charging effects.

4.4.1 Effect of Paramagnetic Impurities at the Interface on Magnetoresistance

According to our earlier discussion, magnetization of the ferromagnetic electrodes is aligned when the external magnetic field is applied. However, because of the existence of isolated paramagnetic impurities at the interface, a much higher external magnetic field is required to attain such parallel alignment of magnetization of the ferromagnetic electrodes. Presence of such misaligned impurities can lead to diversified transport processes.

Let us suppose ***resonant tunnelling*** through a single paramagnetic impurity found at the barrier between two ferromagnetic electrodes, as shown in the illustration (Fig. 4.8). In this case, electrons are supposed to transit elastically from one ferromagnetic electrode to the impurity at the interface and then to the second ferromagnetic electrode. Obviously, such transit of electrons would take place without losing any coherence.

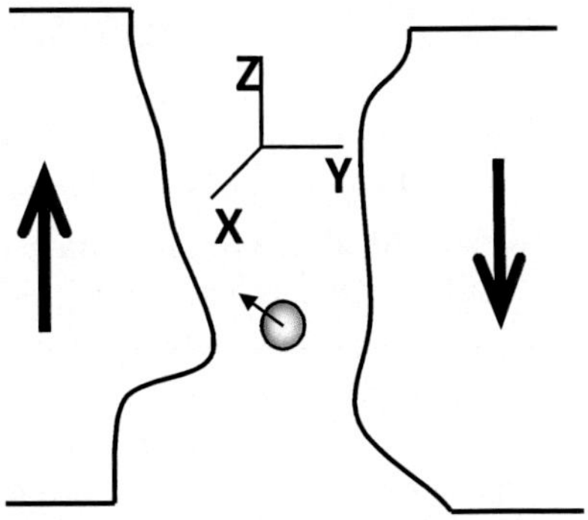

Fig. 4.8 Paramagnetic impurity at the intermediate barrier in a magnetic tunnel junction

Case I: Application of high magnetic field: As stated before, magnetization of the two ferromagnetic electrodes is aligned parallel under the application of high magnetic field. Therefore, ***direct tunnelling process*** occurs only by tunnelling of an electron originating from a state of the majority (minority) spin band of first ferromagnetic electrode to available state in the majority (minority) spin band of the second electrode. Now, considering the magnetization direction along z-axis, probable electronic state of the paramagnetic impurity with its magnetic moment aligned along x-axis becomes

$$|\Psi\rangle = \frac{1}{\sqrt{2}}(|\uparrow\rangle + |\downarrow\rangle). \tag{4.32}$$

Thus, it is quite suggestive that an electron with a given spin orientation, deriving from the first ferromagnetic electrode can undergo tunnelling via the paramagnetic impurity, ***with equal probability***, into both up- and down-spin polarization states of the second ferromagnetic electrode. This is coherent tunnelling of electrons.

Case II: Application of low magnetic field: On the other hand, application of a low magnetic field can align the magnetization of the ferromagnetic electrodes. However, such low magnetic field will have a negligible influence on the magnetization alignment of paramagnetic impurities at the interface. Therefore, a contribution to the conductance of the junction, insensitive to applied magnetic field, also comes from the tunnelling through the magnetic impurities at the interface. This in turn results in reduced MR. Obviously, application of a strong magnetic field aligns the magnetic moments associated with those impurities at the interface. As a result, MR is recovered close to that of a clean junction.

Now, considering ***incoherent tunnelling*** through such impurities at the interface, tunnelling electrons lose their spin orientation, i.e., their memory of spin during their passage through the impurity. This in turn also leads to reduction of MR. Incoherent tunnelling might be important and needs to be considered if magnetic clusters play the role of impurity. Such impurity is actually having several internal degrees of freedom.

4.4.2 Effect of Magnetic Excitations on the MR

Up to this point of discussion, we have only considered the elastic tunnelling process that does not lead to creation or absorption of excitations at the junction. This is an oversimplified assumption, since in this case different kinds of interactions between quasiparticles were not considered. However, in realistic scenario, there are a number of inelastic processes feasible in a junction. Such inelastic processes can of course influence MR.

An evident signature of inelastic process is their distinctive temperature dependences. Increase in temperature means more available thermal energy, $k_B T$, i.e., an increased number of excitations in the system. An alternative way to provide energy to a junction is through the applied voltage to the system. In case of applied bias, the associated energy scale is eV. Therefore, it comes out that the contribution to the conductance of inelastic processes scales in the similar manner with temperature or voltage.

4.4.3 Effect of Magnetic Properties of the Interface on MR

Recalling the very basic features of interfaces, it can be stated that atoms at the interfaces of a given junction are surrounded by an environment that is significantly different from that at the bulk of each electrode. Consequently, this may lead to the change in the magnetic structure of the interface (Mattis 1988). Indeed, the characterization of the interfacial magnetic properties is pretty difficult. Also, no unique formulation can offer solution to the problem. A simple estimation of the changes of the interfacial magnetic structure can be done by analysing a semi-infinite chain of ferromagnetically coupled spins. For instance, reduction in the number of nearest neighbours causes a shift of the density of magnetic excitations towards lower energies. Thus, at low temperatures the spins at the surface undergo more fluctuations compared to than in the bulk. Such fluctuations at the interface is detrimental to MR. Additionally, the surface excitations can also mediate *spin-flip* processes and thereby may cause decrease in MR of the junction.

4.4.4 Effect of Charging in Granular Systems on MR

In case of magnetic granular systems, metallic junctions mainly determine its transport properties. MR of such magnetic grains is usually determined by the tunnelling of conduction electrons across the interfaces between neighbouring grains or grain boundaries. Now, addition of one electron charge in a given grain increase its charge and thereby causes change in its electrostatic energy, which is approximately equal to $[e^2/(\varepsilon R)]$, where R is the average radius of the grain and ε the dielectric constant of the surrounding material. In case of a larger electrostatic energy than the thermal energy, $k_B T$, tunnelling phenomenon gets suppressed. This is referred to as Coulomb blockade (Devoret and Grabert 1992).

In this context, we may mention that the interplay of charging effects and magnetism results in a wide variety of phenomena. For instance, the resistance of granular materials exhibits an upturn at low temperatures. This feature can be explained considering average charging energy, assuming a given grain sizes and barrier heights distribution (Coey et al. 1998; Balcells et al. 1998; Helman and Abeles 1976). Now, application of a magnetic field causes decrease in the effective charging energy due to increase in intergrain conductance. Obviously, tunnelling between grains would be more efficient during parallel alignment of magnetization of the grains.

4.5 Measurement of Spin Relaxation Length and Time in the Spacer Layer

In an attempt to extract spin relaxation length (L_S) and spin relaxation time (τ_s) in a paramagnetic material, a standard method is to carry out ***spin valve experiment*** (discussed earlier). The change in resistance with the corresponding change in magnetization alignment from parallel to antiparallel configurations enable one to find out the spin relaxation length and time in the paramagnetic spacer layer. As we already know in case of MTJs, having tunnel barrier as the spacer layer, TMR ratio is given by Jullière formula: TMR $= \frac{2P_1P_1}{(1-P_1P_1)}$, where P_1 and P_2 are the *spin polarizations of the DOS at the Fermi level* of the two ferromagnetic electrodes. More often, P_1 and P_2 are associated with the spin polarizations of the tunnelling current. Such polarizations of the tunnelling current are separately determined via Meservey-Tedrow experiments employing alumina tunnel barrier (Tsymbal et al. 2003). It is noteworthy that the spin polarization of the tunnel current depends on the DOS and also on the probability of tunnelling. Tunnel probability has been found to be dependent on barrier (De Teresa et al. 1999) and may be different for different electronic states in the ferromagnetic material. In a previous study, Co has been found to exhibit a negative spin polarization of tunnelling electrons for $SrTiO_3$ barrier, whereas spin polarization has been found to be positive for alumina barriers.

In order to measure spin relaxation length and time in the spacer layer, further extension of Jullière's formula has been carried out in case of thicker paramagnetic spacers in which spin transport takes place via drift-diffusion or multiple hopping, instead of direct tunnelling between the contacts. Let us assume that the injector and detector interfaces have tunnelling (Schottky) barrier, which occurs in diversified metal/organic interfaces. Thus, in such a scenario, spin-polarized electrons, deriving from one ferromagnetic electrode, are injected through the tunnel barrier into the paramagnetic spacer layer. Let us suppose that the spin polarization of the injected electrons is P_1, which can be determined exactly by the Meservey-Tedrow technique. After entering the paramagnetic spacer layer, these spin-polarized electrons will undergo drift and diffusion (or hop) and thereby proceeding towards the ferromagnetic detector electrode, under the influence of a transport-driving electric field. The spin polarization of the injected spin-polarized electrons (P) has been found to decrease with distance x, measured from the point of injection. Such relaxation process of spin polarization can be modelled as

$$P(x) = P_1 \exp(-x/L_s), \tag{4.33}$$

where L_S is referred to as the spin relaxation or spin diffusion length. There is also experimental confirmation for such exponential decay of spin polarization with distance (Shim et al. 2008). Therefore, when those electrons arrive at the ferromagnetic detector interface, their spin polarization becomes

$$P(d) = P_1 \mathbf{exp}(-d/L_s). \tag{4.34}$$

where d is the distance between the interfaces of the injector/paramagnetic spacer layer and the detector/paramagnetic spacer layer of the spin valve, i.e., between the injection and the detection point. Let us apply Jullière formula on the tunnel barrier of detector/paramagnetic spacer layer interface, which separates the two spin polarizations $P(d)$ and P_2 of the spin valve. In this case, using Eq. (4.31) we have

$$\mathbf{MR} = \frac{2P(d)P_2}{1 - P(d)P_2} = \frac{2P_1P_2\mathbf{exp}[-d/L_s]}{(1 - P_1P_2\mathbf{exp}[-d/L_s])}. \tag{4.35}$$

This is the so-called '***modified Jullière formula***', which is extensively employed to estimate L_s. The physical model described above is depicted in Fig. 4.9.

While applying this formula, the values of P_1 and P_2 are directly adopted from the literature (Tsymbal et al. 2003). Those values may not be the exact values of spin polarization relevant for a particular experiment. Let us mention some cases related to this discussion. For purely organic or organic/inorganic *hybrid* barriers, spin polarizations of the ferromagnetic electrode have been found to be less (Shim et al. 2008; Santos et al. 2007) than the tabulated values (Tsymbal et al. 2003). Again, any surface contamination of the ferromagnetic electrode can also cause reduction of P_1 and P_2. In order to take into account these effects, P_1 and P_2 can be replaced by $\alpha_1 P_1$ and $\alpha_2 P_2$, respectively, where $\alpha_1, \alpha_2 < 1$. $\alpha_1 P_1$ and $\alpha_2 P_2$ can be determined by carrying out spin-dependent tunnelling (Meservey and Tedrow 1994), muon spin rotation (Drew et al. 2009) or two-photon photoemission experiments (Cinchetti et al. 2009). Finally, this model provides valuable insight of the tunnelling process and can be employed to obtain a rough estimate of L_S. Furthermore, dependence of L_S on temperature and bias can be exploited to shed light on the spin dynamics in the paramagnet.

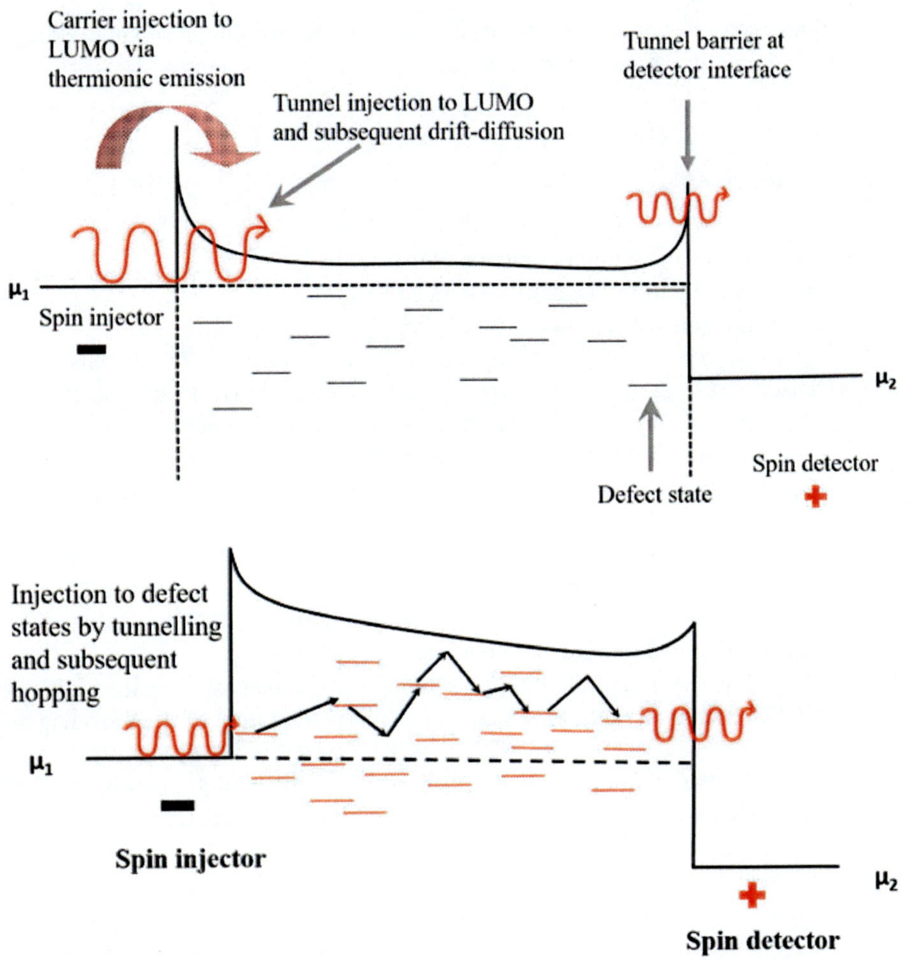

Fig. 4.9 Physical model describing the modified Jullière formula. Tunnel or thermionic injection followed by drift-diffusion (top) and hopping transport (bottom). Spin polarization decreases exponentially with distance from the point of injection

4.6 Conclusions

In this chapter, we have introduced the concept of tunnelling magnetoresistance, magnetic tunnel junction (MTJ) and magnetic junctions. A detailed physical explanation behind the phenomenon has also been presented. In this context, quantum mechanical tunnelling of conduction electrons, which is at the origin of this phenomenon has been discussed in the light of a transfer matrix model. Both the cases, with and without spin-scattering at the interface, have been considered. The Jullière formula has been discussed elaborately. Furthermore, simple description of tunnelling phenomenon has also been presented. The effect of various parameters on tunnel magnetoresistance has also been discussed. Measurement of spin relaxation length and time in the spacer layer is also explained in the chapter.

4.7 Exercises

1. What is magnetic tunnel junction (MTJ)?
2. What is tunnelling magnetoresistance (TMR)?
3. What are contact and tunnel-type junction?
4. What do you mean by spin-dependent conductance of charge carriers?
5. Discuss ***transfer matrix model*** for both cases when (i) no spin-scattering at the interface and (ii) spin scattering takes place at the interface.
6. Formulate ***Jullière model***. Hence write the expression for tunnelling magnetoresistance (TMR) ratio, junction magnetoresistance (JMR) ratio and spin conductance ratio.
7. Explain the conditions required for the formulation of the ***Jullière model***.
8. What is ***two-current model?***
9. Give simple description of tunnelling phenomenon and hence derive the equation related to tunnelling magnetoresistance.
10. Show how with the following Jullière model, one can estimate the magnitude of TMR from the known values of the spin polarization of the ferromagnetic electrodes.
11. What are the main factors that determine the TMR in multilayer films?
12. How does the TMR get affected by the presence of magnetic impurities at the interface?
13. What is the effect of external magnetic field in TMR of the granular system?
14. What is tunnelling magnetoresistance (TMR)? Derive the equation tunnelling magnetoresistance (TMR) $= \frac{2P_L P_R}{1-P_L P_R}$, where P_L and P_R are the polarization of the left and right electrodes, respectively.
15. How the spin relaxation length and time in the spacer layer could be estimated from spin valve experiments?
16. What are the basic differences between GMR and TMR effect?
17. What are the two basic assumptions, based on which ***Jullière*** explained TMR effect?
18. Derive ***modified Jullière formula***. State its significance.
19. Suppose a magnetic tunnel junction with thicker paramagnetic spacer layer of thickness 2 nm, spin polarizations of the DOS at the Fermi level of both the ferromagnetic electrodes are 0.4, tunnelling magnetoresistance obtained at room temperature is 0.2. What is the spin relaxation length in the paramagnetic spacer layer?
20. What do you mean by incoherent tunnelling and resonant tunnelling?

References

R. Meservey, P.M. Tedrow, Spin-polarized electron tunneling. Phys. Rep. **238**, 173 (1994)

L. Ming Loong et al., Flexible MgO barrier magnetic tunnel junctions. Adv. Mater. **28**, 4983–4990 (2016)

S. Bandyopadhyay, M. Cahay, *Introduction to Spintronics* (CRC Press Taylor & Francis, 2008)

M. Julliere, Tunneling between ferromagnetic films. Phys. Lett. **54** A, 225 (1975)

M.B. Stearns, Simple explanation of tunneling spin-polarization of Fe Co, Ni and its alloys. J. Magn. Magn. Mater. **5**, 167 (1977)

D. C. Mattis, *The Theory of Magnetism I: Statics and Dynamics* (Springer, New York, 1988)

M.H. Devoret, H. Grabert, *Single Charge Tunneling* (Plenum Press, New York, 1992)

J.M.D. Coey, A.E. Berkowitz et al., Magnetoresistance of chromium dioxide powder compacts. Phys. Rev. Lett. **80**, 3815 (1998)

L. Balcells, J. Fontcuberta, B. Martinez, X. Obradors, High-field magnetoresistance at interfaces in manganese perovskites. Phys. Rev. B **58**, R14697 (1998)

J.S. Helman, B. Abeles, Tunneling of spin-polarized electrons and magnetoresistance in granular Ni films. Phys. Rev. Lett. **37**, 1429 (1976)

E. Y. Tsymbal, O. N. Mryasov, P. R. LeClair, Spin-dependent tunnelling in magnetic tunnel junctions. J. Phys. Condens. Matter **15**(4), R109–R142 (2003)

J. M. De Teresa, A. Bartnotehelemy, A. Fert, J. P. Contour, R. Lyonnet, F. Montaigne, P. Seneor, A. Vaures, Inverse tunnel magnetoresistance in $Co/SrTiO_3/La_{0.7}Sr_{0.3}MnO_3$: new ideas on spin-polarized tunneling. Phys. Rev. Lett. **82**, 4288–4291 (1999)

J.H. Shim, K.V. Raman, Y.J. Park, T.S. Santos, G.X. Miao, B. Satpati, J.S. Moodera, Large spin diffusion length in an amorphous organic semiconductor. Phys. Rev. Lett. **100**, 226603 (2008)

T.S. Santos, J.S. Lee, P. Migdal, I.C. Lekshmi, B. Satpati, J.S. Moodera, Room-temperature tunnel magnetoresistance and spin-polarized tunneling through an organic semiconductor barrier. Phys. Rev. Lett. **98**, 016601 (2007)

A.J. Drew, J. Hoppler, L. Schulz, F.L. Pratt, P. Desai, P. Shakya, T. Kreouzis, W.P. Gillin, A. Suter, N.A. Morley, V.K. Malik, A. Dubroka, K.W. Kim, H. Bouyanfif, F. Bourqui, C. Bernhard, R. Scheuermann, G.J. Nieuwenhuys, T. Prokscha, E. Morenzoni, Direct measurement of the electronic spin diffusion length in a fully functional organic spin valve by low-energy muon spin rotation. Nat. Mater. **8**, 109–114 (2009)

M. Cinchetti, K. Heimer, J.P. Wustenberg, O. Andreyev, M. Bauer, S. Lach, C. Ziegler, Y.L. Gao, M. Aeschlimann, Determination of spin injection and transport in a ferromagnet/organic semiconductor heterojunction by two-photon photoemission. Nat. Mater. **8**, 115–119 (2009)

Chapter 5
Spin-Transfer Torque

5.1 Introduction on Spin-Transfer Torque

Spin-transfer torque, generally observed in magnetic multilayers, is a very intriguing phenomenon. In fact, magnetic multilayers, exhibiting GMR effect, in which conduction current flows perpendicular to the interfaces, could demonstrate *spin-transfer torque* effect under suitable experimental conditions.

5.1.1 What Is Spin-Transfer Torque (STT)?

At the onset, let us suppose a spin-polarized electron current flows through a ferromagnetic (FM) layer, magnetized to saturation. In this case, a fraction of the spin angular momentum, associated with the spin-polarized electron current, can possibly be *transferred* from the spin polarization of the incident electron current to the magnetization of the FM material. From elementary mechanics, it is well known that the *rate of change of angular momentumwith time* is equal to *torque*, given by the following equation:

$$\tau = \frac{d\vec{L}}{dt}, \tag{5.1}$$

where τ = torque, $\vec{L}$ = angular momentum. Therefore, this phenomenon can be interpreted *as if* spin-polarized electrons applying *torque* directly to the magnetization of the FM layer. Hence, this interaction is referred to as *spin-transfer torque (STT)*.

If the density of the spin-polarized current is sufficiently high, we may obtain precession of the magnetization of the FM materials. An even higher density of the spin-polarized current may result in reversal of the magnetization of the FM materials. This concept of magnetization reversal of a FM layer induced by STT effect was

P. Dey and J. N. Roy, *Spintronics*,
https://doi.org/10.1007/978-981-16-0069-2_5

first proposed by Slonczewskil and Berger (Slonczewski 1989, 1996, 2005) independently in 1996. Important point is that this STT is possibly more effective to modulate the orientation of the magnetization of nanoscale memory devices compared to that of external magnetic fields. In effect, such feasibility of magnetization modulation of ferromagnetic materials employing STT makes it possible to design integrated spintronic devices with reduced dimension and energy consumption than conventional magnetic field actuation. Therefore, STT has attracted great deal of interest for possible implementation in future magneto-electronic devices (Slonczewski 1989, 1996, 2005; Huang et al. 2007; Myers et al. 1999).

5.2 Spin-Transfer Torque in Ferromagnetic Layer Structures

5.2.1 *Single Ferromagnetic (FM) Layer*

Transit of a single electron: Let us first discuss a simplified case of spin-transfer torque observed in a single ferromagnetic (FM) layer. In this case, suppose an electron travels towards a thin FM layer through a non-magnetic (NM) metallic layer, as shown in Fig. 5.1. Let us consider that the magnetic moment associated with the ferromagnetic layer is oriented in the z-direction, whereas the direction of travel of the incoming electron is along the x-direction. Spin magnetic moment of this electron, i.e., its spin polarization makes an angle θ with respect to z-axis in the xz plane.

Region 1: ***Interface between NM metal and the FM layer***

Initially, when the electron enters the FM layer, interface between the NM metal and the FM layer behaves as spin filter, where filtering of the electronic spin states take place. Consequently, relative amplitudes of the spin-up and spin-down components in the transmitted spin wave function of the electron get changed, compared to the incident state at the interface. In this context, we may mention that in copper and cobalt (Cu/Co) interface, the spin-up electrons, which is the majority spin electrons,

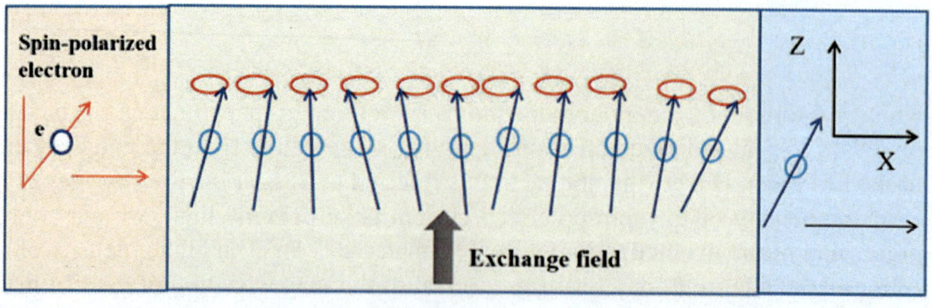

Fig. 5.1 Schematic drawing of a spin-polarized electron travelling through a NM/FM/NM thin film

have larger transmission amplitude through the interface compared to spin-down electrons. As a result, spin polarization of the transmitted electron across the interface between the NM metal and the FM layer is supposed to be oriented at a smaller angle θ with respect to z-axis than its initial incident angle.

Region 2: ***Ferromagnetic (FM) layer***

After entering the FM layer, electron continues to travel through the ferromagnet under proper biasing condition. In this layer, spin magnetic moment of electron experiences strong exchange interaction with the magnetization of the ferromagnet. As already discussed, spin of the incident electron is assumed not to be parallel with the magnetization of the FM. Hence, electron spin is supposed to execute precession about the direction of the exchange field. Considering conservation of angular momentum, the magnetization associated with the FM must also experience precession about the direction of electron spin. This precession has been found to be extremely rapid. As a result, electron traverses only a few lattice spacings per precession period. Thus, even for a 3 or 4 nm thin film, electron may undergo precession several times before leaving the rear side of the FM layer. Finally, whether electron would exit the ferromagnet with either positive or negative spin components in the x and y directions is decided by exactly how many fractions of a turn the electron has precessed before leaving the ferromagnet. Quite expectedly, the z component of the spin should not change during the precessional motion of the electron as the precession occurs around z-axis.

Transit of many electrons: The above discussion is oversimplified since only single electron has been considered. But, in case of a real device, a large number of electrons are travelling simultaneously through the FM layer. Those electrons are propagating with different energies and incident angles in real space. Let us suppose, every electron began with exactly the identical initial angle for its spin polarization. Nevertheless, each electron would undergo a different degree of precession within the FM layer, since each of them is supposed to take a different amount of time to traverse the FM layer. Therefore, even for the fractions of electrons having same initial angle, based on exactly how much fractions of a turn each electron has precessed, it could leave the FM layer with either positive or negative spin components in the x and y directions. Quite expectedly, in this scenario spin component along z direction would not change during the transit of electrons through the ferromagnetic layer. Thus, summing over total spin angular momentum of all the electrons, exiting the FM layer, yields the only non-negligible spin component pointing along z direction, i.e., parallel to the magnetization direction of the FM layer. It can be understood that the spin components along x and y directions should average out to be zero. In this way, by virtue of spin-transfer torque process, spin magnetic moments of the incoming electrons become parallel to the magnetization direction of the FM layer.

5.2.2 Double Ferromagnetic (FM) Layers

Let us consider a device geometry consisting of a NM/FM/NM/FM/NM multilayer nanopillar structure, where NM is a non-magnetic and FM is a ferromagnetic metal as shown in Fig. 5.2. In order to attain spin-transfer torque effect in this multilayer nanopillar, special structural engineering is done on this structure. The special characteristics of this multilayer structure is such that one of its FM layer is deposited as thinner than the consecutive FM layer. Thus, in a multilayer nanopillar structure the successive FM layers are having different thicknesses, i.e., one is thin and other is thick. The thin FM layer is called as a ***free layer*** and the thick layer as a ***fixed layer***. It is understandable that the magnetic moment associated with the free FM layer could be more freely and easily reoriented by spin-transfer torques, whereas magnetic moment of thicker FM layer is more resistant and rigid to be oriented by the torques. The thick FM layer is constructed by high magnetic moment material and thereby serves as the polarizer. This means during transmission through and reflection from this thick FM layer, electron spins can be filtered. Thus, this layer can produce spin-polarized electrons, which eventually reach and act on the magnetization of the free FM layer. Let us consider the thick FM layers are positive polarizers, i.e., thick FM layers preferentially transmit majority-spin electrons. In this context, taking into considerations the thermal fluctuations in the system, let us assume that there is initially a misalignment angle θ, between magnetic moments of those two FM layers. If the magnetic moments in the two FM layers are aligned exactly parallel, then spin-transfer torque would be zero.

Now, two cases can be framed:

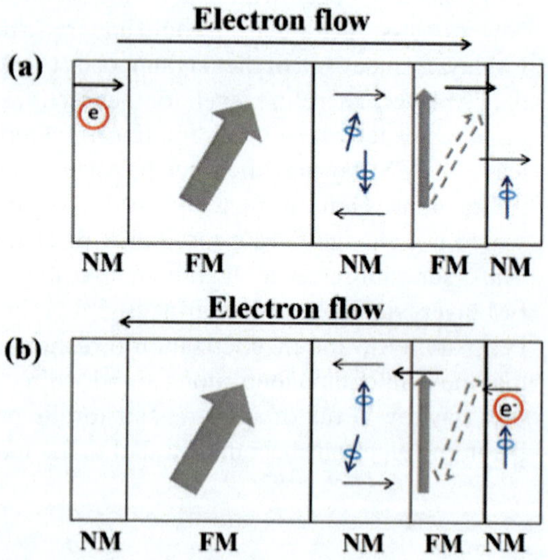

Fig. 5.2 Schematic drawing of spin-transfer torque effect modulating magnetization direction in samples having two ferromagnetic layers. **a** Electrons flowing from the thicker (fixed) ferromagnetic layer to the thinner (free) layer (negative current), **b** electrons flowing from the thinner (free) ferromagnetic layer to the thicker (fixed) layer (positive current)

Case I: Flow of electrons from thicker to thinner layer—In this case, injection of electrons take place from the left side in the multilayer structure, as shown in Fig. 5.2a. This is referred to as ***negative current***. As already mentioned, if unpolarized electrons are injected into the thick fixed layer, then spin-filtering phenomenon would take place. As a result, the spin filtered electrons from the thick fixed layer emerges into the non-magnetic metallic spacer layer having the direction of their average spin magnetic moment parallel to the magnetization direction of the fixed layer. Subsequently, these polarized electrons enter the thin free ferromagnetic layer and interact with the magnetization of this free layer. Consequently, the magnetization of this free layer experiences a torque. This torque in turn tends to orient the associated magnetic moment of the free layer towards the direction of the spin magnetic moment of the incoming polarized electrons from the thick fixed layer. In effect, this torque on the free layer magnetic moment will turn its associated magnetization in the direction of the fixed layer's magnetization. Thus, a parallel orientation between the magnetic moments or magnetizations of these two FM layers can be achieved.

Case II: Flow of electrons from thinner to thicker layer—In this case, injection of electrons take place from the right side in the multilayer structure, as shown in Fig. 5.2b. This is referred to as ***positive current***. This indicates that the net flow of electrons takes place from the free layer to the fixed layer. Similar to *Case I*, unpolarized electrons entering the free FM layer will first experience spin filtering effect and become spin polarized, having average spin magnetic moment parallel to the magnetization of the free FM layer. It is evident that at this step, no net torque is applied to the magnetic moment of the free FM layer by the unpolarized electrons. Subsequently, *spin-polarized* electrons, produced at free FM layer, will flow into the central NM-metallic spacer layer. These spin-polarized electrons will then enter the fixed FM layer. Here, one may expect that spin magnetic moment associated with these spin-polarized electrons would transfer angular momentum and hence apply a torque to the magnetic moment of the fixed FM layer. However, in reality this is not the case. Actually, the magnetic moment of the fixed layer is so rigidly held in place that we cannot expect any torque acting on it. Since the fixed FM layer is made of high magnetic moment material, Fermi level is expected to pass through very little portion of (or may be not at all through) minority spin band. Consequently, the fraction of electrons, having polarization parallel to the magnetization of the fixed FM layer, will be readily transmitted through this layer, whereas the fraction having polarization antiparallel to the fixed layer moment will be reflected back towards the free FM layer from the NM/fixed FM layer interface. Those reflected electrons, having spin magnetic moment antiparallel to the magnetization of the fixed FM layer, now approach to the free FM layer and exert a torque on the magnetic moment of the free FM layer. This, in turn, tends to orient the associated magnetic moment of the free FM layer towards the orientation of the reflected spin moment, i.e., in effect, *away from* the direction of the fixed-layer moment. This implies that a large-enough positive current might destabilize the parallel orientation of the magnetization of those two FM layers and might establish an antiparallel alignment among them.

5.3 Spin-Transfer Torque (STT)-Driven Magnetization Dynamics

Different kinds of magnetization dynamics have been observed for the free layer magnetic moment induced by spin-transfer torque (STT) effect. In order to describe such dynamics, along with spin-transfer torque, it is essential to consider some additional torques acting on the free layer moment including, (i) torque due to the applied magnetic field; (ii) torque due to magnetic anisotropies; (iii) damping torque, which bring the moment at its lowest energy configuration, and (iv) torque due to possible thermal fluctuations. Schematic representation of those torques acting on an arbitrary moment (M) is given below:

Let us assume an external magnetic field is applied in the *z*-direction and that the free FM layer is having a purely uniaxial magnetic anisotropy with easy axis also directed along *z* direction. Thus, both (i) torque due to the applied magnetic field and (ii) torque due to magnetic anisotropy are acting along *z* direction. Let us also suppose that the magnetization of the fixed FM layer direct along the same *z*-direction. Following the general convention, the free FM layer can be considered as a single-domain magnet. To simplify the description, let us ignore the effects of thermal fluctuations. Also, the magnetization corresponding to the fixed and free FM layers are considered not to be aligned exactly parallel to each other at the beginning.

Let us discuss two cases separately, in one ***STT is present***, whereas in another ***STT is absent***.

5.3.1 *Magnetization Dynamics in Absence of STT*

As already mentioned, there are combined effect of two torques, arising from applied magnetic field and magnetic anisotropy, on the magnetic moment of free FM layer along the *z*-direction. The effect of this torque causes the precession of this free-layer moment about *z*-direction. Let us consider that at some instant the magnetic moment associated with the free FM layer is oriented at an angle θ with respect to *z*-direction, as depicted in Fig. 5.3. As generally understood, damping torque accounts for loss of energy from the magnetization of the free FM layer to its environment.

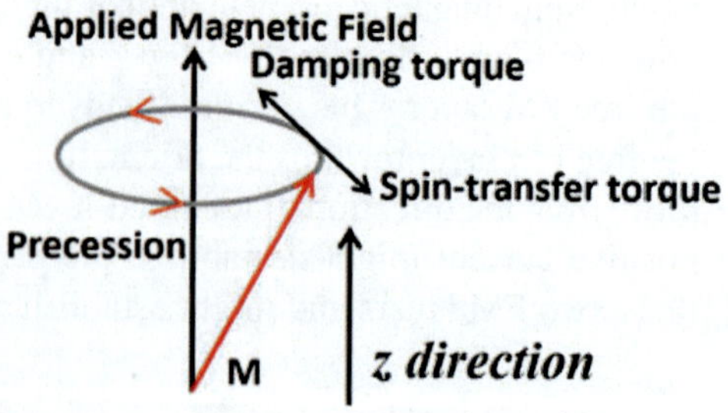

Fig. 5.3 Schematic demonstration of various kinds of torques acting on a single-domain nanomagnet. External magnetic field is applied along z-direction

Consequently, the associated magnetic moment of the free FM layer would tend to attain its lowest energy. Now, the lowest energy corresponds to the alignment of the magnetic moment along the direction of the applied magnetic field, i.e., along z-axis. Therefore, it is evident that the damping torque should point towards the z-axis in a manner, as shown in Fig. 5.3. Hence, it comes out that in absence of any term related to spin-transfer torque, the magnetization associated with the free FM layer will spiral towards the z-axis and eventually finish up by attaining the lowest-energy configuration directing along z-direction.

5.3.2 *Magnetization Dynamics in Presence of STT*

This case corresponds to the action of an additional torque that can point either in the same or opposite direction to the damping torque, depending on the sign of the current that produces this spin-transfer torque (with reference to Sect. 5.3.2). Recalling Sect. 5.3.2, under the application of negative current, the spin-transfer torque would be so generated that it will simply strengthen the damping torque. Consequently, the magnetization associated with the free FM layer would spiral or relax even faster towards the z-direction, compared to the case when there is no STT.

On the other hand, with the application of small positive currents, the generated spin-transfer torque would be such that it will oppose the effective damping, i.e., spin-transfer torque will be acted opposite to the direction of damping torque, as shown in Fig. 5.3. As a result, the magnetization of the free FM layer would spiral or relax rather slowly towards the z-direction, compared to the case when there is no STT.

Furthermore, depending on the magnitude of the applied magnetic field and the detailed mutual angular dependence of the spin-transfer and damping torques, magnetization dynamics can be of diversified forms. Let us discuss these cases separately.

(a) Under the application of appropriate magnitude of positive currents, it is feasible for the magnetization of the free FM layer to spiral up to ever-increasing values of θ all the way to $\theta = \pi$. This implies that the magnetic moment associated with the free FM layer can attain a stable static state with magnetization antiparallel to the fixed FM layer. Experimentally, such *simple spin-transfer torque-driven magnetization reversal* is achievable under the application of low applied magnetic fields. Noteworthy, the magnetization of this free FM layer can be controllably switched back to the parallel alignment with the fixed layer moment by applying a sufficiently large negative current. This phenomenon and the related device are under current investigation for applications in magnetic memory devices (Myers et al. 1999; Katine et al. 2000).

(b) In a similar scenario as described above, with the optimization of parameter values in some different ways, the magnetization of free FM layer may not spiral all the way to $\theta = \pi$. Rather, it might attain some dynamical equilibrium state at an arbitrary intermediate angle, where it perform precession about z-direction. In this case, the energy gained from spin-transfer torque and the energy lost due to damping torque are balanced over each cycle of precession of free layer magnetic moment. Significantly, this experimental observation explores an intriguing fact, in which spin-transfer torque exploits a DC applied current to produce steady-state magnetic precession oscillations at GHz or tens of GHz frequencies (Kiselev et al. 2003; Slonczewski 2002).

5.4 Experimental Results

In order to realize experimentally detectable spin-transfer torque effect in a device, the prime condition is that the flow of electronic current should be restricted to a small diameter. This in turn implies that to attain spin-transfer torque, device fabrication would be very much critical.

To summarize some crucial conditions

1. Following the basic idea of spin-transfer torque mechanism, the amount of current required to induce magnetic excitations, employing spin-transfer torque, scales with the total magnetic moment associated with the free FM layer. This is realizable by constructing the free FM layer a few nanometres thick and a few hundreds of nanometres in diameter. Because of this nanometric dimension of the device, free layer magnetic moment will be pretty less and correspondingly less current is needed to induce magnetic excitations by spin-transfer torque mechanism.
2. Another crucial point of concern for employing nanoscopic devices is that in this case spin-transfer torque effect dominate over the effects of the magnetic field, which is supposed to be pretty small because of very less current has produced it.
3. Spin-transfer torque effects are very easily attainable in devices, which are small enough that the free FM layer becomes single magnetic domain. It is well known that in case of single magnetic domain, movement of magnetic moment is favourable for the spin-transfer torque effect to be observed. It should be mentioned that the dimension of the devices should be close to 100 nm scale. Here we present some illustrations (Fig. 5.4) of various kinds of magnetic-multilayer devices that can demonstrate spin-transfer torque effects.

5.4.1 *Point Contact Device*

Point contact device is a mechanical device, in which a magnetic multilayer is contacted by a sharp metal tip. Such contact region is formed on the scale of few tens of nanometres. Point contact devices, produced lithographically, have also been employed to produce similar device geometry. In these both types of point contact

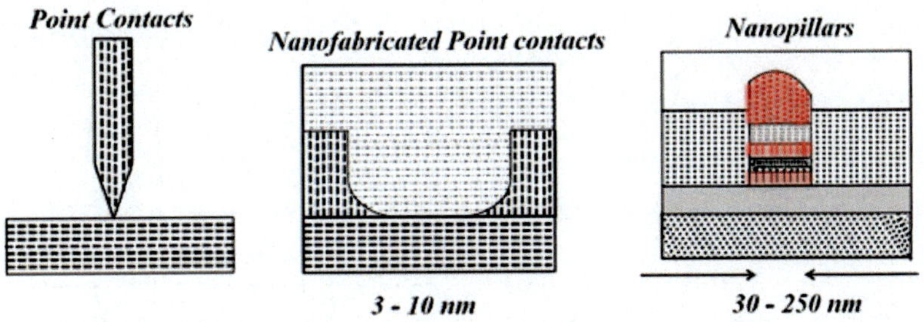

Fig. 5.4 Schematic demonstration of some magnetic-multilayer devices, having typical sample geometries that can demonstrate spin-transfer torque effects

devices, excitations of the ferromagnetic layers can be produced within a nanoscale region in the vicinity of the point contact.

5.4.2 *Multilayer Nanopillar*

As already discussed in Chap. 2, when spins are injected from one FM (FM1) layer into another FM layer (FM2), the spin injection results in a torque on the magnetic moment of the FM layers. This phenomenon results in several diversified effects. As an example, magnetic moment associated with one of the FM layers may even reverse its orientation, which is referred to as ***spin-injected magnetization switching*** (SIMS). In another case, magnetic moment associated with one of the FM layers may perform continuous oscillations, which is called as ***spin-torque oscillation*** (STO). Generally, SIMS and STO are obtained in magnetic nanopillars, which are made up of magnetic multilayers, as demonstrated in Fig. 5.5.

Figure 5.5 exhibits a typical device structure, consisting of two FM layers, FM1 and FM2, constructed by Co and one NM layer, made up of, e.g., Cu or MgO, inserted between them. When electronic current flows through this device structure, initially electrons are polarized by FM1, then passing through NM1 these polarized electrons are injected into the FM2 layer. The net spin magnetic moment of the polarized injected electrons from FM1 interact with the magnetic moment of the host ferromagnetic material (FM2) through exchange interaction. As a result, net spin moment of the polarized electrons, originated from FM1, exerts torque on the magnetic moment of FM2. If this exerted torque is large and strong enough, magnetization associated with FM2 may get reversed or execute continuous precession along a particular direction, decided by the mutual orientation of the net spin magnetic moment of the incoming polarized electrons and the magnetization of FM2 layer.

Furthermore, we have presented here a study on the magnetization switching behaviour of the free FM layer in a Py (2 nm)/ Cu (6 nm)/ Py (20 nm) spin valve nanopillar at 4.2 K for two distinct cases. In one case, switching of the resistance, i.e., resistive hysteresis of the free FM layer is excited by an external magnetic

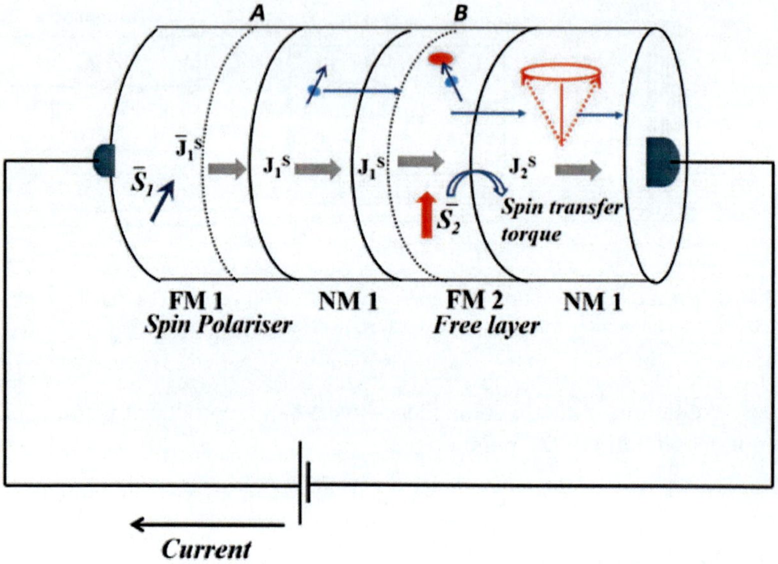

Fig. 5.5 Schematic illustration of a magnetic nanopillar that can demonstrate SIMS and STO

field applied in the plane of the sample along the easy axis (Fig. 5.6a), whereas in another case it is driven by a current (Fig. 5.6b). The resistive hysteresis loop in Fig. 5.6a shows regular switching behaviour of the free-layer magnetic moment between parallel (P) and antiparallel (AP) alignment with the magnetic moment of the fixed FM layer.

Figure 5.6b confirms the potential of the current to switch the magnetic moment of the free FM layer between parallel and antiparallel states with the magnetic moment of the fixed FM layer, similar to that of applied magnetic field as shown in Fig. 5.6a. Following the same sign convention of exciting current (as discussed in Sect. 5.2.2),

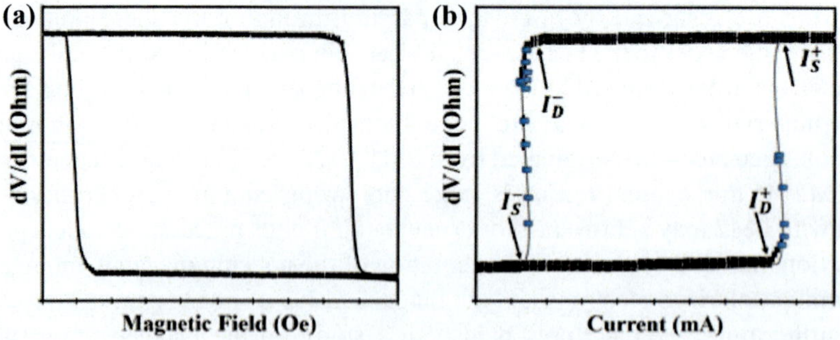

Fig. 5.6 **a** Switching of free ferromagnetic layers in a nanopillar device by a magnetic field at 4.2 K. **b** Switching in the same device obtained by spin-transfer torque effect, arising from an applied current at 4.2 K

a sufficiently large positive current results in a sudden rise of resistance from lower-resistance to higher-resistance state. Noteworthy, lower- and higher-resistance states correspond to the parallel and antiparallel alignments, respectively, of free layer magnetic moment with respect to the fixed layer moment. It has already been discussed that positive current corresponds to the flow of electronic current from the free to the fixed FM layer. As is evident from the above-made discussion, in this case, parallel orientation between free and fixed layer magnetic moment gets destabilize and antiparallel alignment between them gets stabilize. Thus, magnetic moment associated with the free FM layer, in this high-resistance antiparallel configuration, can be reversed back to the parallel orientation by applying a sufficiently large negative current (Myers et al. 1999). It should be clearly mentioned that such switching of free layer magnetic moment between parallel and antiparallel configurations with fixed layer moment can be justified and understood with the mechanism of spin-transfer torques. More subtle point is that such switching is not a mere consequence of current-induced magnetic fields. This can be confirmed from the fact that in nanoscopic devices the requisite current levels are so small that it cannot produce magnetic fields of the desired magnitude to switch the magnetic moment of free layers. Noteworthy observations in the differential resistance are small peaks or shoulders (Fig. 5.6b), appearing at the currents I_D^+ and I_D^-, before the resistance exhibits large jumps in its value (Urazhdin et al. 2003). This experimental feature can be attributed to the turn-on of a dynamical state, in which the magnetic moment associated with free FM layer executes small-angle precession. This subtle experimental feature further assures us about the origin of magnetization switching in this present case. It has already been discussed that at the onset, spin-transfer torque drives the free layer moment into a precessional mode, which is then followed by reversing of the magnetic moment to attain the final static state.

5.5 Spin-Transfer Torque in Magnetic Multilayer Nanopillar

Let us imagine an electron system in which exchange interactions take place between the conduction *s*-electrons and the magnetic *d*-electrons, holding local magnetic moments, as schematically shown in Fig. 5.7a. It is known that the total spin angular momentum would be conserved during *s–d* exchange interaction. Thus, it is straightforward to state that a reduction in the sub-total angular momentum associated with the conduction electrons would be equal to the enhancement of the sub-total angular momentum corresponding to the magnetic *d*-electrons. Suppose during transport through the FM2 layer of a magnetic pillar, spin angular momentum of a conduction electron changes owing to the *s–d* interaction. Now, the same amount of angular momentum, equal to the change experienced by conduction electrons, should be transferred to the *d*-electrons in the FM2 layer. Considering the rule for conserving the spin angular momentum

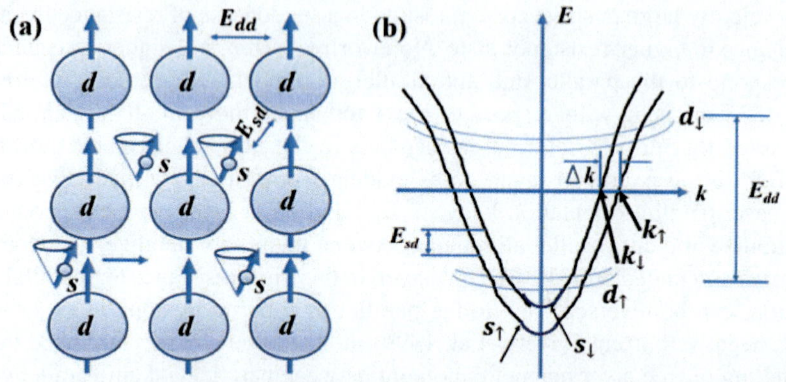

Fig. 5.7 **a** A simplified representation of *s*–*d* model to describe spin-transfer torque effect. Picture depicts flow of *s*-electrons among localized *d*-electrons. *s*–*d* exchange interaction results in precession of *s* and *d* electrons. Precession angle of the *d*-electrons is considerably smaller than that of *s*-electrons because *d*-electrons produces a single large local spin magnetic moment. **b** Schematic band structure of a ferromagnetic 3D transition metal. *s*-bands are *free electron like* having small narrow spin splitting, whereas *d*-bands have large spin splitting

$$\frac{\partial \vec{s}}{\partial t} + \nabla.\hat{\vec{j}}^{s} = 0. \tag{5.2}$$

From Eq. (5.2),

$$\frac{d\vec{S}_2}{dt} = \vec{J}_1^{S} - \vec{J}_2^{S}, \tag{5.3}$$

where $\vec{S}_2$ represents the total angular momentum, associated with the magnetic moment of the FM2 layer. We obtain spin currents $\vec{J}_1^{S}$ and $\vec{J}_2^{S}$ by integrating the spin current density that flows in NM1 and NM2 layer, respectively, over the cross-sectional area of the pillar. Spin–orbit interaction in FM2 can be neglected because of very thin dimension of FM2. Equation 5.3 clearly shows that a torque can indeed be exerted on the local angular momentum as a consequence of transfer of spin from the conduction electrons. Hence, this type of torque is referred to as the ‘spin-transfer torque’.

5.5.1 *Spin-Transfer Torque Exerted in Metallic Junctions*

Let us consider that the thickness of the FM1 layer is larger than its spin diffusion length. Thus, conduction electrons become spin-polarized after passing through the FM1 layer along the direction of the total angular momentum, i.e., $\vec{S}_1$ of this layer.

Subsequently, these spin-polarized electrons are injected into the adjacent NM1 layer. Let us now suppose that in polar coordinate system, the orientation of $\vec{S}_1$ is along (θ, ϕ). Accordingly, the spin wave function of the injected electrons spin can be expressed as

$$|(\theta, \phi)\rangle \cos\frac{\theta}{2}|\uparrow\rangle + e^{i\phi}\sin\frac{\theta}{2}|\downarrow\rangle$$
$$\text{or,} \quad = \begin{pmatrix} \cos\frac{\theta}{2} \\ e^{i\phi}\sin\frac{\theta}{2} \end{pmatrix}. \tag{5.4}$$

Conventionally, $|\uparrow\rangle$ and $|\downarrow\rangle$ corresponds to the spin eigenstates along the $+z$ and $-z$ directions, respectively. Considering FM2 layer is magnetized along the +z direction, energy bands of the conduction electrons, i.e., *s*-electrons are split into $s_\uparrow$ and $s_\downarrow$ bands (Fig. 5.7b). Accordingly, the spin wave function of the injected *s*-electrons into FM2 layer is also divided into $s_\uparrow$ and $s_\downarrow$ partial waves. This means that the associated Bloch states of those injected *s*-electrons correspond to different wavevectors, say, $k_\uparrow$ and $k_\downarrow$. Consequently, during their travel through the FM2 layer of thickness d_2, each of these partial waves having wavevectors $k_\uparrow$ and $k_\downarrow$ have attained the phase equal to $d_2k_\uparrow$ and $d_2k_\downarrow$, respectively. Hence, after travelling ballistically through a very thin FM2 layer, the spin wave functions associated with these outgoing electrons will become

$$\begin{pmatrix} e^{ik_\uparrow d_2} & 0 \\ 0 & e^{ik_\downarrow d_2} \end{pmatrix} \begin{pmatrix} \cos\frac{\theta}{2} \\ e^{i\phi}\sin\frac{\theta}{2} \end{pmatrix} e^{ik_\uparrow d_2} \begin{pmatrix} \cos\frac{\theta}{2} \\ e^{i(\phi+(k_\downarrow - k_\uparrow)d2)}\sin\frac{\theta}{2} \end{pmatrix}. \tag{5.5}$$

It is evident from Eq. 5.5 that ϕ is modified by $\left(\phi + \left(k_\downarrow - k_\uparrow\right)d_2\right)$. Thus, it can be readily understood that the spins of the conduction electrons undergo precession around $\vec{S}_2$ by $\left(k_\downarrow - k_\uparrow\right)d_2$ (rad).

Considering practical situations, most of the films are polycrystalline in nature. Thus, each conduction electron is supposed to travel along different crystal orientations, which in turn results in different phases and hence precession angles for different electrons. Therefore, on taking average over ensemble of electrons, the transverse components, i.e., *x* and *y* components of the injected spins cancel each other and hence disappear. In this case, following Eq. 5.3, the change in the spin current producing spin-transfer torque would take the form

$$\frac{d\vec{S}_2}{dt} = g(\theta)\frac{J^Q}{-e}\left(\frac{\hbar}{2}\begin{pmatrix} \cos\phi\sin\theta \\ \sin\phi\sin\theta \\ \cos\theta \end{pmatrix} - \frac{\hbar}{2}\begin{pmatrix} 0 \\ 0 \\ \cos\theta \end{pmatrix}\right) = g(\theta)\frac{J^Q}{-e}\frac{\hbar}{2}\vec{e}_2 \times \left(\vec{e}_1 \times \vec{e}_2\right), \tag{5.6}$$

where $\vec{e}_1 = (\cos\phi \sin\theta, \sin\phi \sin\theta, \cos\theta)$ and $\vec{e}_2 = (0, 0, 1)$ are unit vectors along the direction of angular momentum of FM1 and FM2 layers, respectively. It is noteworthy that for $\vec{S}_1$ the expectation values of the Pauli spin matrices, i.e., $\sigma_x, \sigma_y, \sigma_z$, are $\cos\phi \sin\theta, \sin\phi \sin\theta, \cos\theta$. J^Q is the charge current and $\frac{J^Q}{-e}$ corresponds to the number of electrons flowing per unit time. $g(\theta)$ represents the efficiency of spin-transfer process. Moreover, $g(\theta)$ has been found to be dependent on the spin polarization (P) of the conduction electrons flowing through the ferromagnetic layers and on the relative angle, θ between $\vec{S}_1$ and$\vec{S}_2$. Slonczewski (1996) has proposed a formula for this parameter $g(\theta)$ by considering a free electron model. The formula is suitable for the CPP-GMR junctions as follows:

$$g(\theta) = \left[-4 + \left(P^{-1/2} + P^{+1/2}\right)^3 (3 + \cos\theta)/4\right]^{-1}. \tag{5.7}$$

In this formula, the effects of electron reflection at the NM1/FM2 interface have also been taken into account.

5.5.2 *Spin-Transfer Torque Exerted in Magnetic Tunnel Junctions (MTJs)*

Let us derive the spin-transfer torque exerted in magnetic tunnel junctions (MTJs) according to the derivation by Slonczewski (1989, 2005). In this case, NM1 can be considered as a barrier layer constructed by MgO or Al_2O_3. FM1 layer has been assumed to be thick enough that at any arbitrary point inside FM1 layer, the conduction electron spins are relaxed and directed parallel to $\vec{S}_1$. Again, following the above discussed mechanism of spin decoherence, transverse spin components of the conduction electrons disappear at any arbitrary point inside FM2 layer. As a result, conduction electron spins align parallel to $\vec{S}_2$ inside FM2 layer. It can be understood that the spins of the conduction electrons are either the majority or minority spins of the host FM materials depending on their magnetization orientations. Hence, the total charge current in the MTJ can be expressed as a sum of the following four components:

$$J^Q = J^Q_{\uparrow\uparrow} + J^Q_{\uparrow\downarrow} + J^Q_{\downarrow\uparrow} + J^Q_{\downarrow\downarrow}. \tag{5.8}$$

Here, suffixes $\uparrow$ and $\downarrow$ are used to indicate the majority and minority spin channels, respectively. For instance, $J^Q_{\uparrow\downarrow}$ signifies flow of spin-polarized conduction electrons from majority spin band of FM1 layer into that of minority spin band of FM2 layer. In order to characterize such current corresponding to each spin sub-channel, the general approach is to use *conductance* (as discussed in Chap. 4) as given by $G_{\substack{\uparrow\uparrow \\ \downarrow\downarrow}}$.

Let us now suppose that V is the applied voltage and the form of the spin wave functions in the FM1 layer is given by

$$\text{wavefunctions corresponding to majority spin electrons: } \cos(\theta/2)|\uparrow\rangle + \sin(\theta/2)|\downarrow\rangle;$$
$$\text{wavefunctions corresponding to minority spin electrons: } \sin(\theta/2)|\uparrow\rangle - \cos(\theta/2)|\downarrow\rangle. \tag{5.9}$$

On the other hand, majority and minority spin wavefunctions in the FM2 layer are given by $|\uparrow\rangle$ and $|\downarrow\rangle$, respectively. Therefore, currents can be expressed as

$$J^{Q}_{\substack{\uparrow\uparrow\\ \downarrow\downarrow}} = VG_{\substack{\uparrow\uparrow\\ \downarrow\downarrow}}\cos^2\frac{\theta}{2}$$
$$J^{Q}_{\substack{\downarrow\uparrow\\ \uparrow\downarrow}} = VG_{\substack{\downarrow\uparrow\\ \uparrow\downarrow}}\sin^2\frac{\theta}{2}. \tag{5.10}$$

Thus, spin currents at any two arbitrary points at FM1 and FM2 layers can be obtained as

$$\vec{J}_1^{S} = \frac{\hbar}{2}\frac{1}{-e}\left(J^Q_{\uparrow\uparrow} + J^Q_{\uparrow\downarrow} - J^Q_{\downarrow\uparrow} - J^Q_{\downarrow\downarrow}\right)\vec{e}_1$$
$$\vec{J}_2^{S} = \frac{\hbar}{2}\frac{1}{-e}\left(J^Q_{\uparrow\uparrow} - J^Q_{\uparrow\downarrow} + J^Q_{\downarrow\uparrow} - J^Q_{\downarrow\downarrow}\right)\vec{e}_2. \tag{5.11}$$

Therefore, total current is obtained as follows:

$$J^Q = \frac{1}{2}V\left[\left(G_{\uparrow\uparrow} + G_{\downarrow\downarrow} + G_{\uparrow\downarrow} + G_{\downarrow\uparrow}\right) + \left(G_{\uparrow\uparrow} + G_{\downarrow\downarrow} - G_{\uparrow\downarrow} - G_{\downarrow\uparrow}\right)\vec{e}_2.\vec{e}_1\right]. \tag{5.12}$$

Therefore, total spin-transfer torqueis obtained as follows:

$$\frac{d\vec{S}_2}{dt} = \frac{\hbar}{2}\frac{1}{-e}\frac{1}{2}V\left[\left(G_{\uparrow\uparrow} - G_{\downarrow\downarrow}\right) + \left(G_{\uparrow\downarrow} - G_{\downarrow\uparrow}\right)\right]\left(\vec{e}_2 \times \left(\vec{e}_1 \times \vec{e}_2\right)\right). \tag{5.13}$$

It is evident from Eq. 5.12 that tunnel conductance depends on $\cos\theta$, whereas Eq. 5.13 exhibits $\sin\theta$ dependence of the spin-transfer torque, which is analogous to the case as given by Eq. 5.6.

5.6 Possible Applications of Spin-Transfer Torques

A plethora of research activities are presently being carried out on the application of STT (Krivorotov et al. 2005; Zhu and Zhu 2004; Nazarov et al. 2002; Kato et al. 2004; Kikkawa and Awschalom 1999; Weber et al. 2001; Joly et al. 2006a, b).

5.6.1 Magnetic Random Access Memory

It is now evident from our discussion that spin-transfer torque effect has potentials of controllable switching of magnetic moments of two ferromagnetic layers back and forth between a high-resistance antiparallel and a low-resistance parallel state. This in turn proposes that STT might be employed to write information within non-volatile magnetic random access memories. In fact, switching produced by spin-transfer torque mechanism has been found to be more effective than that obtained by current-induced magnetic fields to control magnetic bits. Other potential advantages of STT-driven magnetization switching over magnetic field induced switching are as follows:

1. Since there is no need of magnetic field to realize spin-transfer torque effect, accordingly in this case, there is no requirement for designing and fabrication of extra bit lines;
2. Again absence of any magnetic field causes decrease of perturbation, coming from neighbouring magnetic elements while writing of bits in array;
3. While independently maintaining the magnetic-anisotropy barriers needed for thermal stability, it is possible to minimize the currents responsible for spin-transfer torque switching. Therefore, there is feasibility to continuously scale the size of the magnetic bit to the lithography limits of silicon processing;
4. Compared to magnetic field switching, in case of spin-transfer architectures, the switching characteristic is not necessarily required to be so uniform. This in turn leads to much less demanding device tolerances.

Challenges for applications of spin-transfer torques

The two main challenges that restricted the applications of spin-transfer torque in memory technologies are summarized below:

1. *All*-metal spin valve devices are characterized by very low resistances in the range of 1–10 Ω. This is pretty less than the range 1–10 kΩ, which is required to yield reasonable *signal-to-noise* ratio in case a silicon circuit is employed to read the magnetic configuration (Zhu and Zhu 2004; Nazarov et al. 2002).
2. It should be noted that critical currents, needed to excite magnetic switching, have decreased steadily as a result of improvements in device processing. For example, critical currents have reduced from 5 mA in Co devices (used earlier) to approximately 0.3 mA in the Py samples (existing devices) (Katine et al. 2000).

Nevertheless, even smaller values of critical currents are desirable. For instance, critical currents of 0.1 mA could possibly control magnetic bits employing minimum-area silicon CMOS transistors. This in turn could feasibly materialize the realization of very dense memory circuits.

5.6.2 Spin-Transfer Torque-Driven Microwave Sources and Oscillators

As already mentioned in Sect. 5.3, dynamical magnetic modes, corresponding to microwave frequency, can be excited by DC spin-transfer torques. Recently, such experiments are under rigorous investigations for the use of high-speed signal processing, as nanoscale microwave sources, oscillators and amplifiers. The spin-transfer torque devices, executing precession motion, may be implemented as sources and detectors for wireless chip-to-chip communications (Kiselev et al. 2003; Krivorotov et al. 2005).

5.7 Conclusions

In this chapter, we have introduced the concept and definition of spin-transfer torque. Furthermore, we have discussed spin-transfer torque effect in layer structures, both for *single* and *double* ferromagnetic (FM) layers. We have also addressed interesting topic like spin-transfer torque-driven *magnetization dynamics*, both in absence and presence of spin-transfer torque. Few experimental results, such as point contact device, multilayer nanopillar etc., have also been discussed in this case. We have also taken into account some possible applications of spin-transfer torques, such as magnetic random access memory, spin-transfer torque-driven microwave sources and oscillators etc., in this chapter. However, in an attempt to employ spin-transfer torque-based magnetic sensor in current-perpendicular-to-the-plane GMR magnetic read heads, spin-transfer torques has been found to generate noise and reduce the effectiveness of the sensor. Therefore, a detailed understanding of spin-transfer torque excitations is indeed important for reduction of these hazardous effects. Presently, a lot of research is actively going on in this field of STT.

5.8 Exercises

1. What is spin-transfer torque process?
2. Why is the spin-transfer torque limited to nanoscale samples?
3. What are the different processes used to identify the mechanism of spin-transfer torque in the nanoscale samples?

4. What are the applications of spin-transfer torque process?
5. What are the limitations in using transfer torque mechanism in spintronic devices?
6. How STT process can excite parallel orientation between the magnetic moments of two ferromagnetic layers? Hence, mention the condition for exciting antiparallel orientation of magnetic moments of the same ferromagnetic layers.
7. Define Magnetization Reversal Process.

References

B. Huang, D.J. Monsma, I. Appelbaum, Phys. Rev. Lett. **99**, 177209 (2007); X. Waintal, E.B. Myers, P.W. Brouwer, D.C. Ralph, Phys. Rev. B **62** 12317 (2000)
L. Joly, J.K. Ha, M. Alouani, J. Kortus, W. Weber, Phys. Rev. Lett. **96**, 137206 (2006a)
L. Joly, L. Tati Bismaths, W. Weber, Phys. Rev. Lett. **97**, 187404 (2006b)
J.A. Katine, F.J. Albert, R.A. Buhrman et al., Phys. Rev. Lett. **84**, 3149 (2000)
Y. Kato, R.C. Meyers, A.C. Gossard, D.D. Awschalom, Nature (London) **427**, 50 (2004)
J.M. Kikkawa, D.D. Awschalom, Nature (London) **397**, 139 (1999)
S.I. Kiselev, J.C. Sankey, I.N. Krivorotov et al., Nature **425**, 380 (2003)
I.N. Krivorotov, N.C. Emley, J.C. Sankey, S.I. Kiselev et al., Science **307**, 228 (2005)
E.B. Myers, D.C. Ralph, J.A. Katine, R.N. Louie, R.A. Buhrman, Science **285**, 867 (1999)
A.V. Nazarov, H.S. Cho, J. Nowak, S. Stokes, N. Tabat, Appl. Phys. Lett. **81**, 4559 (2002)
J.C. Slonczewski, Phys. Rev. B **39**, 6995–7002 (1989)
J.C. Slonczewski, J. Magn. Magn. Mater. **159** L1–L7 (1996); S. Datta, D. Das, Appl. Phys. Lett. **56**, 665 (1990)
J.C. Slonczewski, J. Magn. Magn. Mater. **247**, 324 (2002)
J.C. Slonczewski, Phys. Rev. B **71**, 024411(1–10) (2005)
S. Urazhdin, N.O. Birge, W.P. Pratt Jr., J. Bass, Phys. Rev. Lett. **91**, 146803 (2003)
W. Weber, S. Riesen, H.C. Siegmann, Science **291**, 1015 (2001)
J.G. Zhu, X.C. Zhu, IEEE Trans. Magn. **40**, 182 (2004)

Chapter 6
Magnetic Domain Wall Motion

6.1 Introduction on Magnetic Domain Wall Motion

6.1.1 *What Is Magnetic Domain Wall?*

In nature, often we found ferromagnetic materials, i.e., 3d transition metals, Fe, Co and Ni, in unmagnetized condition. In this demagnetized state, i.e., $M = 0$, ferromagnetic materials are divided into a number of small regions. Each of those small regions is magnetized to saturation and is called magnetic domains (Weiss 1907). More subtle point is that within a single magnetic domain magnetization is saturated, but magnetizations of different magnetic domains are oriented in different directions (Fig. 6.1). This implies that the magnetization orientations of neighbouring domains are not parallel. Consequently, in this demagnetized or $\mathrm{M} = 0$ state of ferromagnetic materials, any two neighbouring domains are separated by a region in which the direction of magnetic moments gradually changes from one direction to another. This transition region of the orientation of magnetic moments is called magnetic domain wall.

6.1.2 *Why Do Domains Exist?*

Figure 6.2 exhibits two different hypothetical configurations of domain. Figure 6.2a represents the configuration where both the exchange and the anisotropy energy are minimal. The reason behind minimal exchange energy is that all the magnetic moments are aligning parallel in this configuration. On the other hand, minimal anisotropy energy is attained because in this configuration, the magnetization axis is an easy axis. But, as it is evident from Fig. 6.2a, numerous uncompensated magnetic 'poles' appear at the ferromagnetic sample surface that gives rise to magnetostatic energy deriving from dipole–dipole interaction. The situation arises where at one surface there is magnetic North Pole, whereas at another there is magnetic South

P. Dey and J. N. Roy, *Spintronics*,
https://doi.org/10.1007/978-981-16-0069-2_6

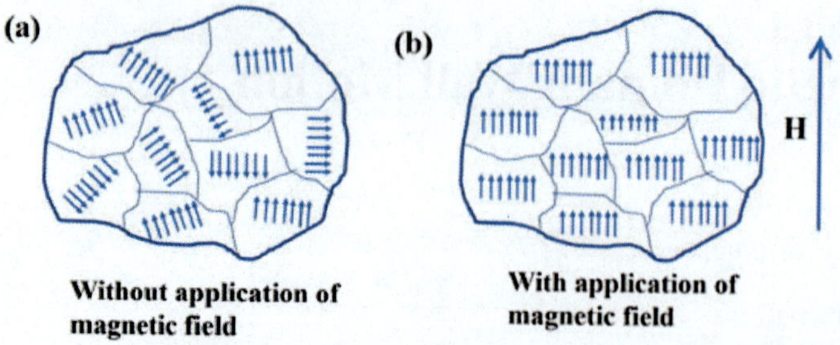

Fig. 6.1 Schematic configuration of magnetic domain of a piece of ferromagnetic material, **a** without application of any magnetic field, i.e., in a demagnetized state and **b** with application of a sufficiently strong magnetic field leads to saturation

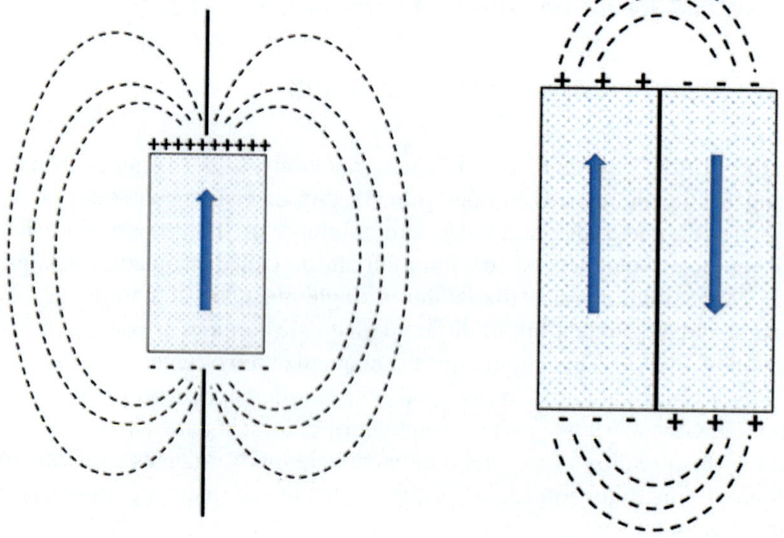

Fig. 6.2 Schematic demonstration of the creation of magnetic domains leading to a decrease of the demagnetizing energy

Pole. This, in turn, indicates a very large demagnetizing energy associated with the sample, i.e., similar to exchange and anisotropy energy, demagnetizing energy is not minimal in this case.

An oversimplified approach to reduce this large demagnetizing energy might be to divide the whole sample into two regions having opposing magnetization direction, i.e., the so-called magnetic domains, as shown in Fig. 6.2b. This, in turn, would result in less magnetostatic energy, as can be clearly understood from Fig. 6.2b. Thus, it comes out that demagnetizing energy contribution can be reduced significantly by the introduction of magnetic domains in the ferromagnetic sample. However, the creation of magnetic domains in a ferromagnetic sample yields transition region,

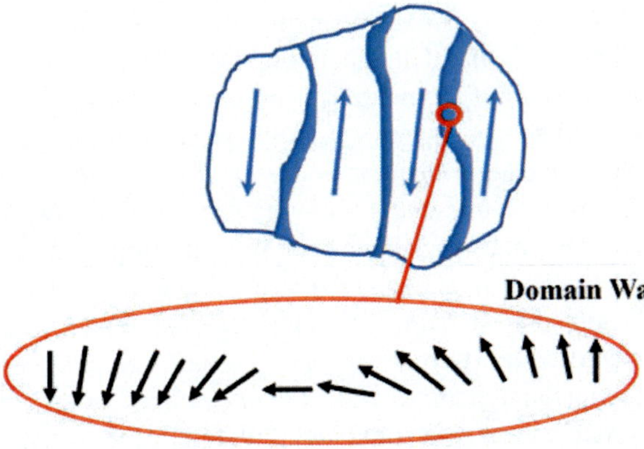

Fig. 6.3 Schematic representation of magnetic domain wall exhibiting a transition region, in which the spin direction gradually varies

named *magnetic domain wall*, between them. Evidently, gradual change in orientation of magnetic moments takes place at this *magnetic domain wall* region, in order to attain the opposing orientation of magnetizations in those two domains. Therefore, magnetic moments at *magnetic domain wall* are neither parallel to each other nor anymore parallel to the easy axis. Consequently, both the contribution of the exchange energy and the anisotropy energy is larger in this configuration than that shown in Fig. 6.2a. However, in this configuration (Fig. 6.2b), a comparatively smaller number of magnetic moments in the magnetic domain wall region are involved in enhancing both the exchange and the anisotropy energy. Therefore, for a ferromagnetic sample of macroscopic dimension, this domain configuration leads to smaller total energy than in the single magnetic domain state.

6.1.3 *What Is Domain Wall Width?*

The transition region between two domains, i.e., magnetic domain wall has a finite width '*d*', which is decided by the exchange and the anisotropy energy (Fig. 6.3).

6.1.4 *Why Small Particles Are Always Mono-domain?*

The introduction of domains and hence of domain walls causes the reduction of demagnetizing energy and the enhancement of domain wall energy. Let us consider a ferromagnet having linear dimension l. In this case, expression[1] for demagnetizing energy can be well approximated as $E_d = Al^3$ and that of magnetic domain wall energy as $E_w = Bl^2$, with A and B are constants. Following our conjecture of a

[1] ***'Introduction to Magnetic Materials' by*** B. D. Cullity and C. D. Graham, Wiley, 10.1002/9780470386323, Institute of Electrical and Electronics Engineers, Inc.

critical value of the linear dimension, $l_c = B = A$, it can be well understood that when $l > l_c$, the demagnetizing energy is larger than the domain wall energy. Thus, this condition seems to favour a *multidomain* state of a ferromagnetic material. On the other hand, when $l < l_c$, the domain wall energy of a single magnetic domain would be larger than the demagnetizing energy of this system. Hence, this condition seems to favour a *mono-domain* state of a ferromagnetic material. Noteworthy, such critical values of l_c have been typically found in the nm regime.

As already stated, in bulk ferromagnetic materials, the magnetic domains are arranged in such a way that the vector summation of all the magnetic moments associated with different magnetic domains comes out to be zero. This, in turn, leaves the ferromagnetic materials *unmagnetized* in the virgin state. With the reduction of dimension of a ferromagnetic material, the number of domains and hence domain walls can be controlled in a systematic manner. Therefore, in case of nanoscopic materials, domain wall motion can be controlled in any particular direction. This kind of tuning feasibility of magnetic domain wall motion allows us to explore the associated domain wall dynamics for possible implementation in the futuristic novel magnetic storage devices. Consequently, a new avenue of the study of dynamical behaviour of magnetic domain wall has opened up in the field of spintronics. In this direction, a number of experimental studies on domain wall dynamics in nanomagnetic systems have already been demonstrated at the laboratory level.

6.2 Magnetic Domain Wall Motion in Spintronics

Advancement in lithography techniques facilitates the fabrication of the nanoscale magnets possessing simple magnetic domain structure. Owing to their simplified domain configuration, such nanodimensional magnets are suitable for carrying out basic studies on the magnetization reversal process. For instance, in a ferromagnetic (FM) nanowire [definition of nanowire] with submicron width two important processes such as nucleation and propagation of a magnetic domain could be effectively realized. Figure 6.4i shows an FM nanowire, where the magnetization is

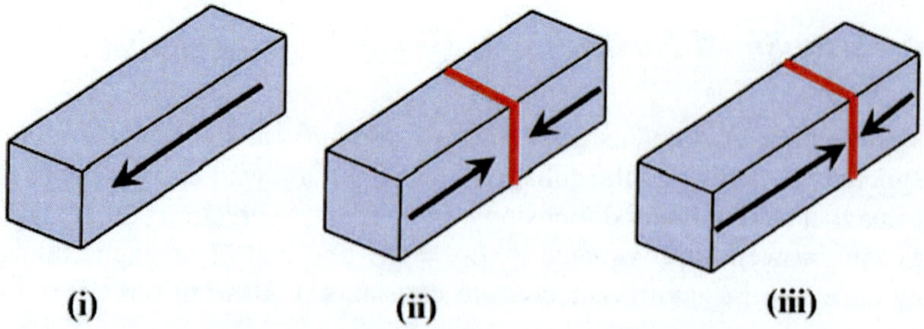

Fig. 6.4 A simplified schematic representation of the magnetization reversal process in a ferromagnetic nanodimensional wire

restricted to be oriented parallel to the axis of the wire because of its shape anisotropy. Suppose an external magnetic field is applied against the magnetization direction, subsequently, nucleation of a magnetic domain wall takes place at one end of the wire and eventually propagates through the wire to the other end. This is clearly demonstrated in Fig. 6.4ii, iii.

6.2.1 *Detection of Domain-Wall (DW) Propagation*

Investigation of nucleation and propagation of a magnetic domain wall (DW) has been conveniently and suitably carried out on a magnetic nanowire, where only one dimension is being *macroscopically* realizable and other two dimensions are nanoscopic. The idea behind the choice of such nanoscopic wire, for this kind of study, lies in its reduced dimension. Since two out of three dimensions are already being reduced, DW motion is realizable and measurable in only one dimension and hence, monitoring of DW motion in that dimension is sufficient for extracting necessary information at any point of time. However, owing to the minute volume of the magnetic nanowire, detection of the small variation in the magnetic moments, induced by the propagation of the DW, is quite difficult. In this attempt, a highly sensitive technique such as GMR has been proposed for the detection of DW propagation in magnetic nanowires (Baibich et al. 1988; Himeno et al. 2005a). In Fig. 6.5, we present a schematic representation for the detection of DW propagation in magnetic wires by using the GMR effect. For this purpose, let us consider a nanowire, composed of trilayer GMR structure consisting of a ferromagnetic (FM), non-magnetic (NM) and FM layers, as shown in Fig. 6.5. The well-known fact in case of any kind of GMR structure is that the resistance of the device structure is the largest for antiparallel magnetization configuration between two ferromagnetic layers (Fig. 6.5a), whereas it is the smallest for the parallel magnetization configuration (Fig. 6.5c). During the magnetization reversal process from Fig. 6.5a–c, i.e., at any intermediate steps as shown in Fig. 6.5b, the total resistance of the device structure can be written as the sum of the resistances of the parallel and antiparallel magnetization parts. Thus, the

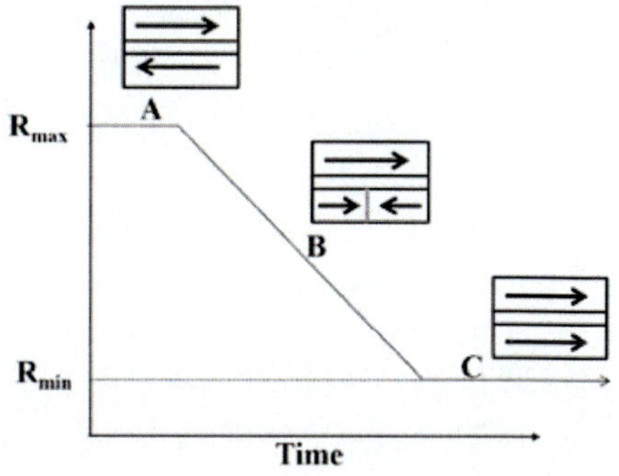

Fig. 6.5 Schematic variation of resistance of a trilayer spin valve structure as a function of time with the application of magnetic field. This enables us the detection of DW propagation in magnetic wires by using the GMR effect

resistance, R of this system is given by

$$R = \frac{x}{L}R_{\uparrow\uparrow} + \frac{L-x}{L}R_{\uparrow\downarrow} \tag{6.1}$$

where, x is the position of the domain wall, L the length of the film, $R_{\uparrow\uparrow}$ and $R_{\uparrow\downarrow}$ are the resistances for parallel and antiparallel magnetization configurations, respectively.

From Eq. 6.1, we understand that the determination of the position of the DW in the magnetic nanowire is feasible by '***Simple Resistance Measurement***'. In our supposed nanowire having FM/NM/FM trilayer device structure, the bottom FM layer is considered to be thin, whereas the top FM layer is thick. Figure 6.6a exhibits the intriguing variation of the resistance of this trilayer nanowire system as a function of the external magnetic field.

In this case, a small magnetic field is applied at the beginning of the measurement to achieve magnetization alignment in one ferromagnetic film in the direction of the applied field. Subsequently, the value of the magnetic field is increased, and the corresponding resistance has been measured accordingly. Magnetoresistance measurement evidence essentially four very sharp leaps. The first and second leaps, observed at low applied magnetic fields, correspond to the magnetization reversal of the thin top ferromagnetic layer, whereas third and fourth leaps, obtained at high applied

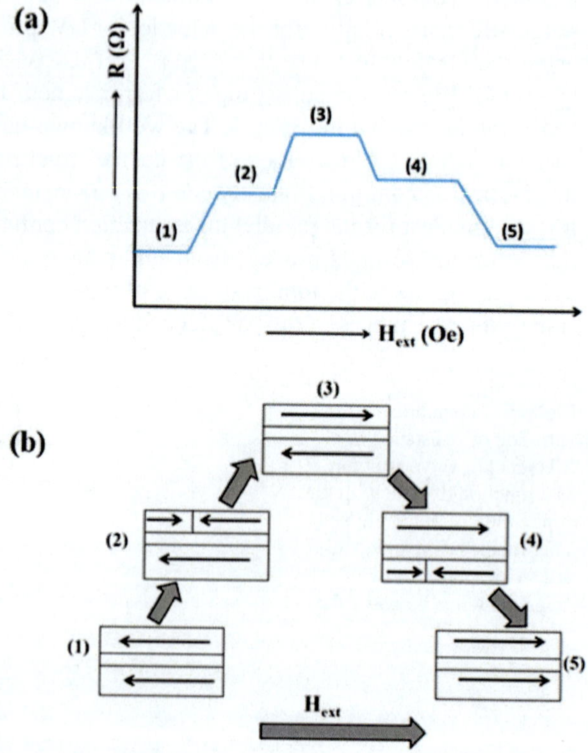

Fig. 6.6 **a** Schematic variation of resistance of a trilayer spin valve structure with applied magnetic field. **b** Magnetic domain structures can be inferred from the resistance measurement. Direction of the external magnetic field is also shown

magnetic fields, correspond to the magnetization reversal of the thick bottom ferromagnetic layer. A schematic representation of the magnetization reversal behaviour of this kind of trilayer GMR nanowire structure is also given in Fig. 6.6b.

6.3 Ratchet Effect in Magnetic Domain Wall Motion

6.3.1 What Is Rachet Effect?

Ratchet effect implies the restricted feature of *reversal* of any sort of human processes once it has already been executed. A realistic example is the mechanical ratchet that holds a spring tightly as a clock is wound up. In a simpler way, we may define ***Ratchet as an effect that can limit the motion to one particular direction.*** In a macroscopic scale, such an effect can be realized by employing a pawl and a wheel with asymmetric-shaped teeth. In effect, the pawl constrains the wheel to rotate in one particular direction.

6.3.2 Rachet Effect in Magnetic Domain Wall Motion

In this context, let us consider a magnetic nanowire having asymmetric artificial structure, as shown in Fig. 6.7. It can be understood that any artificial neck in a magnetic nanowire, as represented in Fig. 6.7, acts as a pinning potential for the magnetic DW motion. It is well known that energy of a DW is proportional to its area. Accordingly, in such kind of a wire having an artificial neck, DW has larger energy at wider position. This energy continues to vary along the axis of the wire and finally generates the pinning potential for the DW at the artificial neck. This is the way a DW can be trapped in such artificial neck. Now, the force required to shift the DW against that pinning potential is given by the derivative of the DW energy with respect to the DW position. This derivative, in turn, is proportional to the slope of the artificial neck. Consequently, different depinning fields are expected subject to the propagation direction of a DW. Because of this, discrepancy in depinning fields between rightward and leftward propagation directions of DWs results in unidirectional motion of a DW. This is referred as *ratchet effect* in magnetic DW motion.

Fig. 6.7 Schematic representation of a special kind of magnetic nanowire structure with an artificial neck

6.3.3 *Rachet Effect in Spintronics*

Structure

In order to discuss rachet effect in spintronics, let us consider a trilayer magnetic nanowire, consisting of $Ni_{81}Fe_{19}$ (5 nm)/Cu (20 nm)/$Ni_{81}Fe_{19}$ (20 nm). Figure 6.8 illustrates the schematic representation of the top view of the nanowire (Himeno et al. 2005b). The completed structure is fabricated onto thermally oxidized Si substrates by *e*-beam lithography and lift-off method. The uniqueness of this nanowire structure is that its main body consists of four asymmetric shaped notches. In order to measure current and voltages, four probes have been constructed by a non-magnetic material (Cu). Furthermore, two narrow Cu wires are attached crossing the ends of the magnetic wire and there is a wide Cu wire protecting the notched part of the magnetic nanowire. Finally, these Cu wires are electrically insulated from the magnetic nanowire by SiO_2 layers.

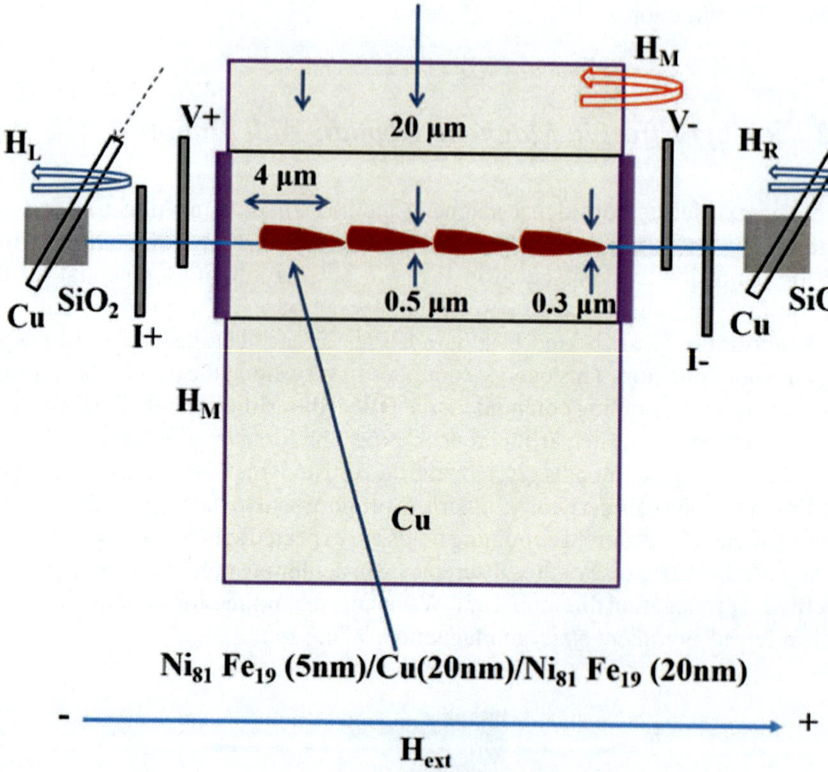

Fig. 6.8 Schematic representation of top view of a trilayer special kind of magnetic wire having NiFe (5 nm)/Cu (20 nm)/NiFe (20 nm) spin valve structure for demonstrating ratchet effect

Operation

When an electric current is passed through each Cu wire, local magnetic fields, H_L, H_M or H_R are generated. These local magnetic fields act on the left end, the main part of the wire with notches and on the right end of the magnetic wire, respectively. As is evident from Fig. 6.8, H_L and H_R can trigger the nucleation of a domain wall at the left and right end of the magnetic wire, respectively. Hence, the direction of propagation of a domain wall can be precisely controlled by the local fields, H_L and H_R.

Figure 6.9i exhibits typical variation of resistance with the application of an external magnetic field (H_{ext}) of the trilayered magnetic wire with asymmetric

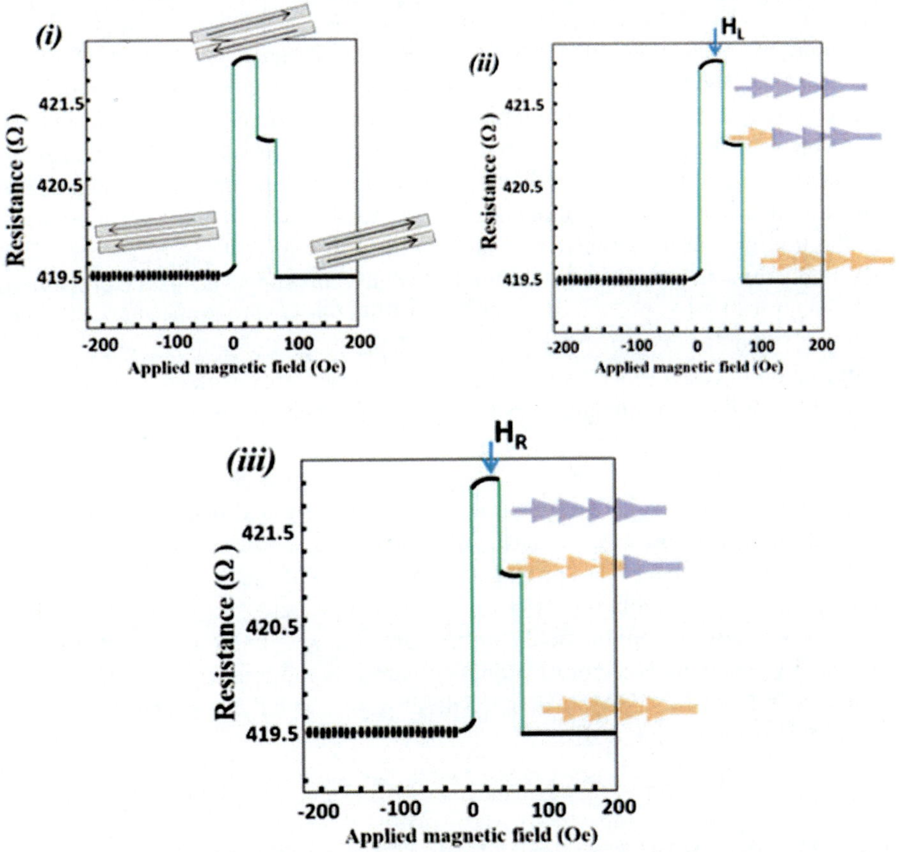

Fig. 6.9 Variation of resistance as a function of magnetic field in a trilayered NiFe (5 nm)/Cu (20 nm)/NiFe (20 nm) spin valve magnetic nanowire structure with asymmetric notches. **i** Magnetoresistance curve of the spin valve system, **ii** magnetoresistance curve when a magnetic domain wall is injected into the NiFe (20 nm) layer from the left end of the magnetic wire by the pulsed H_L, **iii** magnetoresistance curve when a magnetic domain wall is injected into the NiFe (20 nm) layer from the right end of the magnetic wire by pulsed H_R (Himeno et al. 2005)

notches. Initially, the resistance has been found to increase at lower applied magnetic fields, which corresponds to the magnetization reversal of the thin NiFe (5 nm) layer. Successively, the decrease in resistance observed at higher value of H_{ext} is correlated with the magnetization reversal of the thick NiFe (20 nm) layer. It should be noted that our experimental signature does not evidence the pinning of magnetic domain wall at any of the asymmetric notches during the magnetization reversals of NiFe (20 nm) layer because the associated nucleation field is quite large.

In an attempt to nucleate a domain wall in the NiFe (20 nm) layer at smaller value of applied H_{ext} and further to pin the domain wall at the notch, one can adopt the technique of producing a ***pulsed local magnetic field*** at the terminal of this magnetic wire. Figure 6.9ii presents the consequence of the injection of domain wall into the NiFe (20 nm) layer of the magnetic wire by H_L. This means that in addition to the normal measurements, a pulsed magnetic field H_L is applied at the left end of the magnetic wire. The magnitude and the duration of the pulsed H_L were 200 Oe and 100 ns, respectively. As a result of the application of this pulsed magnetic field H_L, the resistance has been found to abruptly decrease and remained at an intermediate value between the largest and the smallest values, as shown in Fig. 6.9ii. This, in turn, implies that a magnetic domain wall, injected from the left end of the magnetic wire by the application of H_L, gets pinned at the first notch as shown in Fig. 6.9ii. Followed by the injection of the magnetic domain wall, further increase in H_{ext} results in an abrupt decrease in resistance to the smallest value. This suggests that the propagation of domain wall takes place to the right end of the wire overcoming the asymmetric notches. From this experimental feature, the depinning field for the magnetic domain wall motion in case of its rightward propagation can be determined.

Figure 6.9iii exhibits magnetoresistance measurement when the magnetic domain wall is injected from the right end of the wire by the application of H_R. Similar to Fig. 6.9ii, in this case also the depinning field corresponding to the leftward propagation of the magnetic domain wall can be determined from the experimental result as shown in Fig. 6.9iii. As is evident from Fig. 6.9ii, iii, the depinning field corresponding to the leftward propagation is found to be much lower than the rightward propagation. In this scenario, an AC magnetic field, having amplitude lying between those two depinning fields for both propagation directions, is applied. Quite expectedly, this induces a unidirectional domain wall motion, which is referred to as 'magnetic ratchet effect'. More detailed description about the ratchet effect has been given in Himeno et al. (2005b).

6.4 Domain Wall Motion Velocity Measurements

Another important parameter that is essential to know for the implementation of magnetic domain wall motion in spintronics is the velocity of domain wall motion. In fact, experimental research on magnetic domain wall motion and its possible implementation for technological applications has been going on intensively (Sixtus and Tonks 1931; Ono et al. 1999; Yamaguchi et al. 2004; Allwood et al. 2002; Versluijs et al. 2001; Parkin 2004; Numata et al. 2007; Levy and Zhang 1997; Lepadatu and Xu 2004; Versluijs et al. 2001; García et al. 1999; Lepadatu et al. 2005; Himeno et al. 2004; Atkinson et al. 2003). In order to determine domain wall velocity in 1931, Sixtus and Tonks (1931) have proposed a circuit diagram (Fig. 6.10). In their arrangement, a homogeneous magnetic field is generated by employing a main coil. Nucleation of magnetic domain wall takes place by a local magnetic field, produced by an additional coil. Two search coils are positioned around the wire at a known separation, as shown in Fig. 6.10. Because of the motion of the domain wall along the wire from left- to right-hand side, successive voltage surges have been produced in those two search coils. As these coils are placed at a known distance, the velocity of the domain wall can easily be calculated.

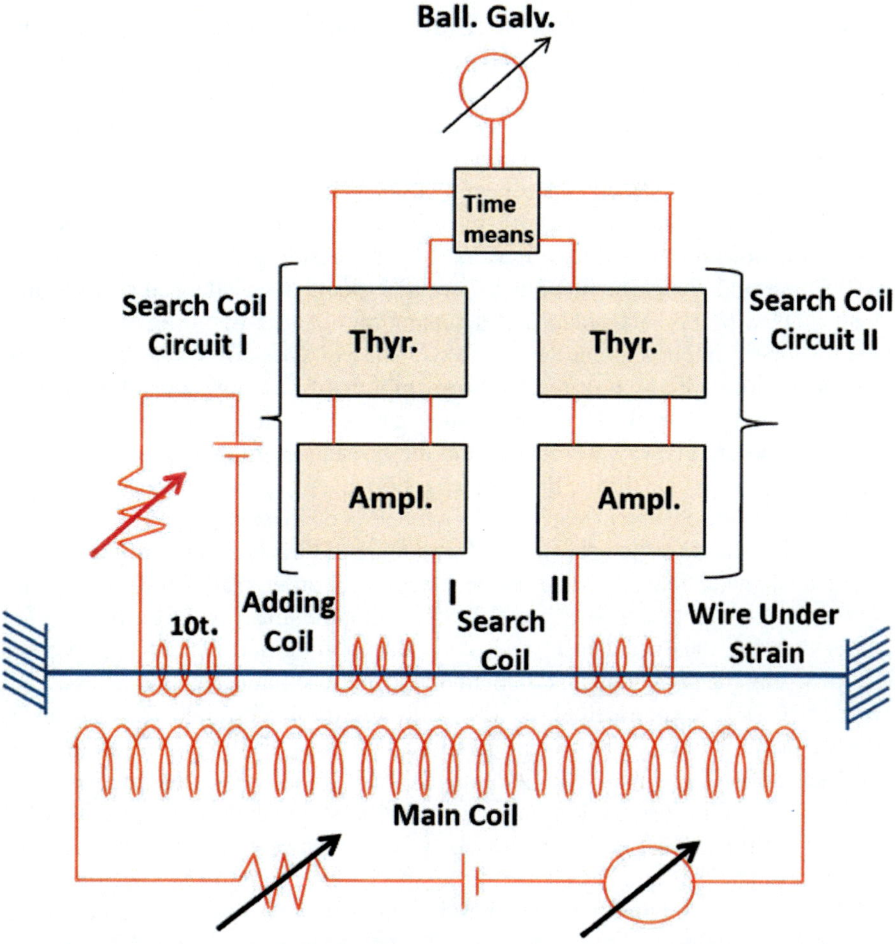

Fig. 6.10 Schematic circuit diagram for DW velocity measurements, proposed by Sixtus and Tonks (1931)

It should be noted that measurement of domain wall velocity using GMR method has an advantage over dynamical measurements because of its simplicity (Ono et al. 1999). As discussed in Sect. 6.2.1, GMR detection method gives us information about domain wall position [Eq. (6.1)]. Hence, domain wall velocity, $v = dx/dt$, can also be determined by the time-domain measurements. Rearranging the terms in Eq. (6.1), we obtain

$$x = \frac{L(R_{\uparrow\downarrow} - R)}{(R_{\uparrow\downarrow} - R_{\uparrow\uparrow})} \tag{6.2}$$

$$v = \frac{dx}{dt} = -\frac{L}{(R_{\uparrow\downarrow} - R_{\uparrow\uparrow})}\frac{dR}{dt}. \tag{6.3}$$

Equation (6.3) clearly shows that GMR method can provide us information on the time variation of the domain wall position, i.e., domain wall velocity. This technique offers an advantage over conventional experimental methods (Sixtus and Tonks 1931), where domain wall velocity measurements, employing Kerr microscopy, can provide only the average velocity of a domain wall.

Let us give a simplified experimental description of DW velocity measurement. In order to carry out DW velocity measurements, trilayer structures of $Ni_{81}Fe_{19}$ (40 nm)/Cu(20 nm)/$Ni_{81}Fe_{19}$(5 nm) have been considered. The width of the wire is 0.5 mm and four current–voltage terminals have been attached to the sample with voltage probe being placed at a separation distance of 2 mm. The magnetic field has been applied along the wire axis. The voltage across two voltage probes has been monitored by a differential pre-amplifier and a digital oscilloscope. The current flowing through the electromagnet has also been monitored by the digital oscilloscope. In this way, we obtain both the resistance of the trilayer structure and applied magnetic field during the magnetization reversal simultaneously (Fig. 6.11).

It exhibits the change in resistance as a function of an externally applied magnetic field of the trilayer structure at 77 K temperature. At the beginning of the measurement, a magnetic field of 500 Oe has been applied in order to align the magnetization of the device structure in one direction, i.e., along the direction of the applied field. After that, the measurement of resistance has been carried out at 10-ms intervals by sweeping the applied magnetic field towards the counter direction at a sweeping rate of 20 Oe/s. Experimental results, exhibiting the largest value of resistance in the magnetic field range between 80 and 120 Oe, suggest antiparallel alignment of magnetization in that magnetic field range. The appreciable change in resistance at 80 and 120 Oe is attributed to the magnetization reversals of the 5-nm-thick NiFe and 40-nm-thick NiFe layers, respectively. The absence of any measured point in our experimental results, as shown in Fig. 6.11, during the magnetization reversal of the 40-nm-thick NiFe indicates that the magnetization reversal is accomplished within 10 ms. However, magnetization reversal of the 5-nm-thick NiFe takes place gradually with increasing the applied magnetic field. This, in turn, indicates that the magnetization reversal of this 5 nm thin NiFe layer takes place by the successive pinning and depinning of a magnetic DW. Let us focus on the magnetization reversal

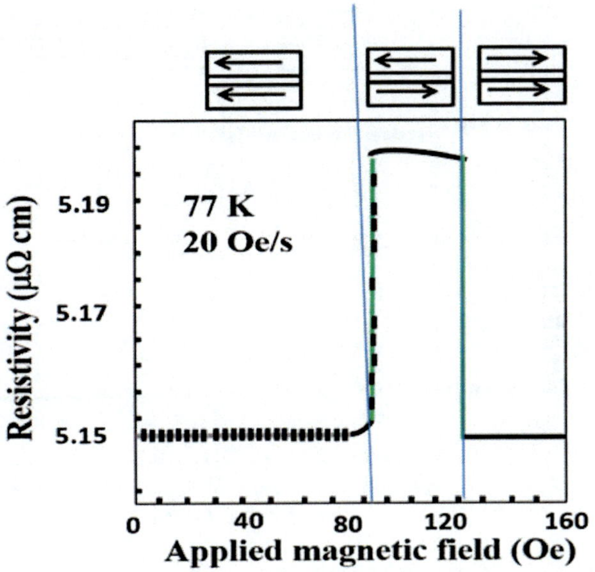

Fig. 6.11 Schematic variation of resistance as a function of the external magnetic field at 77 K of a $Ni_{81}Fe_{19}$ (40 nm)/Cu(20 nm)/$Ni_{81}Fe_{19}$(5 nm) trilayer structures (Figure adapted and redrawn from Ono et al. 1999)

of the 40-nm-thick NiFe. Figure 6.12 exhibits the variation of resistance as a function of time during the magnetization reversal of 40-nm-thick NiFe layer. The data have been collected at 40-ns intervals. The linear portion of the resistance *versus* time curve, as shown in Fig. 6.12, suggests that the propagation velocity associated with the magnetic DW remains constant during the magnetization reversal process of the 40-nm-thick NiFe layer. For instance, the propagation velocity of the magnetic DW at an applied magnetic field of 121 Oe is estimated to be 182 m/s. This has been calculated from the 2 mm separation of the two voltage probes and 11 μs time of travel of the domain wall across that separation. Considering the sweeping rate of the magnetic field to be 20 Oe/s, the variation of the magnetic field during the magnetization reversal is less than 2×10^5 Oe, that is, the external magnetic field is regarded as constant during the measurements.

Nowadays, spintronic devices, whose operation is based on the motion of magnetic domain wall, are gaining widespread of attention (Yamaguchi et al. 2004; Allwood et al. 2002; Versluijs et al. 2001; Parkin 2004; Numata et al. 2007). Although, direct observation of *domain wall motion*, induced by current was made feasible by magnetic force microscopy technique (Yamaguchi et al. 2004), a more sophisticated experiment to measure domain wall velocity was proposed and performed by Himeno et al. (2004). In their experiment, two Cu wires are set crossing the magnetic wire at the ends of the wire. Such Cu wires can generate pulsed local magnetic fields, arising due to the flow of pulsed electric current through those Cu wires. Because of this pulsed local magnetic field, nucleation of magnetic DW takes place at the end of

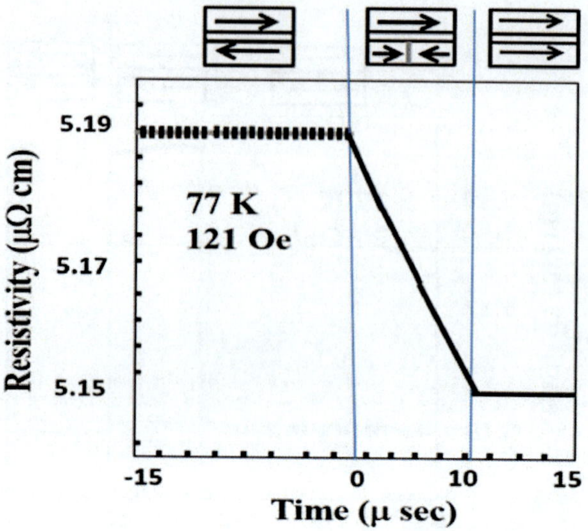

Fig. 6.12 Schematic variation of resistance as a function of time during the magnetization reversal of the 40-nm-thick NiFe layer at 77 K, which was collected at 40-ns intervals of a $Ni_{81}Fe_{19}$ (40 nm)/Cu(20 nm)/$Ni_{81}Fe_{19}$(5 nm) trilayer structures (Figure adapted and redrawn from Ono et al. 1999)

the magnetic wire under a given external magnetic field. This process is quite similar to the previous one and enables us to determine the DW velocity as a function of the external magnetic field in a controlled manner.

Another approach to measure DW velocity consists of magneto-optic Kerr effect magnetometer, having micron-scale spatial resolution, along with the pulsed magnetic field. An appreciably high DW velocity over 1000 m/s with high mobility of 30 m/s Oe has been realized for a single-layer 5-nm-thick $Ni_{80}Fe_{20}$ wire with 200 nm in width (Atkinson et al. 2003).

6.5 Current-Driven Domain Wall Motion

Now, we will introduce a brief idea about *current-driven* domain wall motion. Let us consider a magnetic wire where two magnetic domains are separated by a domain wall as shown in Fig. 6.13i. Magnetic moments, as shown by arrows, associated with those two magnetic domains are oriented at an angular displacement of 180°. According to the definition, magnetic domain wall represents the transition region of the magnetic moments between neighbouring domains, i.e., the direction of moments gradually changes in the domain wall.

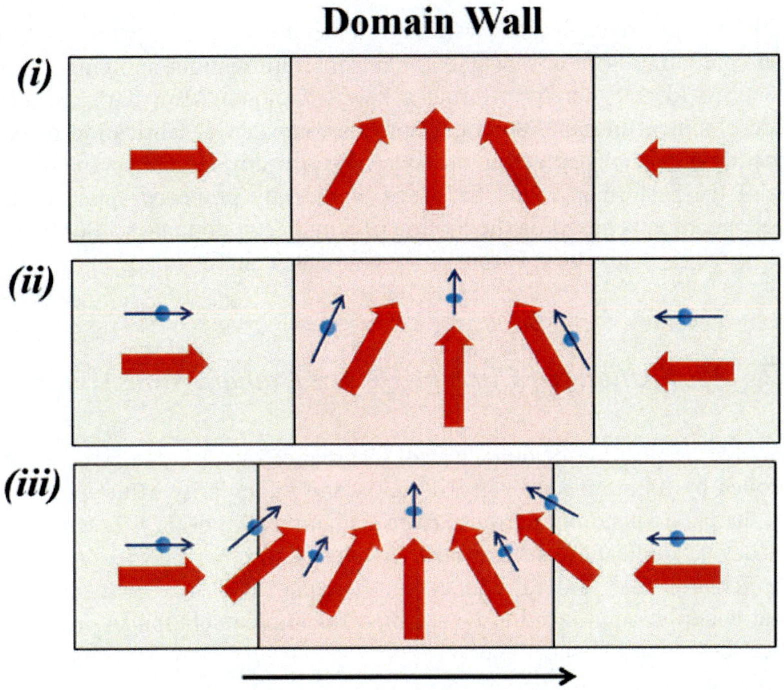

Fig. 6.13 Schematic illustration of current-driven domain wall motion

6.5.1 What Will Happen if Electric Current Flow Through This Domain Wall?

Assume that a conduction electron passes through the domain wall from left- to right-hand side. During this passage of conduction electrons through the domain wall, *s–d* interaction between the spin magnetic moment of conduction electrons and magnetic moments at the domain wall is expected to take place. Consequently, the spin magnetic moments of conduction electrons tend to follow the orientation of local magnetic moments at domain wall, as demonstrated in Fig. 6.13ii. Correspondingly, the local magnetic moments at the domain wall rotate in opposite direction (Fig. 6.13iii). Thus, it comes out that a flow of electric current through the domain wall can cause its displacement and thereby yields domain wall motion. As already mentioned, direct observation of such *current-driven domain wall motion* was possible utilizing magnetic force microscopy technique (Yamaguchi et al. 2004). Furthermore, propagation of the domain wall has been found to be in opposite direction to that of current direction. Moreover, the propagation direction can be reversed by switching the polarity of this current. It is quite intriguing to point out that this observation of *current-driven domain wall motion* is consistence with the spin transfer torque mechanism.

This current-driven domain wall motion has potential to modulate the magnetization configuration of any magnetic nanostructure without applying any external magnetic field and thereby offering a new technique. Also, with the technological development in nanolithography process, nowadays, fabrication of nanoscale magnetic wires has become quite easier. This, in turn, inspires dramatic improvement of the performance and functions of recently proposed spintronic devices, whose operation is based on the motion of a magnetic domain wall (Allwood et al. 2002; Versluijs et al. 2001; Parkin 2004; Numata et al. 2007).

6.5.2 *Applications of Current-Driven Domain Wall Motion*

Evidently, the position of domain wall in a nanostructured magnetic wire can be controlled by tuning the intensity, duration and the polarity of the pulsed current. Thus, the phenomenon of current-driven domain wall motion is potential for spintronic device applications such as novel memory and storage devices (Allwood et al. 2002; Versluijs et al. 2001; Parkin 2004; Numata et al. 2007). In this context, some critical conditions that need to be satisfied for implementation in practical device applications should be mentioned:

1. Low threshold current density;
2. High domain wall velocity;
3. Stability and controllability of domain wall position.

These three conditions should be satisfied simultaneously for the operations of real devices.

6.6 Conclusions

In this chapter, we have presented an introduction on magnetic domain walls motion in magnetic materials. We have discussed several factors, like the origin of magnetic domain wall, domain wall width, the reason for small particles to be mono-domain, etc. Furthermore, we have addressed the relevance of magnetic domain walls motion in spintronics. In this direction, we have first focussed our discussion on the detection of domain-wall propagation. Moreover, we have also reviewed ratchet effect in magnetic domain wall motion and its applicability in the field of spintronics. Other important issue that has been addressed is the measurements of domain wall motion velocity. Furthermore, in the context of spintronics application, we have discussed current-driven domain wall motion.

6.7 Exercises

1. What is magnetic domain wall?
2. How does the resistance of a multilayer film changes with the motion of a domain wall?
3. What is Ratchet effect?
4. Which is the better technique to determine the domain wall velocity and explain how?
5. How can one obtain higher domain wall velocity in nanoscale devices?
6. What are the advantages of current-driven domain wall motion over the external applied field one?
7. Why do domains exist? Why small particles are always mono-domain?
8. How domain-wall propagation can be detected in a magnetic nanowire?
9. Describe Ratchet effect in a trilayer magnetic wire with schematic representation.
10. What is current-driven domain wall motion?
11. What will happen if an electric current flows through this domain wall?
12. What are the applications of current-driven domain wall motion?

References

D.A. Allwood et al., Science **296**, 2003 (2002)
D. Atkinson, D.A. Allwood, G. Xiong, M.D. Cooke, C.C. Faulkner, R.P. Cowburn, Nat. Mater. **2**, 85 (2003)
M.N. Baibich et al., Phys. Rev. Lett. **61**, 2472 (1988)
N. García, M. Munoz, Y.W. Zhao, Phys. Rev. Lett. **82**, 2923 (1999)
A. Himeno, T. Ono, S. Nasu, T. Okuno, K. Mibu, T. Shinjo, Dynamics of magnetic domain walls in nanomagnetic systems. J. Magn. Magn. Mater. **272–276**, 1577 (2004)
A. Himeno et al., J. Magn. Magn. Mater. **286**, 167 (2005a)
A. Himeno et al., J. Appl. Phys. **97**, 066101 (2005b)
S. Lepadatu, Y.B. Xu, Phys. Rev. Lett. **92**, 127201 (2004)
S. Lepadatu, Y.B. Xu, E. Ahmad, J. Appl. Phys. **97**, 10J708 (2005)
P.M. Levy, S. Zhang, Phys. Rev. Lett. **79**, 5110 (1997)
H. Numata et al., *Symposium on VLSI Technical Digest* (2007), pp. 232
T. Ono, H. Miyajima, K. Shigeto, K. Mibu, N. Hosoito, T. Shinjo, Science **284**, 468 (1999)
S.S.P. Parkin, U.S. Patent No. 6834005 (2004)
K.J. Sixtus, L. Tonks, Phys. Rev. **37**, 930 (1931)
J.J. Versluijs, M.A. Bari, J.M.D. Coey, Phys. Rev. Lett. **87**, 026601 (2001)
P. Weiss, J. Phys. **6**, 661 (1907)
A. Yamaguchi et al., Phys. Rev. Lett. **92** 077205 (2004); **96** 179904(E) (2006)

Chapter 7
Opto-spintronics

7.1 Introduction: What Is Opto-spintronics?

Opto-spintronics is an emerging and fascinating branch of spintronics where light is used for the study and/or control of electron spin. Opto-spintronics combines the effect of light with the spin of charge carriers and hence, fast and accurate control of electron spins is expected. We can say that opto-spintronics is an inter-disciplinary field of overlapping region of optics, electronics and magnetics (shown in Fig. 7.1).

- Spintronics is a spin-dependent phenomenon applied to electronic devices.
- Optoelectronics is the study and application of electronic devices that interact with light.
- Magnonics is the study of spin waves and magnetism. Opto-magnonics is manipulating spin waves with light pulses.

This emerging field of spintronics has received tremendous momentum as femtosecond laser has the capability of fastest change in magnetic state of matter. Now let us see what is femtosecond laser and why it is so important?

7.2 What Is so Special About Femtosecond Laser?

A femtosecond laser is one kind of ultrashort laser where light is emitted in femtosecond domain, i.e., of the order of 10^{-15} s (see Fig. 7.2). Femtosecond laser is technologically important for its wide use in automobile, medical instrumentation and micro-electronics industries. It has also triggered the ultrafast controlling of magnetic materials.

Some special features of this laser are mentioned below

- Being very short optical pulse, it bears the ability of instant excitation.
- Monitoring of excited state can be done with greater temporal resolution (≈12–50 fs).

P. Dey and J. N. Roy, *Spintronics*,
https://doi.org/10.1007/978-981-16-0069-2_7

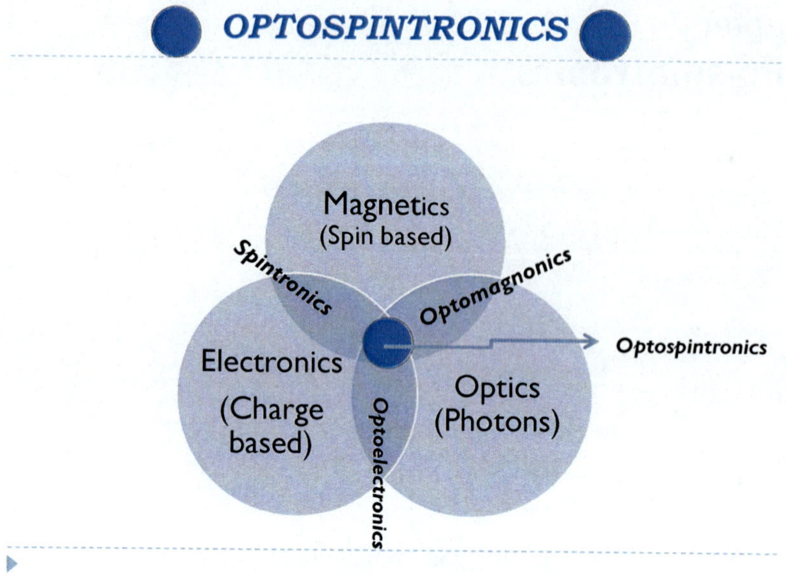

Fig. 7.1 Opto-spintronics: an inter-disciplinary field

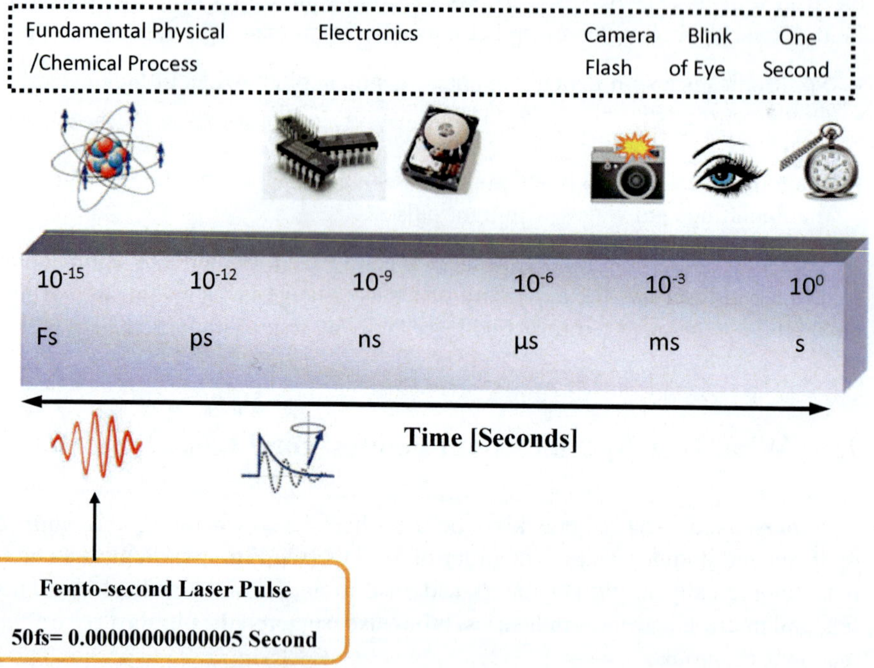

Fig. 7.2 Examples of operating timeline of different devices and phenomena

- Dissipation and transfer of energy take place on a time scale greater than 100 femtosecond.
- Can provide very high peak power. Laser with 10 mJ pulse and having hundred femtosecond duration can produce peak power 100 GW (relation: I ~ J/τ,where I denotes the produced peak power, J and τ represent the pulse energy and pulse duration, respectively.)
- Femtosecond laser pulse can create an ultraintense magnetic field strength of the order of hundreds of Tesla.

7.3 Issues to Be Considered: Why Do We Need Optical Manipulation?

To get preferred result in spintronics, many issues are to be considered and major challenges are to be faced and overcome. Some of them are electron's spin lifetime optimization, spin coherence detection in nanoscale structures, spin-polarized carriers transport, etc. Challenges also include skilful management of electron and nuclear spin in ultrafast time scale. Efficient controlling of magnetic states is the key towards the development of high speed information storage device in magnetic media. Magnetization manipulation requires alternative ways other than magnetic field that will be effective at or above room temperature.

The development of femtosecond lasers, as external stimuli, has triggered the path of ultrafast manipulation of the magnetic order (Beaurepaire et al. 1996; Kirilyuk et al. 2010; Satoh et al. 2010, 2009, 2016; Chen et al. 2019; Wolf et al. 2001; Pershan 1963). Thanks to Bigot and coworkers for their milestone work on ultrafast switching in Nickel at Strasbourg in the year of 1996. Femtosecond light pulses offer the challenging but interesting possibility of probing a magnetic system on a time scale much faster than the time scale of spin–orbit interaction (SOI) or magnetic precession (see Fig. 7.3). Generally the time scale for SOI is around 1–10 ps whereas for magnetic precession, it is 100 ps–1000 ps. Ultrashort light pulse-based magnetic

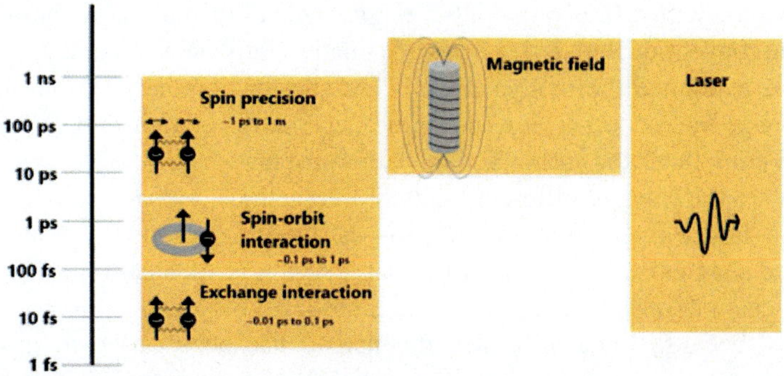

Fig. 7.3 Time scales in magnetism (adapted and redrawn from Ref. (Kirilyuk et al. 2010))

manipulation has created high impact for future spin-based electronics devices and quantum computation. Femtosecond optical excitation gives raise a fundamental question. The question is whether magnetization reversal would be possible, which is faster than within half a precessional period? These questions may not play a decisive role at larger time scale and for equilibrium states, but it becomes very important when magnetically ordered spins, the electron system and the lattice become dynamically isolated.

Such investigations on ultrafast magnetization dynamics are technologically relevant where there is a lag between manipulating speed and data storing, creating a so-called ultrafast technology gap. The same thing is also applicable in spintronics as in, for example, MRAM devices. Therefore, the investigation of the fundamental as well as practical speed limits of magnetization direction are surely of great importance. Logical bits designated by 'ones' and 'zeros' are stored in magnetic memory devices by assigning the magnetization vector either 'up' or 'down'. Conventionally, magnetic bit is recorded by reversing the magnetization on the application of magnetic field. Intuitively, one would expect that switching could be infinitely fast, limited only by the attainable strength and shortness of the magnetic field pulse.

However, recent experiments show that deterministic magnetization reversal does not take place if the magnetic field pulse is shorter than two picoseconds. Could optical pulses be an alternative?

7.4 Laser Pulse and Its Impact on a Magnetic System

Demonstration of ultrafast demagnetization of a Ni film by a 60 femtosecond optical laser pulse triggers the emergence of laser-based controlling of magnetization (Beaurepaire et al. 1996). Findings were also confirmed by subsequent experiments and opened the possibility of light generated coherent magnetic precession, laser-induced spin reorientation or even modification of magnetic structure on a time scale of one picosecond or less. However, despite the reporting of many interesting experimental results, the ultrafast optical manipulation of magnetism is still not clearly understood. The excitation with fs laser pulse puts a magnetic medium into a non-equilibrium state where the conventional macro-spin approximation fails and the magnetic phenomena cannot be explained thermodynamically. At shorter time scales (sub-picosecond), the exchange interaction is viewed as time-dependent phenomenon. All these issues seriously complicate the understanding of this problem and give raise many questions.

What are the roles of different kinds of interactions, namely, spin–orbit interaction, spin–lattice interaction and electron–lattice interactions in the ultrafast light controlled magnetism? How does the band structure of electron affect the laser-assisted changes in magnetic order? A thorough investigation may answer these questions, because light controlled magnetism has been observed in dielectrics, metals and also in semiconductors. As magnetism is related to angular momentum, then demagnetization is concerned with fundamental question of the conservation and transfer of angular momentum. Different types of angular momentum transfer

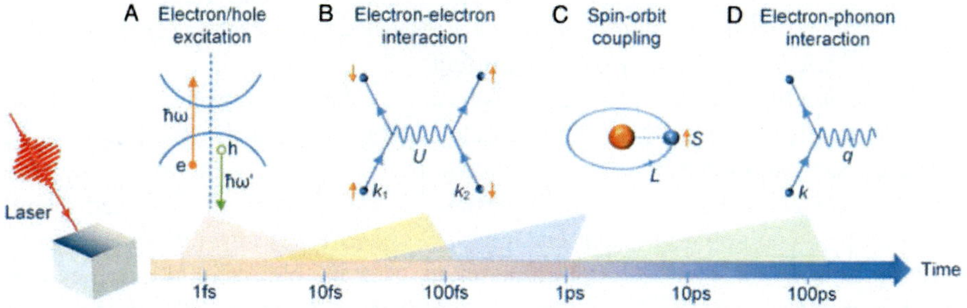

Fig. 7.4 Schematic diagram of various interactions involved in the laser-induced ultrafast demagnetization process. The effective time scale given is an estimated illustration; however, its actual value is material dependent. (Ref. (Chen et al. 2019))

channels are involved in the demagnetization process (Fig. 7.4): Light-induced electron/hole excitation, electron–electron interaction with spin exchange, spin–orbit coupling and electron–phonon interaction (Chen et al. 2019).

The effects of a laser pulse on a magnetic system could broadly be classified into two categories (Kirilyuk et al. 2010; Zvezdin and Kotov 1997; Pavlov et al. 1997; Eremenko et al. 1992; Kimel et al. 2005). The first category is concerned with thermal effects where the change of magnetiozation is due to change in temperature. The second category is related with non-thermal effect.

7.4.1 Thermal Effects

In this case, photon absorption pumps the energy into the medium. The change in the magnetization is related with spin temperature. The direct energy transfer is not effective from light to spins as spin-flip transitions are not allowed in the electric dipole approximation. Rather, light transfers the energy into the system of electron and phonon. Internal equilibration processes like electron–electron, electron–phonon and electron–spin interactions determines the time scale of the subsequent change of magnetization. This time scale is very short for iron or nickel ferromagnets and can be down to 50 femtosecond. In contrast, this time scale founds to be around nanosecond for dielectric magnets as direct electron-spin processes are absent.

7.4.2 Non-thermal Effects

The possibility of ultrafast non-thermal control of magnetization by light is much more interesting. Here, the change in the magnetization is not simply due to an increase in temperature. In one hand, it gives greater freedom for the manipulation of the magnetization, and on the other hand, it prevents unwanted heating and possible damage in devices. Basically, two types of non-thermal effect are seen: (i) photo-magnetic effect and (ii) opto-magnetic effect.

Photo-magnetic Effect: This effect is based on photon absorption and deals with the direct influence of electromagnetic radiation on magnetic properties. Photomagnetic effects are observed in ferrimagnetic garnet films where the linearly polarized laser pulses change magnetocrystalline anisotropy via optically induced electron transfer between non-equivalent crystal sites. The effect is ultrafast and can induce magnetization to start precessing immediately after the photoexcitation. It can even switch the magnetization. Though the excitation process appears to be instantaneous, the relaxation time of the photo-induced anisotropy is several nanoseconds.

Opto-magnetic Effect: This process does not need any absorption of photon. It is one type of impulsive effect and just opposite to magneto-optic effect. In magneto-optics, the influence of the magnetization is generally observed as a change in the light's polarization caused by a magnetically polarized material (Faraday Effect) (see Fig. 7.5a), whereas in opto-magnetics, circularly polarized light induces a magnetic polarization in the material (Inverse Faraday Effect) (see Fig. 7.5b). Opto-magnetic interaction of femtosecond laser pulses with magnetic medium has the base on non-dissipative Raman scattering mechanisms.

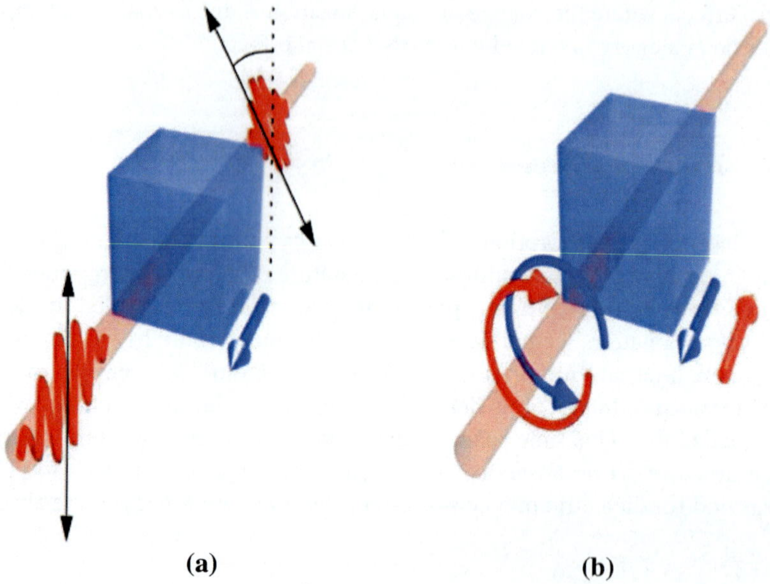

Fig. 7.5 **a** Magneto-optics (Faraday Effect) **b** Opto-magnetics (inverse Faraday effect)

7.5 Interaction of Photons and Spins

Polarization rotation of a linearly polarized light takes place when it travels through a magnetized medium (Faraday effect). This effect shows that a magnetically ordered medium affects photons by changing its state of polarization. Now the question arises whether the inverse phenomenon is also feasible where polarized photons can indeed affect the magnetization?

Let us discuss the possibility of photon–spin direct interaction using energy considerations. It is known that for an isotropic, non-absorbing, magnetically ordered medium having static magnetization **M**(0) in a monochromatic light field **E**(ω), the thermo dynamical potential Φ can be written as

$$\Phi = \alpha_{ijk} E_i(\omega) E_j(\omega)^* M_k(0) \tag{7.1}$$

α_{ijk} is the magneto-optical susceptibility (see Refs. Beaurepaire et al. (1996) and Kirilyuk et al. (2010) and Ref. Wolf et al. (2001) for details.)

Optical polarization can be expressed as

$$\mathbf{P}(\omega) = \partial \Phi / \partial \mathbf{E}(\omega)^* \tag{7.2}$$

Equation (7.1) shows that **P**(ω) should have a contribution $\mathbf{P}^{(m)}$ proportional to the magnetization **M**, and hence,

$$P_i^{(m)} = \alpha_{ijk} E_j(\omega) M_k(0). \tag{7.3}$$

Equation (7.3) can be used to determine the rotation of plane of polarization given by

$$\Theta_F = \alpha_{ijk} M_k(0) \omega L / cn \tag{7.4}$$

where, c is the speed of light in vacuum, n is the refraction coefficient of the medium, ω is the light frequency and L is the propagation distance.

Now from Eq. (7.1), one can also find that an electric field of light at frequency ω will act on the magnetization as an effective magnetic field directed along the wave vector of the light **k** and can be written as

$$H_k = -\partial \Phi / \partial M_k = \alpha_{ijk} E_i(\omega) E_j(\omega)^* \tag{7.5}$$

The magneto-optical susceptibility α_{ijk} is a fully antisymmetric tensor having a single independent element α in an isotropic media. Therefore, Eq. (7.5) can be rewritten as

$$H_k = \alpha \left[E_i(\omega) \times E_j(\omega)^* \right] \tag{7.6}$$

From this, it becomes clear that right- and left-handed circularly polarized waves should act as magnetic fields of opposite sign. Hence, from Eq. (7.6), we can say that, in addition to magneto-optical Faraday effect where the polarization of light is affected by the magnetization M, the same susceptibility α also determines the inverse opto-magnetic effect where circularly polarized light modifies the magnetization via inverse Faraday effect (IFE). In recent years, IFE has acquired renewed interest in solid-state physics, especially in magnetic material. It is recognized that IFE could offer an alternative and new path to ultrafast, all-optical magnetization reversal. The situation may be different and interesting when the product **E(ω) x E(ω)*** is changed much faster than the fundamental time scales in a magnetically ordered material, given by the spin precession period and the spin–lattice relaxation time.

Let us consider the excitation of spins by a laser pulse with duration $\Delta t = 100$ fs (spectrally broad $\Delta\omega$ ~5 THz). Initially the electron is in the ground state E_1 and its spin is up. If photon is acted on this electron, there will be an increase in orbital momentum. This effectively increases SOI and hence the probability of a spin-flip process. If the photon energy is less than the band gap energy (difference between the ground state E_1 and the nearest excited state E_2) then the electron will not jump to the excited state. Instead, the electron will have spin flip in its ground state.

This process is associated with the coherent reemission of a photon of energy $\hbar\omega_2 = \hbar(\omega_1 - \Omega_m)$. Here, $\hbar\Omega_m$ corresponds to the energy of a magnon in magnetically ordered materials. The presence of frequencies ω_1 and ω_2 in the laser pulse can stimulate coherent spin-flip process (see Fig. 7.6). Materials having large magneto-optical susceptibility can show ultrafast spin-flip process, which is around 20 femtosecond. This mechanism does not require annihilation of a photon and hence much more effective than magnetic dipole transitions.

7.6 Experimental Techniques

The study of ultrafast spin dynamics needs methods to detect the changes that occurred in given magnetic medium with high temporal resolution. Several methods are used in this regard like.

- Pump and probe
- Optical probe
- Far infrared probe
- X-ray probe.

A brief description is given below about the above-mentioned probing techniques.

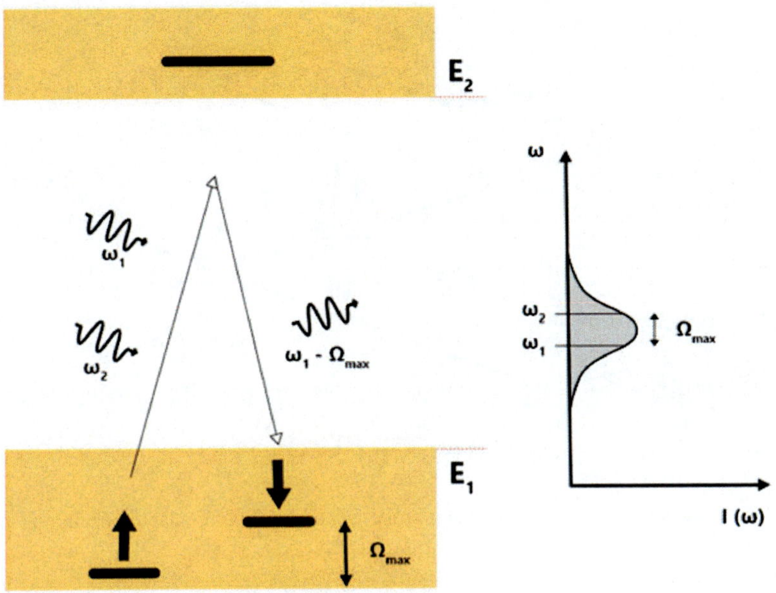

Fig. 7.6 With strong spin–orbit coupling, the light frequency ω_1 causes a transition into a virtual state. Light frequency ω_2 stimulates the relaxation back to the ground state and produce a magnon. (Adapted and redrawn from Ref. (Kirilyuk et al. 2010))

7.6.1 Pump and Probe Method

Pump and probe method is an effective tool for the investigation of ultrafast light triggered magnetization (see Fig. 7.7). In this time-resolved technique, a first light pulse is the 'pump' pulse, which triggers a photoinduced process. The delayed second pulse (probe) detects the corresponding changes. The temporal resolution of the experiment is determined by the duration of pump and probe pulses. We need very short pulses to observe fast processes. Hence, in an ideal system, pulses should be as short as possible and broadly tunable**.** The probe pulse should be as short as possible and can be chosen in the far-infrared, optical, ultraviolet or X-ray spectral regions.

7.6.2 Optical Probe

Electric dipole approximation can describe the interaction between medium and probe pulse in the spectral range of light 0.4 μm–10 μm. Such interaction can be analysed with the help of thermodynamical potential Φ. For an isotropic non-dissipating, magnetically ordered medium with static magnetization **M**(0) or antiferromagnetic vector **l**(0) in a monochromatic light field **E**(ω), neglecting terms of order higher than 3 in **E**(ω), Φ can be written as

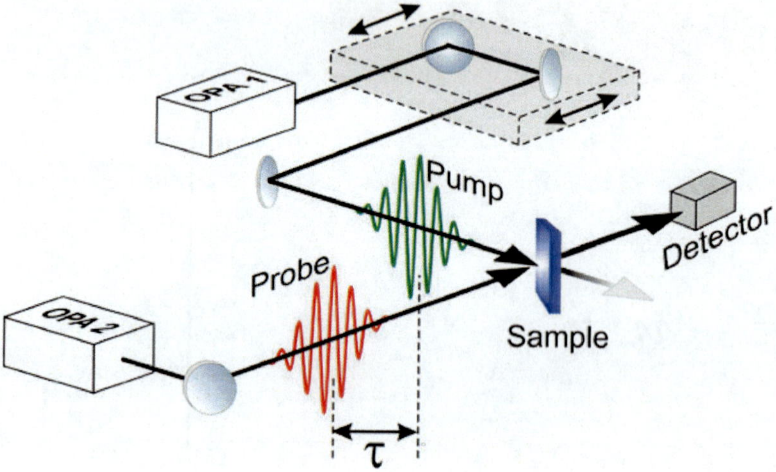

Fig. 7.7 Schematic diagram of pump and probe technique (Figure adapted from https://polli.faculty.polimi.it/spectroscopy/)

$$
\begin{aligned}
\Phi = {} & \chi_{ij}^{(l)} E_i(\omega)^* E_j(\omega) + \alpha_{ijk}^{(l)} E_i(\omega)^* E_j(\omega) M(0)_k \\
& + \beta_{ijk}^{(l)} E_i(\omega)^* E_j(\omega) I(0)_k + \chi_{ijk}^{(nl)} E_i(2\omega)^* E_j(\omega)\, E_k(\omega) \\
& + \alpha_{ijkl}^{(nl)} E_i(2\omega)^* E_j(\omega)\, E_k(\omega) M(0)_l \\
& + \beta_{ijkl}^{(nl)} E_i(2\omega)^* E_j(\omega)\, E_k(\omega) I(0)_l
\end{aligned}
\quad (7.7)
$$

where $\chi_{ij}{}^{(l)}$, $\alpha_{ijk}{}^{(l)}$, $\beta_{ijk}{}^{(l)}$, $\chi_{ijk}{}^{(nl)}$, $\alpha_{ijkl}{}^{(nl)}$ and $\beta_{ijkl}{}^{(nl)}$ are tensors that define the optical properties of the medium, and the superscript *l* indicates linear and *nl* indicates the non-linear response, respectively. Equation (7.7) is a generalization of Eq. (7.1) from which the Faraday effect was derived.

Similarly, polarization rotation of light can be obtained upon reflection from a magnetized medium because of the term $\mathbf{P}^{(m)}$. [The phenomenon of magnetization-induced polarization rotation upon reflection of light is known as the magneto-optical Kerr effect (MOKE)]. Hence, the linear magneto-optical effects can be utilized as a probe of the magnetization of a medium.

The linear terms in antiferromagnetic vector **I**(0), as given Eq. (7.7), show that linear optics can also serve as a probe of magnetic order in geometries where Faraday or Kerr effect is absent or in a material with no net magnetization, such as antiferromagnets or ferrimagnets. It is important to note that all linear M–O phenomena are sensitive to certain projections of the magnetic vectors **(M)** and **(I)** and can be observed in all media, irrespective of their crystal symmetry or crystallographic orientation.

In non-linear optical approximation of Eq. (7.7), the terms of third-order are to be taken into account. In that approximation, the optical field $\mathbf{E}(\omega)$ can induce a polarization in the medium at the double frequency $\mathbf{P}$ (2 ω). This phenomenon is known as second-harmonic generation (SHG) where the medium is excited by the optical field **E** (ω) and generates light with a frequency double of initial one, i.e., 2ω. The intensity of the second harmonic light in ferromagnets or antiferromagnets can be found as

$$I(2\omega) \sim \mathbf{P}|(2\omega)|^2$$
$$= \left[\chi_{ijk}{}^{(nl)} E_j(\omega)\, E_k(\omega) + \alpha_{ijkl}{}^{(nl)} E_j(\omega)\, E_k(\omega) M(0)_l\right]^2$$

Or $[\chi_{ijk}{}^{(nl)} E_j(\omega)\, E_k\, (\omega) + \alpha_{ijkl}{}^{(nl)} E_j\, (\omega)\, E_k\, (\omega) I(0)_l]^2$ respectively.

Hence, the equation tells that the SHG can also be a measure of the magnetization **M**(0) or antiferromagnetic vector **l**(0). Compared to linear magneto-optical effects, the second harmonic is generated only in centrosymmetric media (in electric dipole approximation). Consequently, the magnetic second-harmonic generation (MSHG) technique can be used as a unique tool for probing surface and interface magnetism.

In the visible spectral range, the optical response of media is dominated by electric dipole transitions. Spin flip and the selection rules for such transitions are not allowed. A strong spin–orbit interaction is required for the sensitivity of light to magnetic order in case of both linear and non-linear magneto-optical (M–O) effects. A strong coupling between spins and orbitals results in substantial values of α and β components in Eq. (7.7). Hence, the magneto-optical effects can provide information about the magnetic state of a medium.

The analysis of results obtained from M–O studies of solids is troublesome as the optical transition in visible spectral region is relatively broad. Besides that, magneto-optical effects are helpful only as indirect probes of spin ordering as they are proportional to **M** and **l**. Therefore, the interpretations of time-resolved M–O measurements bear a number of uncertainties. Laser-induced excitation can alter the populations of the excited states and change the symmetry of the ground state. As a result, the M–O response may be changed even without affecting its magnetic order and magnetization vector. Thermo dynamical description is unsuitable for an explanation of femtosecond laser-induced effects in magnets in terms of time-dependent tensors. Laser-induced M–O Kerr or Faraday effects should be described considering non-linear effect.

7.6.3 Far-Infrared (F-IR) Probe

Far-infrared spectroscopy is used to measure low-energy optical excitations in high magnetic fields such as various electron magnetic resonances (ESR, cyclotron resonance, antiferromagnetic resonance). The low excitations in the F-IR or THz region are point of concern since the electronic properties of quantum materials are usually determined by the low lying charge and magnetic excitations. Significantly, the frequencies of magnetic resonance of most antiferromagnets and spin-flip transition of many rare-earth ions lie in the frequency range from 100 GHz to 3 THz. A combination of magnetic field and optical probe enables us to retrieve a wealth of information in magnetic materials.

7.6.4 X-ray Probe

Dichroism is the property of a material in which different absorption coefficients are found when light with different polarization states travels through the medium. Nowadays, the term dichroism refers to polarization-dependent photon absorption. Anisotropies in the charge or the spin in the material could be the origin of dichroism. Latter case is related to magnetic dichroism. Two types of magnetic dichroism techniques, namely linear and circular, are observed in X-ray region. X-ray magnetic linear dichroism (XMLD) refers the difference in absorption in linearly polarized perpendicular and parallel to the quantization axis. X-ray magnetic circular dichroism (XMCD) refers to the difference in absorption for left- and right circularly polarized light. Magnetic properties of ferromagnets (usually metal) are suitably studied with XMCD whereas antiferromagnets (usually oxides) are studied with XMLD techniques. Polarized soft X-rays play a pivotal role on studying the origin of fundamental effects such as exchange bias and magnetic anisotropy. The integrated circular dichroism signal can be used to measure the magnitude of the orbital and spin components of the magnetic moment using sum rules. Time-resolved XMCD measurement with temporal resolution in sub-picosecond order can provide unique information into ultrafast light-induced magnetic changes.

7.7 Demagnetization of Metallic Ferromagnets: 3TM Model

Laser-induced magnetization dynamics experiments confirm that the demagnetization is very complicated and involves different relaxation processes. The interactions among lattice, electrons and spins can qualitatively be explained with the phenomenological three temperature model (3TM) as given in Fig. 7.8. The said model can be described by three coupled differential equations assuming certain equilibrium within the subsystem,

$$C_e d(T_e)/dt = -G_{el}(T_e - T_l) - G_{es}(T_e - T_s) + P(t),$$

$$C_s d(T_s)/dt = -G_{es}(T_s - T_e) - G_{sl}(T_s - T_l),$$

$$C_l d(T_l)/dt = -G_{el}(T_l - T_e) - G_{sl}(T_l - T_s),$$

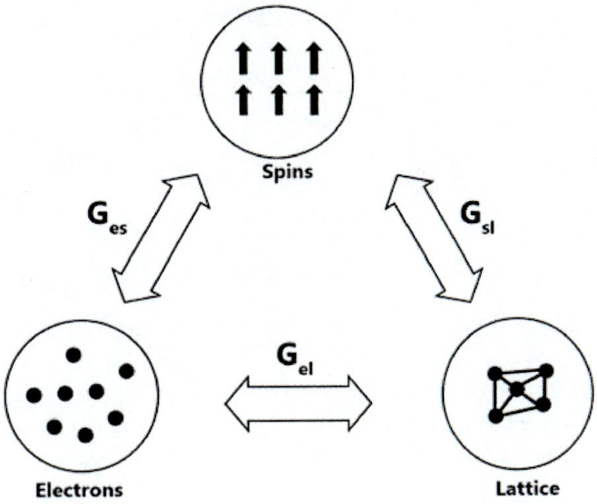

Fig. 7.8 Interaction among electrons, spins and lattice in 3TM (Adapted and redrawn from Ref. (Kirilyuk et al. 2010))

where the coupling between the *i*th and *j*th baths is described by *Gij*, heat capacity *is* represented by *Ci*, temperature of the corresponding system temperature is *i* and the optical input is P(t). The phenomenological parameters G_{ij} coefficients give us the information regarding the strength of a particular link only, not about the nature of the interaction. Equilibration process among different reservoirs (electrons, spins and lattice) upon the excitation of a femtosecond pulsed laser takes place at different stages.

The electrons respond practically instantaneously with an electromagnetic excitation. Laser-induced demagnetization process can be explained in step by step as mentioned below.

Step-1: Laser pulse is incident on the sample. Then, electron–hole pairs are created within very short time scale of ~ 1 fs.

Step-2: Within 50 fs to 500 fs, the electronic system is elevated at temperatures *Te* by electron–electron interactions and reaches equilibrium.

Step-3: The equilibrated electron heats up the lattice through the electron–phonon interaction within 100 femtosecond to 1 picosecond for metals. Consequently, T_l is increased and finally, electron and lattice systems come in thermal equilibrium at the end of a picosecond.

It is remembered that the angular momentum conservation is also one important part in any magnetic system. Therefore, in the process of ultrafast demagnetization, a certain amount of angular momentum should be taken away from the spin system. The laser-induced demagnetization of Ni could be related to a direct coupling between photons and spins.

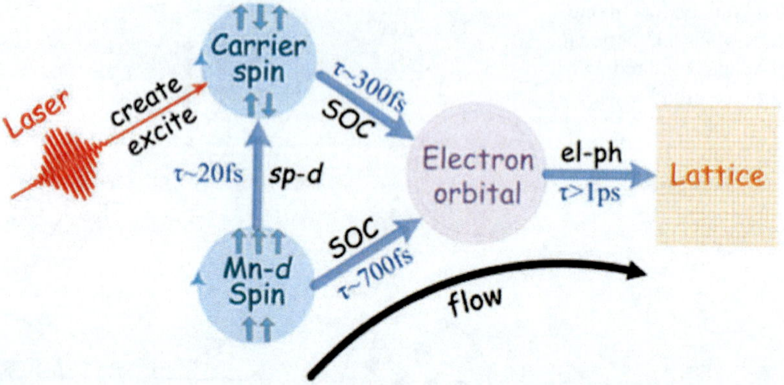

Fig. 7.9 The flow of angular momentum in GaMnAs. τ represents the angular momentum relaxation time. (Ref. (Chen et al. 2019))

7.8 Demagnetization of Ferromagnetic Semiconductor: GaMnAs

A wide range of materials like ferromagnetic metals, antiferromagnetic dielectrics and semiconductors show light-induced ultrafast demagnetization process. Ferromagnetic (III, Mn)V semiconductors are the most favourable and encouraging candidates towards realization of next-generation multifunctional spintronic devices. In (III, Mn)V semiconductors, Mn–Mn spin exchange interaction is mediated by the hole carriers. This has helped to modify the magnetic dynamics by changing the itinerant carrier density with femtosecond laser (Chen et al. 2019; Tesařová et al. 2014).

Four types of angular momentum present in GaMnA systems whose total value is to be conserved once light is turned off. They are: (i) localized Mn-d spin, (ii) delocalized carrier spin, (iii) electron orbital angular momentum and (iv) lattice phonon angular momentum. Their angular momentum flow in the demagnetization process is schematically illustrated in Fig. 7.9. Initially through the strong sp–d interaction, the laser induces electron excitation and transfers a major part of Mn–d spin into itinerant carriers in a very short time. Then, through spin–orbit coupling, the spin is relaxed for both Mn–d and itinerant carriers. In about 1 ps, the spin angular momentum is transferred to the orbital degree of freedom. Then, on the time scale of a few picoseconds, the angular momentum is transferred to the lattice through the slow electron–phonon coupling.

7.9 Antiferromagnetic Opto-spintronics

Conventional ferromagnetism-based spintronic devices depend on the controlling and modification of magnetic moments. However, antiferromagnetic materials have drawn much more attention in recent time due to its intriguing features compared to ferromagnets: they.

- are robust against perturbation,
- produce no stray fields,
- exhibit ultrafast dynamics.
- generate large magneto-transport effects.

7.9.1 *Brief History of the Emergence of Antiferromagnetic Spintronics*

Antiferromagnetic spintronics has been formally emerged inspired by the discovery of a spin-valve-like magneto-resistance of an antiferromagnet based tunnel junction in 2011. Extensive researches have been carried out to study the roles of antiferromagnets in spintronics devices. In 2013, the room temperature tunnelling anisotropic magnetoresistances have been realized in antiferromagnetic material. The emergence of a room temperature antiferromagnetic memory resistor is witnessed in 2014. The reversible electrical switching in antiferromagnets is observed in 2016. Of late, many fascinating phenomena like large anomalous Hall effect, spin Hall magnetoresistance and skyrmions have been noticed with antiferromagnets. They have accelerated the research for the development of antiferromagnetic spintronics. Field-free switching of magnetization in antiferromagnets through spin–orbit torque has also been realized. Therefore, it becomes very important to know the physics and engineering for skilful control of magnetic states of antiferromagnets. Efficient controlling of magnetization can be done through magnetic, electrical, optical and strain manipulation. Here we will discuss about optical manipulation only.

7.9.2 *Probing and Optical Manipulation of Antiferromagnets*

In antiferromagnetic materials, the elementary magnetic moments are spontaneously long-range ordered. However, the net magnetic moment is zero or small compared to the sum of the magnitudes of the participating magnetic moments. In compared to ferromagnetism, there are different numbers of arrangement of magnetic moments on a lattice to give net moment zero (Kimel et al. 2004; Němec et al. 2018; Feng et al. 2015; Saidl et al. 2017; Nishitani et al. 2012; Manz et al. 2016; Radu et al. 2010). All nearest-neighbour magnetic atoms are collinear and aligned antiferromagnetically in perovskite $LaFeO_3$. Alternatively aligned ferromagnetic planes are present in $LaMnO_3$. In $YMnO_3$, three magnetic sublattices are rotated by 120°. Helical order is found in rare-earth metals and multiferroics like $TbMnO_3$. Electromagnetic radiation faces a strong challenge towards reorientation of spins in antiferromagnets. The required field strength may be of the order of tens or hundreds of Tesla. Fortunately, laser light has proved him as a powerful tool to control and detect spins and their ultrafast dynamics in magnetic materials. Based on interaction of AFs with electromagnetic radiation, there are three detection techniques, namely LO studies, NLO studies and THz spectroscopy.

Magneto-optics (MO) has been established as an efficient probe of magnetic order with good temporal and spatial resolutions. Let us say that an AF has only two equivalent magnetic sublattices. If their magnetization is represented by $\mathbf{M_1}$ and $\mathbf{M_2}$, then we can define two orthogonal vectors of magnetization $\mathbf{M} = \mathbf{M_1} + \mathbf{M_2}$ and antiferromagnetism $\mathbf{M} = \mathbf{M_1} - \mathbf{M_2}$ (Fig. 7.10a–b).

Many time-resolved MO studies have been performed in canted antiferromagnets. Large magneto optic Kerr effect was predicted also for non-collinear antiferromagnets Mn_3X (where, X = Rh, Ir, Pt) (Feng et al. 2015). Let us consider a cartesian coordinate system with the x- and y-axes in the sample plane and the z-axis along

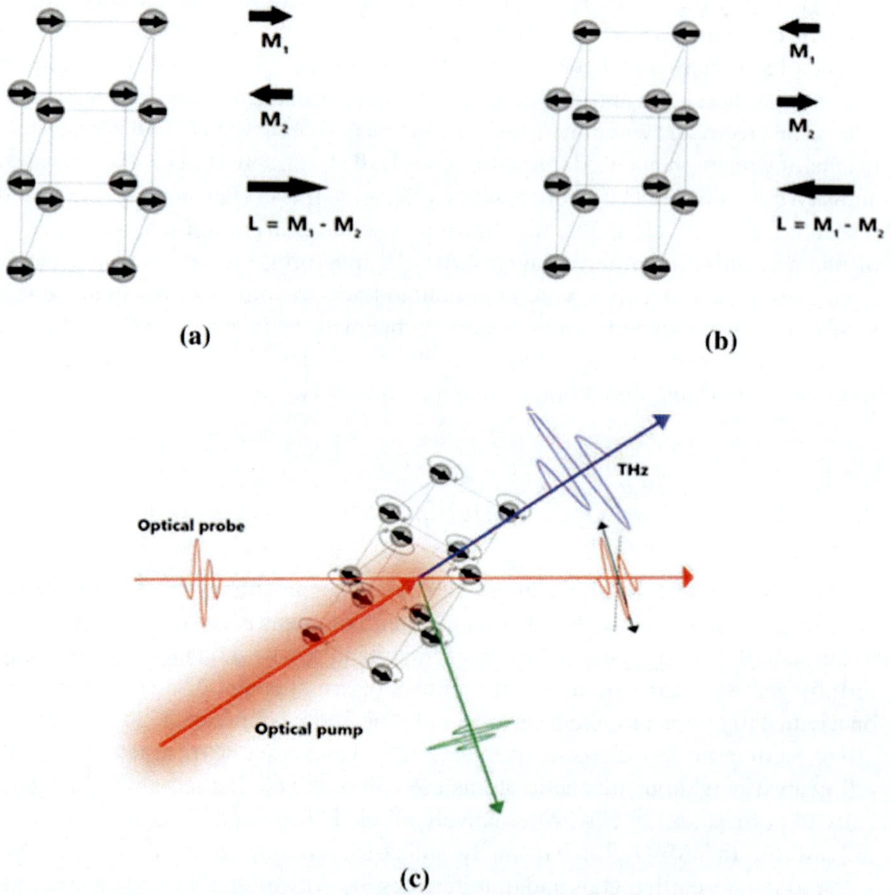

Fig. 7.10 Investigation of electromagnetic radiation-induced antiferromagnetism [adapted and redrawn from Ref.15, Němec, *et.al, Nature Physics (2018)*]

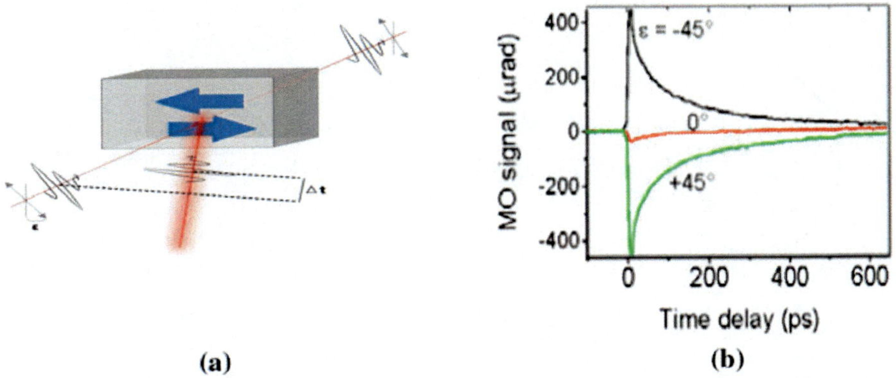

Fig. 7.11 **a** and **b** reduction of rotation of plane of polarization in CuMnAs film. (Adapted and redrawn from Ref. (Němec et al. 2018))

the normal to the sample, where **M** is oriented. An optical strong pulsed pump laser, preferably in red or IR region, is incident and breaks equilibrium in antiferromagnets. Measurement can be done by detecting the rotation of polarization state of optically weaker probe pulse. Magnetization-assisted second harmonic generation or precessing spin emitted electromagnetic radiation in the THz range can be utilized to monitor the static and/or dynamic properties antiferromagnets, (see Fig. 7.10c).

Antiferromagnetic order detection is a tough task. Laser-induced experimental findings of change of state from antiferromagnetic to paramagnetic were found in $FeBO_3$. Quenching of magnetic order (demagnetization) in $FeBO_3$ is a result of increase in magnon temperature. This happens due to transfer of energy from heated lattice. This phenomenon can also be helpful for magnetic characterization of antiferromagnets as demonstrated in a thin film CuMnAs metal (as given in Fig. 7.11). Schematic representation of reduction of the rotation of plane of polarization due to the Voigt effect is shown in Fig. 7.11a. Figure 7.11b gives the change of MO signal as a function of time delay(Δt) between pump and probe pulses.

The antiferromagnetic order is changed to ferromagnetic order in metallic FeRh. Around 380 K, ultrashort external excitations induce a first-order magneto-structural transition from an AF to ferromagnetic phase (as given Fig. 7.12a). Figure 7.12b shows that at low temperatures, FeRh is antiferromagnetic with local iron moments $m_{Fe} = 3\mu B$ and no reasonable moment on rhodium.

At elevated temperatures, the system is ferromagnetic with local iron and rhodium moments. Figure 7.12c shows the local and areal growth of magnetization that generates net magnetization by alignment of individual domains. Growing of demagnetizing field that leads to canting total effective field is depicted in Fig. 7.12d. The homogeneous magnetization starts precessing around the new effective field as shown in Fig. 7.12e.

Laser-assisted ultrafast reorientation of spins in antiferromagnets has also been reported in the dielectric orthoferrite $TmFeO_3$, which observed temperature-dependent magnetic anisotropy at 80–91 K.

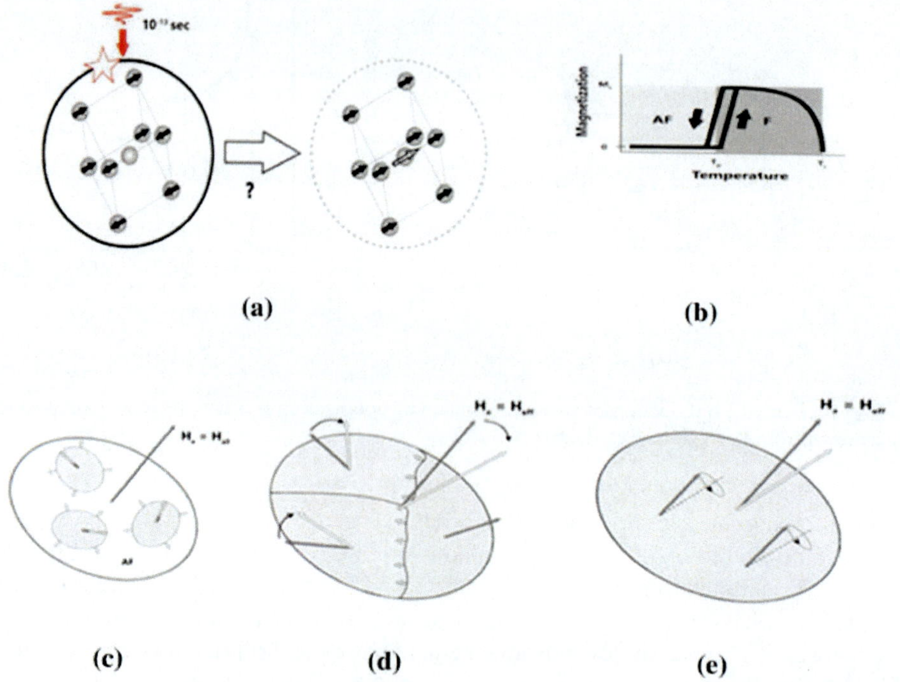

Fig. 7.12 Ultrafast modification of magnetic order in FeRh (**a**–**d**) (adapted and redrawn from Ref. (Němec et al. 2018))

7.10 All-Optical Spintronic Switching

The prevailing notable magnetization switching technologies (see Fig. 7.13) are based on spin transfer torque (STT), spin–orbit torque (SOT) and magnetoelectric/voltage-controlled (ME) switching. Speed is the major hurdle towards the extensive selection of spintronic devices. The switching time of spintronic devices is still on the order of hundreds of picoseconds, which is much greater than that of the silicon field-effect transistors where it less than 5 ps.We have already observed that ultrafast femtosecond laser can control and demagnetize the magnetic medium on sub-picosecond time scales even in the absence of external magnetic field. Now the question is whether we can switch the magnetization without magnetic field? The answer is yes, we can. Femtosecond laser pulse can also switch the magnetization, i.e., can reverse the spins from one direction to other without magnetic field. This effect is known as all optical spin switching (AOS). It bears tremendous potential in the field of magnetic data storage, memory and logic based devices.

AOS combines two important characteristics, i.e., (i) strong storage capability of spintronic media and (ii) efficient speed delivery power of fs laser, yet free from magnetic field. All optical switching can broadly be classified into two categories. One category is based on helicity of polarized light and another one is based on

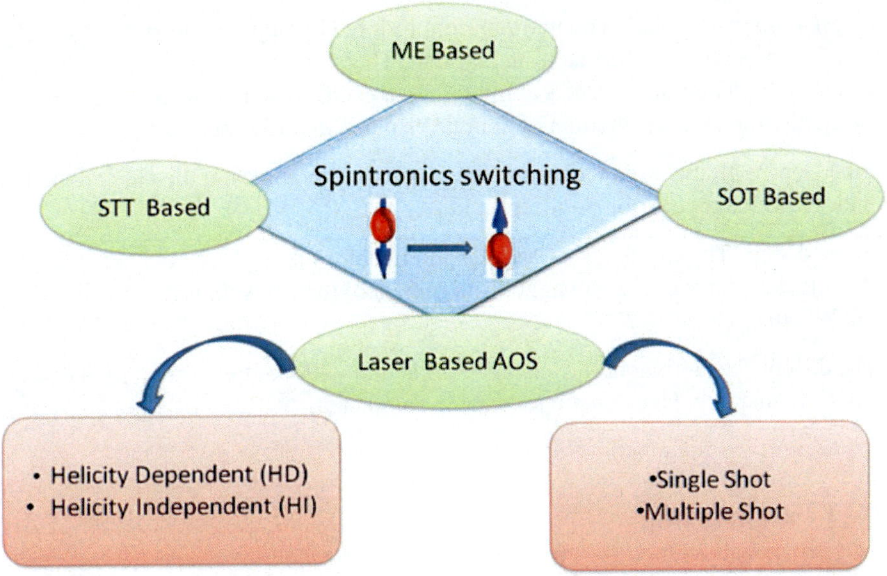

Fig. 7.13 Schematic diagram of different spintronic switching. STT: Spin Transfer Torque, SOT: Spin Orbit Torque, ME: Magneto Electric, AOS: All-Optical Switch

number of pulses needed to switch the spin. The first category refers to HD-AOS where spin switching is determined by helicity of circularly polarized light pulse. Spin reversal by right-circularly polarized light (RCPL) pulse with helicity σ + is just opposite to that of by left-circularly polarized light (LCPL) having helicity σ-. That means, if spin is switched from 'down' state to 'up' state by RCPL pulse then, switching from 'up' state to 'down' state is done by LCPL pulse. RCPL will have no effect on up spin, i.e., if the magnetic domain is in 'up' state, it will remain in 'up' state. Similarly, LCPL will not have any effect on down spin. The linearly polarized light (π) does not switch the spin, rather it breaks original domain in to multidomain consisting of randomly oriented 'up' and 'down' spin. In case of helicity independent all optical (HI-AOS) switching scheme, spin state can be switched by all σ + , σ− and π. Operations of AOS largely depend on material and type of laser pulse. Based on number of pulses required to switch the spin, all-optical switching may be divided into single-shot switching and multishot switching. Multipulses are required to switch spins in most of the materials like CoAgPt and FePt. The magnetization reversal from fully 'up' to fully 'down' requires at least 10^2–10^3 light pulses. Materials that show single shot switching are GdFeCo, Pt/Co/Gd, Co/Pt/Co/GdFeCo and Pt/Co/Pt. All AOS materials can be grouped into three: Feromagnetic, weak and strong ferrimagnetic. One spin orientation is dominant in both feromegnetic and weak ferrimagnetic materials. But in strong ferrimagnetic materials, spin can be switched by CPL with different helicities. Circularly polarized femtosecond laser pulses act as equally short magnetic field pulses via the inverse Faraday effect. Single 40 fs CPL pulse can fully reverse the magnetization in GdFeCo ferromagnetic materials without external magnetic field. The main reasoning force behind this all-optical switching is ultrafast heating of the electronic system. There are two distinct spin sub-lattices in GdFeCo. They are aligned antiparallel to each other and bear uncompensated magnetic moments. Switching of spin takes place in different steps.

First step: With the laser excitation, the magnetic sublattices of the Gd and the FeCo undergo demagnetization at different rates.

Second step: The exchange of angular momentum takes place between sublattices that induce a transient ferromagnetic alignment of the two sublattices by flipping the FeCo spins.

Third step: The antiparallel exchange interaction with the FeCo sublattice causes to flip Gd sublattice. Hence, spin reversal is completed.

7.11 Conclusion

Opto-spintronics, in which the electronic spin polarization is controlled by light, is an emerging and fascinating branch of spintronics. Furthermore, optical approach enables us to manipulate the spin with or without magnetic materials. Development and optimization of future high-speed information storage devices need ultrafast controlling of magnetic order. In spite of the extensive research in last few decades, there are growing interests in exploring and understanding of femtosecond laser-assisted spin dynamics. A wide range of materials starting from ferromagnetic metals, semiconductors and clusters to antiferromagnetic dielectrics responds to light-induced ultrafast demagnetization. In this chapter, we have briefly discussed the challenges to be overcome along with energy-efficient existing and novel methods. We also mentioned the effects of optical laser pulse on a magnetic system. Spin–photon interaction, Faraday effect and inverse Faraday effect have also been discussed. Different methods for the detection of the changes in the magnetization in a medium have been investigated. The processes that lead to an optical laser-induced demagnetization have been explained. Laser-induced demagnetization of GaMnAs, ferromagnetic semiconductor, has been talked about. Special emphasis has been given on *ultrafast* optical controlling of magnetic states of antiferromagnet, i.e., antiferromagnetic opto-spintronics. The outline of different types of all-optical spintronic switching is also given in this chapter. To stimulate continued advancement of the field, more suitable and functional materials should be investigated along with novel methods that are efficient in energy and operation. Technologies associated with computing, communication and control may potentially be revolutionized by the extension of the present state of optical control and modification of magnetic order toward smaller nanoscale dimensions. Rapid developments in integration of spintronics, nano-photonics and plasmonics bring such possibilities in days to come. Expanding research field is fueled by further progress in the laser-controlled magnetism may be associated with the duration down to the time scale of attosecond.

7.12 Exercises

Q.1 Mention some special features of femtosecond laser. Why femtosecond laser pulse is so important in spintronics?

Q.2 Invention of femtosecond laser pulse has bought the revolution in magnetism and in spintronics: Comment on it.

Q.3 What are the advantages of optical manipulation of magnetic order? Explain the thermal and non-thermal effects of a laser pulse on a magnetic system. How photomagnetic effect is different from optomagnetic effect?

Q.4 Explain Faraday and Inverse Faraday effect in magnetism. Briefly discuss the different detection techniques used to study the spin dynamics.

Q.5 With the help of three-temperature model, explain demagnetization of metallic ferromagnets. Mention the process involved in demagnetization of ferromagnetic semiconductor: GaMnAs.

Q.6 Why antiferromagnetic spintronics is getting momentum over ferromagnetic spintronics?

Q.7 Discuss different types of spintronic switching with the help of schematic diagram. What is laser-based all-optical switching?

References

E. Beaurepaire, J. C. Merle, A. Daunois, J. Y. Bigot, Ultrafast spin dynamics in ferromagnetic nickel. Phys. Rev. Lett. **76**, 250 (1996)

D. Bossini, S. Conte Dal, Y. Hashimoto, et al., Macrospin dynamics in antiferromagnets triggered by sub-20 femtosecond injection of nanomagnons. Nat. Commun. **7**, 10645 (2016)

Z. Chen, J.W. Luo, L.W. Wang, Revealing angular momentum transfer channels and timescales in the ultrafast demagnetization process of ferromagnetic semiconductors. PNAS **116**, 19258–19263 (2019)

V.V. Eremenko, N.F. Kharchenko, Y.G. Litvinenko et al., *Magneto-Optics and Spectroscopy of Antiferromagnets* (Spriger, New York, 1992).

W. Feng, G.Y. Guo, J. Zhou et al., Large magneto-optical Kerr effect in noncollinear antiferromagnets Mn3X (X = Rh, Ir, Pt). Phys. Rev. B **92**, 144426 (2015)

A.V. Kimel, B. Ivanov, A, Pisarev, et al., Inertia-driven spin switching in antiferromagnets. Nat. Phys. **5**, 727–731 (2009)

A.V. Kimel, A. Kirilyuk, A. Tsvetkov, V.R. Pisarev, Th. Rasing, Laser-induced ultrafast spin reorientation in the antiferromagnet $TmFeO_3$. Nature **429**, 850–853 (2004)

A.V. Kimel, A. Kirilyuk, P.A. Usachev et al., Ultrafast non-thermal control of magnetization by instantaneous photomagnetic pulses. Nature **435**, 655–657 (2005)

A. Kirilyuk, A.V. Kimel, Th. Rasing, Ultrafast optical manipulation of magnetic order. Rev. Mod. Phys. **82**, 2731–2784 (2010)

S. Manz, M. Matsubara, T. Lottermoser et al., Reversible optical switching of antiferromagnetism in TbMnO3. Nat. Photonics **10**, 453–456 (2016)

P. Němec, M. Fiebig, T. Kmpfrath, A.V. Kimel, Antiferromagnetic opto-spintronics. Nat. Phys. **14**, 229–241 (2018)

J. Nishitani, T. Nagashima, M. Hangyo, Coherent control of terahertz radiation from antiferromagnetic magnons in NiO excited by optical laser pulses. Phys. Rev. B **85**, 174439 (2012)

V.V. Pavlov, R.V. Pisarev, A. Kirilyuk, Th. Rasing, Observation of a transversal nonlinear magneto-optical effect in thin magnetic garnet films. Phys. Rev. Lett. **78**, 2004–2007 (1997)

P.S. Pershan, Nonlinear optical properties of solids: energy considerations. Phys. Rev. **130**, 919–929 (1963)

I. Radu, C. Stamm, N. Pontius et al., Laser-induced generation and quenching of magnetization on FeRh studied with time-resolved x-ray magnetic circular dichroism XE "X-ray Magnetic Circular Dichroism (XMCD)" . Phys. Rev. B **81**, 104415 (2010)

V. Saidl, P. Němec, P. Wadley et al., Optical determination of the Néel vector in a CuMnAs thin-film antiferromagnet. Nat. Photon. **11**, 91–97 (2017)

T. Satoh, S.J. Cho, R. Iida, T. Shimura et al., Spin oscillations in antiferromagnetic NiO triggered by circularly polarized light. Phys. Rev. Lett. **105**, 077402 (2010)

N. Tesařová, T. Ostatnický, V. Novák, K. Olejník et al., Systematic study of magnetic linear dichroism and birefringence in (Ga, Mn)As. Phys. Rev. B **89**, 085203 (2014)

S.A. Wolf, D.D. Awschalom, R.A. Buhrman, J.M. Daughton, et al., Spintronics: a spin-based electronics vision for the future. Science **294**, 1488–1494 (2001)

A. Zvezdin, A.V. Kotov, *Modern Magnetooptics and Magnetooptical Materials* (Institute of Physics Publishing, Bristol, 1997).

Chapter 8
Terahertz Spintronics

8.1 Introduction

The terahertz region occupies the border between the microwave and infrared regions of the electromagnetic spectrum. This region of the electromagnetic spectrum is known as the 'terahertz gap'. Electronics dominates the instrumentation and technology below the THz gap whereas photonic dominates the paradigm, which is above the gap. The convergence between optics and electronics takes place in the THz gap (loosely defined frequency range 0.3–30 THz). Now a day, the THz region of the electromagnetic spectrum is a thrust area for research in physics, chemistry, biology, materials science and medicine (Lee 2009). Despite of some remarkable achievements, there is paucity of devices for the generation, modulation and the detection of THz frequencies. New techniques and technologies are needed for the development of new devices in this area. In this direction, terahertz spintronics can play a pivotal role towards generation and controlling of THz waves utilizing magnetic materials. In this chapter, we will discuss the necessity of terahertz science, its importance in science and technology and finally some spin-dependent phenomena in spintronics structures in the THz range. The chapter will also illustrate different types of spintronic terahertz emitter.

8.2 What Is Terahertz Radiation?

Terahertz radiations are not visible to human eye but we can feel it since a part its spectrum is shared with far-infrared radiation (see Fig. 8.1). Their impact on human body is not harmful as they are low energy radiations. As a result, it does not pose any ionization hazard for biological tissues. Commonly used parameters at 1 THz can be summarized as follows:

Frequency (ν): = 1 Terahertz = 10^3 Gigahertz.

Angular frequency ($\omega = 2\pi\ \nu$): 6.28 Terahertz.

P. Dey and J. N. Roy, *Spintronics*,
https://doi.org/10.1007/978-981-16-0069-2_8

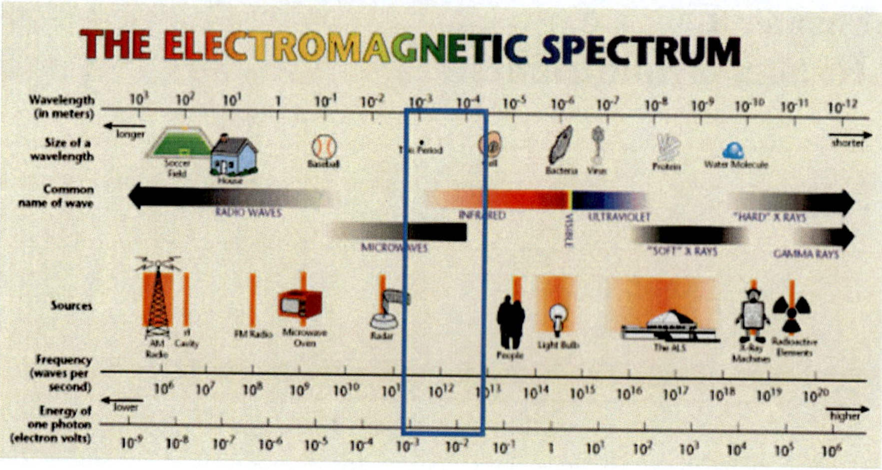

Fig. 8.1 The electromagnetic spectrum. The THz frequency range lies between 'radar' and 'people' and is indicated in blue. Diagram from the LBL Advanced Light Source website (https://www.lbl.gov/MicroWorlds/ALSTool/EMSpec/EMSpec2.html)

Time Period ($T = 1/\nu$): 1 picosecond.
Wavelength ($\lambda = c/\nu$): 300 micrometre [Where, c is the velocity of light].
Photon energy ($E = h\,\nu$): 4.14 meV [h is the Plank's constant].
Temperature ($T = h\,\nu\,/K_B$): 48 K [K_B is Boltzmann's constant].

Due to unavailability of efficient and reliable terahertz source and detector, this portion of E–M spectrum remains underutilized that leads to so-called THz Gap. This gap exists between 0.3THz and 30THz and is being filled up very rapidly. Photonics technologies are advancing from the high-frequency side, while microwave technologies are moving from the low-frequency side.

8.3 Why Terahertz Radiation Is so Important?

Terahertz radiation is very important and significant from both scientific and application point of view. Many fundamental and exciting phenomena are observed in physics, chemistry, material science and biology that occur in picoseconds scales. Some examples of scientific phenomena observed in picosecond scale (THz frequency range) are given below.

- Spin–orbit interaction of electron.
- Rotation of some small molecules.
- Vibration of biologically significant collective modes of proteins.
- Existence of characteristic absorption spectra of many organic substances.
- Absorption by bimolecular water or hydration water.
- In semiconductors, resonance of electrons and their nanostructures.

- Existence of superconducting energy gaps at THz frequencies.
- Oscillation of gaseous and solid-state plasmas.
- Emission of black-body radiation by matter above 10 K.
- Collision time between gas phase molecules at room temperature.

THz technology has a wide range of applications in diversified field such as imaging, sensing, wireless communication, medical science, etc. (see Fig. 8.2). Terahertz waves have the power to penetrate a wide variety of non-conducting materials such as polymers, paper, textiles, ceramics, composite materials, chemical powders. They can also be used for

- Security applications (detection of explosives, drugs, threats and weapons).
- Non-destructive testing (electronics industry, corrosion analysis, agro-food control).
- Medicine and biology diagnosis (e.g., pharmaceutical quality control, protein spectroscopy, tissue characterization, breast tumours, skin cancer and burn depth diagnosis).
- Airport scanning (better than X-rays as it is not harmful to human body).
- Computing and communication (intense THz pulses can induce an ultrafast electric- or magnetic-field switching operation at tens of femtoseconds to

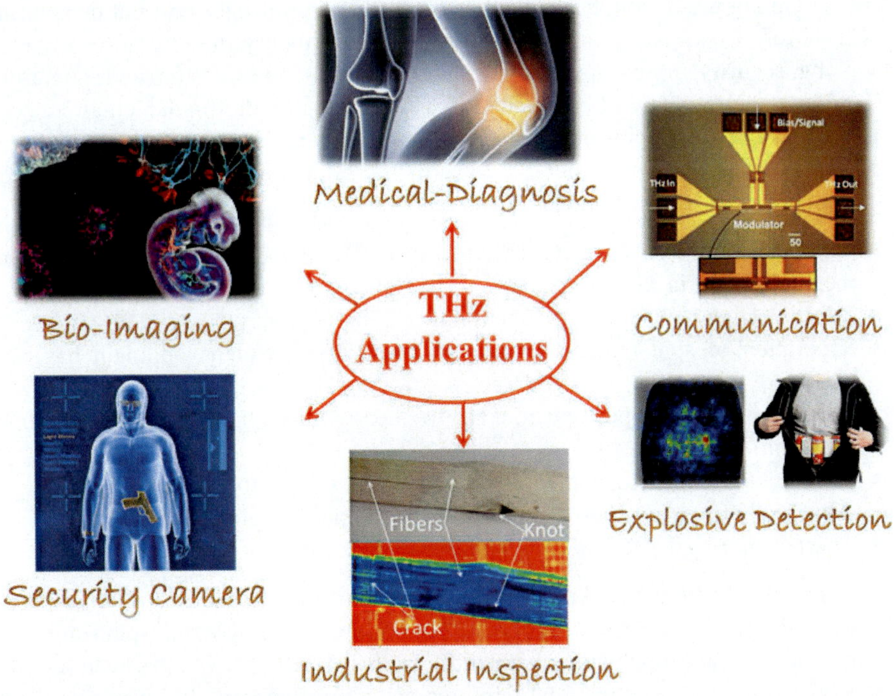

Fig. 8.2 Application of THz radiation

picosecond time scale, which is much faster than what can be achieved through conventional charge-based electronics).

8.4 Why Do We Need Terahertz Spintronics?

Spintronic read head sensors and magnetic random access memory (MRAM) have already impacted on multibillion dollar industry due to its intriguing features like non-volatility, fast switching, low switching energy and capabilities of increased integration densities compared to the conventional semiconductor devices. But recently, the field effect transistor (FET) is reaching its cutoff frequency up to 1 THz. On the other hand, optical fiber communication as well as LAN (local area network) are about to reach terahertz bandwidth. Hence, spintronic technology, operating at gigahertz range, urgently needs to transfer its functionalities from GHz to THz.

The use of spintronics in terahertz range may be of twofold. It may (i) speed up existing computing and (ii) bring a paradigm shift incorporating three-dimensional chip structures or using plasmons or magnons for computing. Though integration densities (number of transistor per unit area) are still increasing, yet frequency clocking remains almost stagnant (at few gigahertz) in last decades. THz spintronics can pave the way of closer synchronization between processing and memory clock by combining ultrafast optics and spintronics. THz spintronics will not be confined with ultrafast computing only; it will extend its periphery in the field of imaging, sensing, security, bio-medical science and many more due to its intriguing features. Terahertz spintronics will be feasible if we observe that the fundamental concepts and phenomena of spintronics also work at terahertz frequency domain. Fortunately, observations are favourable towards the implementation of spin-dependent phenomena in THz scale and are given below.

- Giant magnetoresistance (GMR), for which the Nobel Prize was awarded, has been observed in THz frequency.
- Mott's *two current model* is functional at THz frequency.
- Operation of spin transfer torque (STT) at THz frequency domain has been reported.
- MRAMs are also operative at THz frequency through MTJ (magnetic tunnel junction) based bits.
- THz fields can access elementary spin couplings (e.g., to phonons).
- Spin Hall effect and spin Seebeck effects are operative up to 10 s of THz.
- Inverse spin Hall effect is still operative in THz frequencies.

It has also been reported that the power of spintronic terahertz emitters is at par with GaP- and ZnTe-based electronic THz emitters. However, the spintronic terahertz emitters have much larger bandwidth 1–30 THz. Recently, it has been reported that the magnetization direction of emitter can comfortably modify the polarization of terahertz pulse, which accelerates the development of powerful terahertz near field sources. One important advantage of spintronic-based THz emitters is the possibility

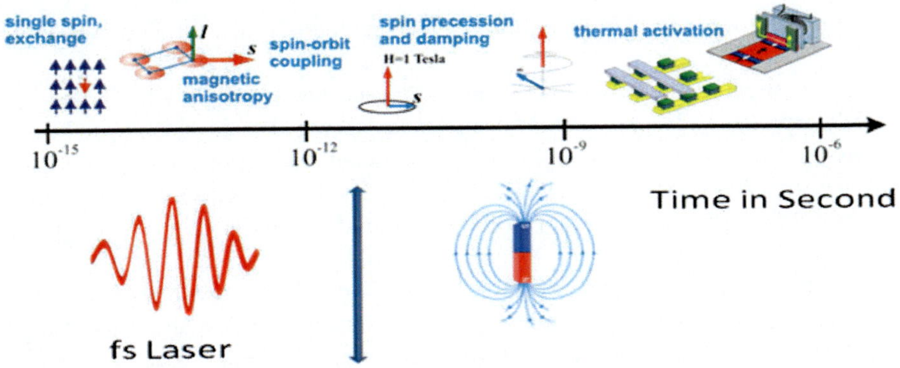

Fig. 8.3 Spintronic phenomena can be observed with external magnetic field or with femtosecond laser pulse

of manipulating emission characteristics by source structure engineering. The invention of ultrafast demagnetization by Bigot and Beaurepaire has paved the way to develop novel and exciting devices exploiting spin-dependent and spin–orbit effect in sub-picoseconds time scale (see Fig. 8.3).

8.5 Spintronic Terahertz Emitter (STE)

Energy-efficient and low-cost terahertz pulses are needed for the complete exploitation of huge potential of THz radiation. Rigorous research efforts have explored various methods and a wide range of materials for the generation of terahertz radiation source (Seifert et al. 2016, 2017, 2016; Seifert 2017; Kimel et al. 2005; Werake et al. 2011; Saitoh et al. 2006; Walowski and Münzenberg 2016; Němec et al. 2018). THz emitters can broadly be classified into two categories (i) charge-based THz emitter and (ii) spin-based THz emitter.

Most of the methods use femtosecond laser pulse to produce THz emission. Methods include photoconductive antennas, optical rectification, photoionization with intense laser pulse, etc. As far as materials are concerned metals, semiconductors and insulators are utilized in emission process. Both polar and non-polar semiconductors have been investigated. Commonly exploited polar semiconductors are ZnTe, GaP, GaSe and GaAs. Unfortunately, they have strong attenuation at terahertz radiation around optical phonon resonance. This restricts the emission between ~1 and 15 THz. The major promising terahertz sources covering wide THz terahertz window, so far, are gas plasmas. However, they usually require amplified laser pulses with high threshold energies of the order of 0.1 mJ. On the other hand, metals can be taken as one of the suitable material classes towards realization of gap free Thz sources. This is because they show wavelength-independent pump absorptivity. They also have short electron lifetimes of ~10 to 50 fs and posses good heat conductivity.

But the main drawback of metal Thz emitter is their poor bandwidth. It is only around 3 THz.

In this situation, it is important to mention that all previously stated THz emitters solely exploit the advantages of the charge but not the spin of the electron and deliver emission spectra with substantial gaps. Low-cost and easy-to-use emitters have to be devised for extensive and extended use of THz spectroscopy. A promising approach is spintronic terahertz emitters, based on a few nanometer thick metallic heterostructure composed of magnetic and non-magnetic materials. The rapid advances in the field of spintronics and femtomagnetism reveal that the electron spin has opened up an entirely new and promising technology for generation, detection and control of spin currents in metals and insulators. In fact, combining the spintronic and photonic properties of magnetic materials, novel THz source has already been reported which is operated by −1 nJ laser pulses from a compact, high-repetition-rate femtosecond laser oscillator. Choosing proper spintronic material and optimizing its geometrical parameters one can realize a single device that may exhibit many advantages like large bandwidth, low pump power, easy operation, scalability and low cost. Spin-based THz emitters can be of two types; (i) femtosecond laser driven and (ii) spin current driven. Now, we will discuss about different laser-driven spin-based THz emitters in detail.

8.5.1 Metallic Spintronic THz Emitter (MSTE)

The new exciting phenomena like spin-dependent Seebeck effect (SDSE), the inverse Spin Hall Effect (ISHE) and the Spin Seebeck effect (SSE) have made spintronic devices more competent towards the efficient generation, transport and detection of spin currents (Werake et al. 2011; Saitoh et al. 2006). These effects have already been discussed in Sect. 2.12. The key role is played by ferromagnetic (FM)/non-magnetic (NM) bilayer for the emission THz radiation. Various types of conventional magnetic materials and binary alloys are used to make THz source (Kampfrath et al. 2013; Li et al. 2017; Wu et al. 2016; Freimuth et al. 2015; Bocklage 2017; Ganichev et al. 2002; Yang et al. 2016; Olejník et al. 2018; Walowski and M. M€unzenberg, 2016; Spintronics for Next Generation Innovative 2015; Kampfrath 2018; Seifert et al. 2017; Lendinez et al. 2019). Some of the promising alloys are: ferrimagnetic magnetite (Fe_3O_4), (anti)ferromagnetic iron rhodium (FeRh) and the ferrimagnetic alloys dysprosium cobalt ($DyCo_5$) and gadolinium iron ($Gd_{24}Fe_{76}$). A schematic structure and basic principle of operation of bilayer THZ emitter are shown in Fig. 8.4.

Basic Principle of Operation:

A femtosecond laser that emits ultrashort pulse hits the ferri-/ferromagnetic (FM) and the non-magnetic (NM) nanostructure. Subsequently, it stimulates the electrons in the ferromagnetic film and brings them in non-equilibrium state. In FM material, majority (spin-up) and minority (spin-down) spin channels have two different transport properties. Hence, a strong spin current is created and flows into the adjacent

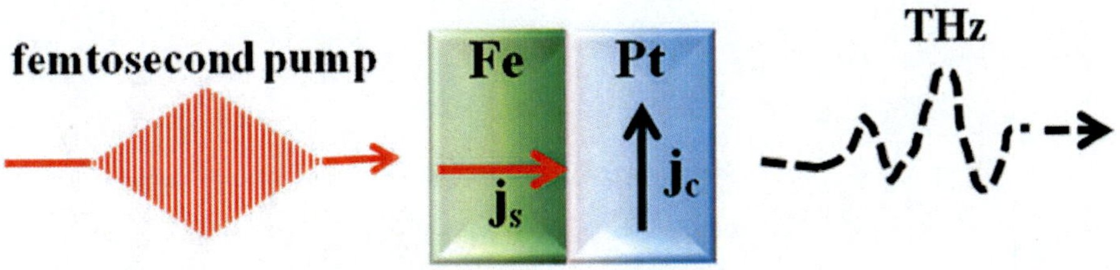

Fig. 8.4 Principle of operation of metallic spintronic terahertz emitter

NM layer. Once, they reach in NM layer, certain physical phenomena come to play, the Inverse Spin Hall Effect (ISHE). Spin-up and spin-down electrons are deflected in opposite directions by spin–orbit interaction (SOI). This leads to transformation of the spin current into ultrafast transient charge current. Finally, this sub-picosecond charge current burst emits a terahertz electromagnetic pulse into the optical far-field. THz signal and the transient THz electric field is shown in Fig. 8.5.

Choice of Materials:

We know that materials matter in device fabrication. The selection of proper ferromagnetic and non-magnetic materials is one of the important aspects of making an efficient THz emitter. The parameter that takes a significant role in this regard is the spin Hall angle. This is related with mean deflection angle of a moving electron. Different materials have different spin Hall angle value and even can show opposite sign. This opens up the possibility of engineering spin Hall currents. The preferences are given to metals having large value of spin Hall angle.

Dependence of terahertz amplitude on non-magnetic material is shown in Fig. 8.6. As an example, Platinum (Pt) achieves larger amplitude than that of Tantalum (Ta) and Iridium (Ir). Whereas, the selection of Tungsten (W) as the non-magnetic material layer yields a comparable magnitude with that of Palladium (Pd) or Platinum, but having an opposite sign. The probable reason for sign reversal is proportional relation of spin Hall angle with the spin–orbital polarization of the electronic states around the Fermi energy, which is opposite for the half-filled d-shell in W and the almost full

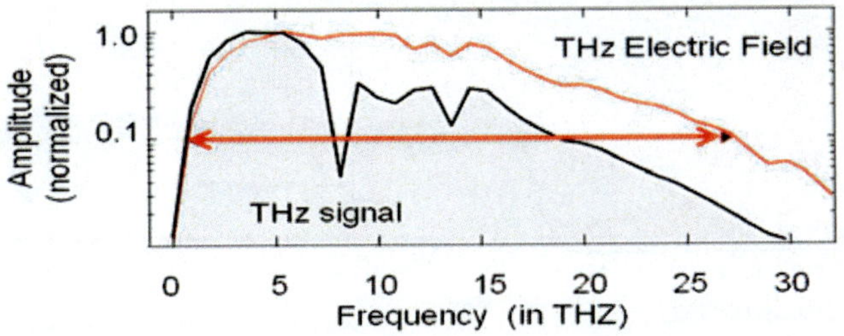

Fig. 8.5 Fourier spectra of the THz signal and the transient THz electric field of metallic spintronic terahertz emitter. Both spectra are normalized to peak amplitude 1. The double arrow illustrates about 30-THz-large bandwidth of the emitter. [Adapted and redrawn from Ph.D. thesis of Seifert (2017)]

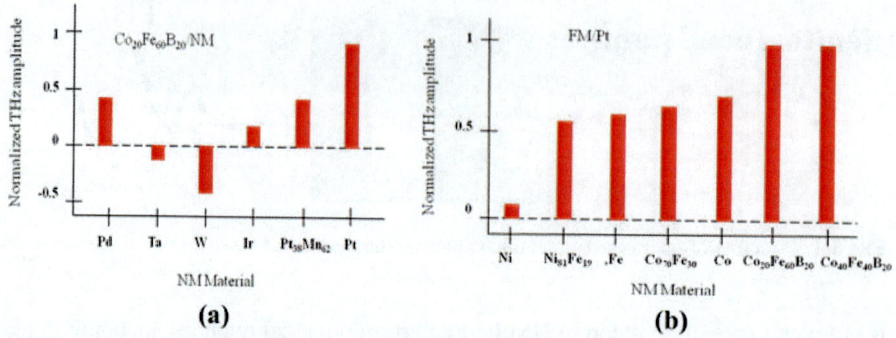

Fig. 8.6 Impact of material choice on emitter performance. **a** THz signal amplitude (RMS) as a function of the NM material used for the $Co_{20}Fe_{60}B_{20}$/NM stack. **b**, THz signal amplitude (RMS) of a FM/Pt heterostructure as a function of the FM material chosen. [Figure is adapted and redrawn from Ref. (Seifert et al. 2016)]

d-shell in Pt. Observations recognize the platinum as promising candidate to be used as non-magnetic material for FM/NM–bilayer THZ emitter. Regarding selection of ferromagnetic material, the conventional materials like iron (Fe), nickel (Ni) and cobalt (Co) are used to make spintronic THz emitter. Maximum amplitude of THz emission is little bit less for nickel compared with Fe or Co. Though the exact reasons are not known, yet it is assumed that the Curie temperature of Nickel, which is 627 K, is less than that of all other FM materials where the value is greater than 1000 K. Furthermore, it has been observed that the amplitude of THz radiation may be increased by adding Boron (B) to Co–Fe alloys. $Co_{40}Fe_{40}B_{20}$/Pt heterostructures may work as one of the powerful THz emission sources. Figure 8.7 depicts the comparative performance of THz signal emitted by the spintronic and other standard ZnTe terahertz emitters. It manifests that spintronic terahertz emitters are better

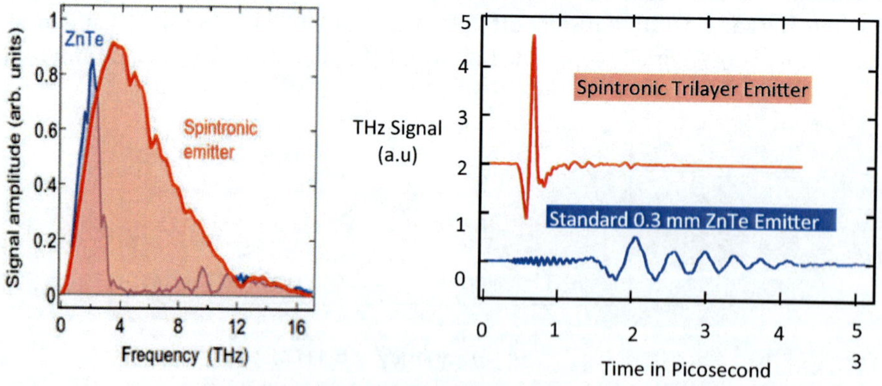

Fig. 8.7 Spintronic emitter performance: THz signal waveforms and resulting amplitude spectra of the spintronic trilayer emitter in comparison to standard THz emitter. Adapted and redrawn from https://magnetism.eu/esm/2018/slides/kampfrath-slides.pdf; Ref. Kampfrath (2018)]

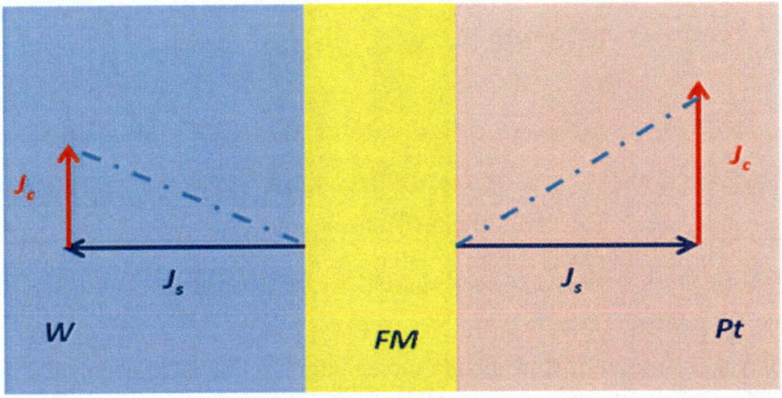

Fig. 8.8 Trilayer structure for THZ emission

performer in terms of its amplitude as well as gap free range coverage (from 1 to 30 THz).

Trilayer

Performance of metallic THZ spintronic emitter can be upgraded by structural engineering of sequence of metallic layers. Figure 8.4 suggests that the spin current moving in forward direction only was considered. That means half of the photo-induced spin current enters into the non-magnetic layer and subsequently converted into charge current. Hence, half of the spin current (backward flowing electrons) remains unutilized. To exploit the full utilization of spin current, another NM layer may be incorporated on the left-hand side of ferromagnetic film (See Fig. 8.8). This structure is a trilayer structure. Tungsten (W) and Platinum (Pt) can be a good choice because they have largest spin Hall angle with opposite sign. Spin current travels in the same direction in W and Pt layer. The amplitude of THz field is increased due to their in-phase radiation. W/$Co_{40}Fe_{40}B_{20}$/Pt is a favourable combination for trilayer terahertz spintronic emitter. The W/CoFeB/Pt trilayer THz emitter can have the capability to emit high energy pulse having duration of 230 femtosecond and a peak field of 300 kV/cm and an energy of 5 nJ.

8.5.2 THz Emitter with Magnetic Insulator (F)/Non-magnetic Metal (N) Layer

Judicious identification of materials with an efficient spin-to-charge conversion is pivotal for future spintronic. The tough task in all-metallic magnetic heterostructures is to find out the involvement of basic process in the ultrafast spin-current generation. Hence, potentially simpler structure is a preferred choice. The easiest way is to achieve it is to use an insulator replacing one of the materials in the FM/NM

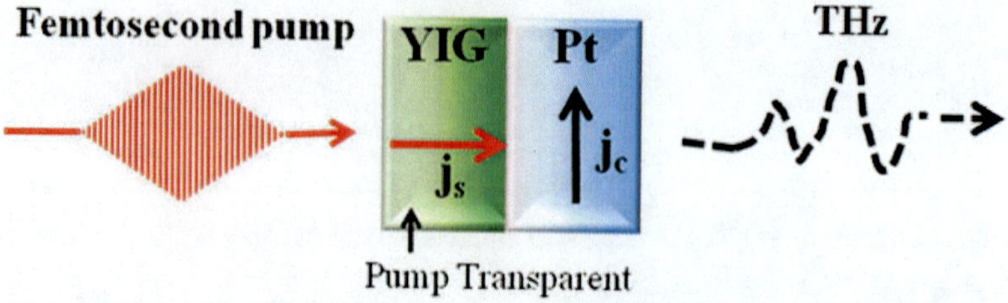

Fig. 8.9 Yttrium–iron–garnet (YIG)/platinum (Pt) bilayer THz emitter

multilayer. Transport of spin angular momentum is a crucial phenomenon for functioning of spintronic devices. Contrast to charge currents, spin currents can also flow in magnetic insulators in the form of spin waves.

Magnons (spin waves) are the spin current carriers in magnetic insulators and can be initiated by heating (i.e., by creating a temperature gradient) an adjacent metal film. In this case, the fundamental effect involved is spin Seebeck effect, which is observed at the interface between a magnetic insulator (F) and a non-magnetic metal (N). Here, a temperature difference induces a spin current density (j_S). Ferrimagnet yttrium–iron–garnet/platinum bilayer is a promising candidate for THz emitter (see Fig. 8.9). A femtosecond laser pulse is incident on a F/N bilayer made of platinum (N = Pt) on top of yttrium iron garnet (F = YIG). Yttrium iron garnet film is transparent to the pump pulse. Homogeneous excitation of platinum film increases its temperature. Any ultrafast spin-current density $\boldsymbol{j_s(t)}$ arising in Pt is converted into a transverse charge current density $\boldsymbol{j_c(t)}$ by the inverse spin Hall effect thereby acting as a source of a THz electromagnetic pulse.

Note: The primary step associated with the formation of the initial spin Seebeck effect (SSE) current is shown in Fig. 8.10 and can be explained as follows:

(i) Optically excited metal electrons make an impact on the interface with the magnetic insulator. Random torque applied by electrons gets rectified through two subsequent interactions. This leads to a net spin current from YIG into the metal. This response is quasi-instantaneous because the YIG spins react without inertia to the impacting metal spins. From application point of view, the observed ultrafast SSE current can be taken as a affirmation of incoherent terahertz spin pumping. Therefore, an instantly heated metal layer is an optimistic transducer for launching ultrashort

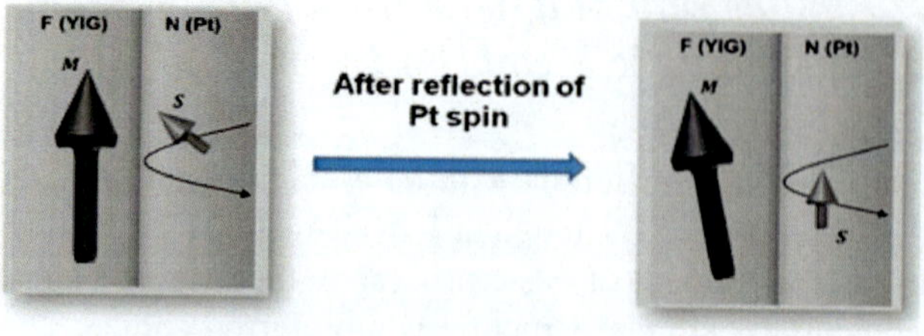

Fig. 8.10 Step linked with the formation of the initial spin SSE current

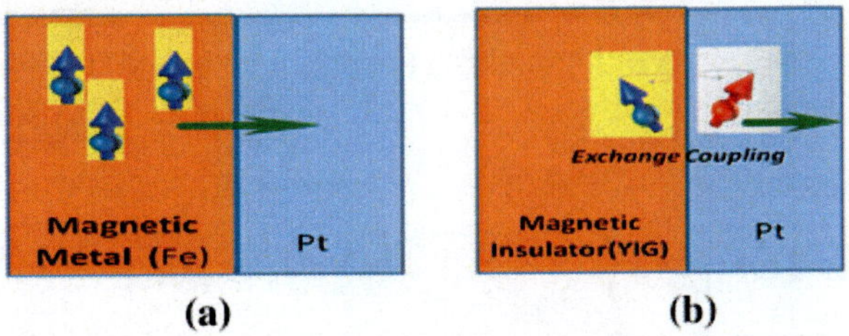

Fig. 8.11 Two types of spin transport due to **a** moving electron **b** spin torque

THz magnon pulses into magnetic insulators. This may be an encouraging aspect for magnon-based information transportation.

Platinum (Pt) spin is incident on interface and reflected spin is aligned more parallel to **M**: ⇈ (similar to STT). The spin current is proportional to (i) rate of reflection events and (ii) number of electron–hole pairs in Pt. The response is quasi-instantaneous since Pt spins traverse the interface region in < 5 fs and YIG spins react without inertia. Thermalized electrons are needed for large spin Seebeck effect.

8.5.3 Comparison Between Metallic Magnet (Fe) and Non-metallic Magnet (YIG)

In Sect. 8.5.1 and Sect. 8.5.2, we have seen that both metallic magnet/non-magnetic metal (i.e. FM/NM) bilayer structure as well as non-metallic magnet/non-magnetic metal (i.e., F/N) bilayer structure can be a powerful spintronic broadband THz source. Two different types of spin transports are observed (see Fig. 8.11) in metallic and non-metallic magnets. One is due to moving electrons, which is possible only in magnetic metals (spin-dependent Seebeck efect). Another one is by torque between adjacent spins, which is even possible for magnetic insulators (magnonic Spin Seebeck Effect). Figure 8.11 depicts the spin current produced in the two above-mentioned structures. Here we have compared the Fe/Pt with YIG/Pt as spintronic THz emitter and has been shown in Fig. 8.12 (Yang et al. 2016; Torosyan et al. 2018).

The Fe to Pt: Spin current has a negligible torque contribution. Spin current is largely due to moving electron.

YIG to Pt: The electron transport is off and the spin current is mainly due to spin torque.

8.5.4 Terahertz Emission by Complex Magnetic Compounds

Spintronic heterostructures have shown their prospect as effective broadband terahertz emitters. Standard FM materials have primarily been investigated with Terahertz emission spectroscopy of FM/NM heterostructures and some of their binary alloys. However, for large-scale application of spintronic THZ emitter, many more

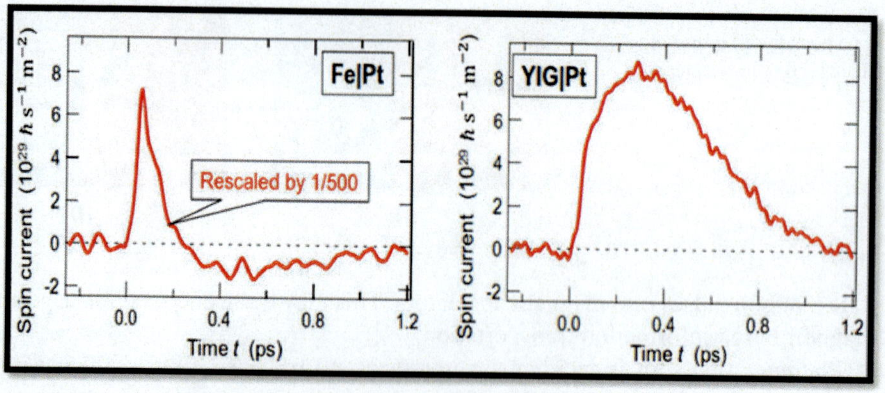

Fig. 8.12 Spin current produced in **a** Fe/Pt and **b** YIG/Pt structure. [Adapted from http://dx.doi.org/10.17169/refubium-214]

magnetic materials need to be explored. In this regard, several complex metallic compounds have been studied in terms of their THz emission (Chau and Elezzabi 2006; Herapath et al. 2019; Nandi et al. 2019). The studied magnetic compounds are ferrimagnetic magnetite (Fe_3O_4), (anti) ferromagnetic iron rhodium (FeRh) and the ferrimagnetic alloys dysprosium cobalt ($DyCo_5$) and gadolinium iron ($Gd_{24}Fe_{76}$).

- **$DyCo_5$and $Gd_{24}Fe_{76}$**: These are RM-TM class alloys consisting of rare earth (RM) and transition metal (TM) elements. $DyCo_5$ and $Gd_{24}Fe_{76}$ have been among the first magnetic media used for high-density magneto optical recording. They bear strong perpendicular magnetic anisotropy, tunable magnetic properties, large magneto optical effects. As a result of that, they were used as first magnetic rewritable memories. Of late, these magnetic materials have shown the all-optical magnetization switching phenomenon in the absence of an external magnetic field. Magnetic heterostructures containing $DyCo_5$ and $Gd_{24}Fe_{76}$ can also emit THz radiation.
- ***Fe_3O_4***: The Fe_3O_4 sample also shows a THz emission but that is about 10 times smaller than from the CoFeB/Pt.
- ***FeRh***: Iron rhodium is a favourable material for heat-assisted magnetic recording. It demonstrates a transition from an antiferromagnetic to feromagnetic phase at a temperature that depends on the actual composition in sample preparation.

It has been found that X = CoFeB is still the most efficient spin current emitter in X/Pt-type bilayers.

8.6 Terahertz Writing

We have already discussed the advantages of antiferomagneic materials in spintronic devices in Chap. 7. Though the antiferromagnetic materials are statically unimpressive, but they are dynamically impressive. Spin precession of electrons

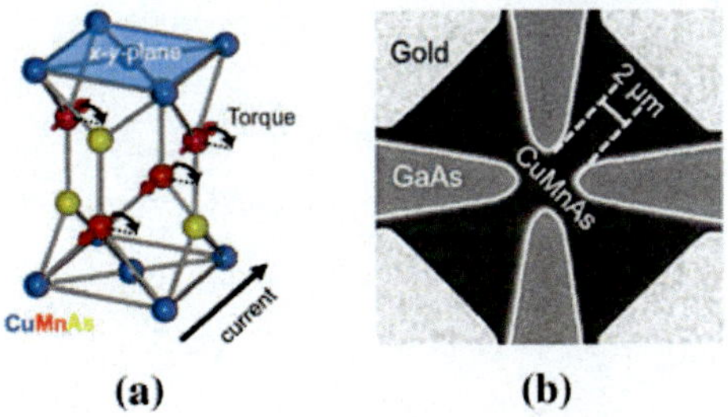

Fig. 8.13 **a** Schematic picture of the CuMnAs crystal. **b** Detailed electron microscope image [Adapted from Olejnik et al., Science Advances (2018) and T. Seifert, Ph.D. thesis Refs. Seifert (2017); Saitoh et al. (2006)]

in antiferromagnets takes place at a much shorter time scale (picosecond scale) than in ferromagnets. This increases the possibility of faster decimation of information. The writing speeds for ferromagnetic memories are restricted to the gigahertz range, whereas it is possible to achieve THz writing speed in antiferromagnets. Besides that, antiferromagnets are more abundant in nature than ferromagnets and many of them are good insulators with low dissipation of energy. Encoding of information is provided by Néel vector in case of antiferromagnets.

The most promising antiferromagnetic material for writing magnetic memory is CuMnAs. Figure 8.13a gives the schematic picture of the CuMnAs crystal. A charge current passing through CuMnAs in the x–y plane imparts a non-equilibrium spin polarization of opposite sign at the two magnetic sublattices due to its crystal symmetry, which is the source of staggered magnetic field on the antiferromagnetic moments. The strength of this magnetic field is proportional to the current-induced polarization and to the exchange coupling between the antiferromagnetic moments and carrier spins. This can be compared with the spin–orbit torque switching mechanism in ferromagnets.

Figure 8.13b shows its microscopy image grown on a gallium arsenide (GaAs) substrate. Four gold contacts are made for electrical connection. The CuMnAs of 50 nm thickness is kept in the middle of the device. Interestingly, the writing can be implemented by both ways (i) by sending trains of MHz current pulses, which can be launched along either x or y direction and (ii) by free-space THz pulses (see Fig. 8.14). The linearly polarized THz electric field drives charge currents in the plane of the antiferromagnetic device whose direction can be conveniently controlled by the THz polarization set by a wire-grid polarizer. Notably, megahertz and terahertz schemes both show very analogous time evolutions of the AMR read-out signal.

8.7 Conclusion

Entirely innovative opportunities are being opened up as more and more powerful terahertz sources become available in science and technology. The progress of generation and detection of strong THz radiation will bring breakthrough in diversified field of technology. It is important to note that the spintronic THz emitters consisting

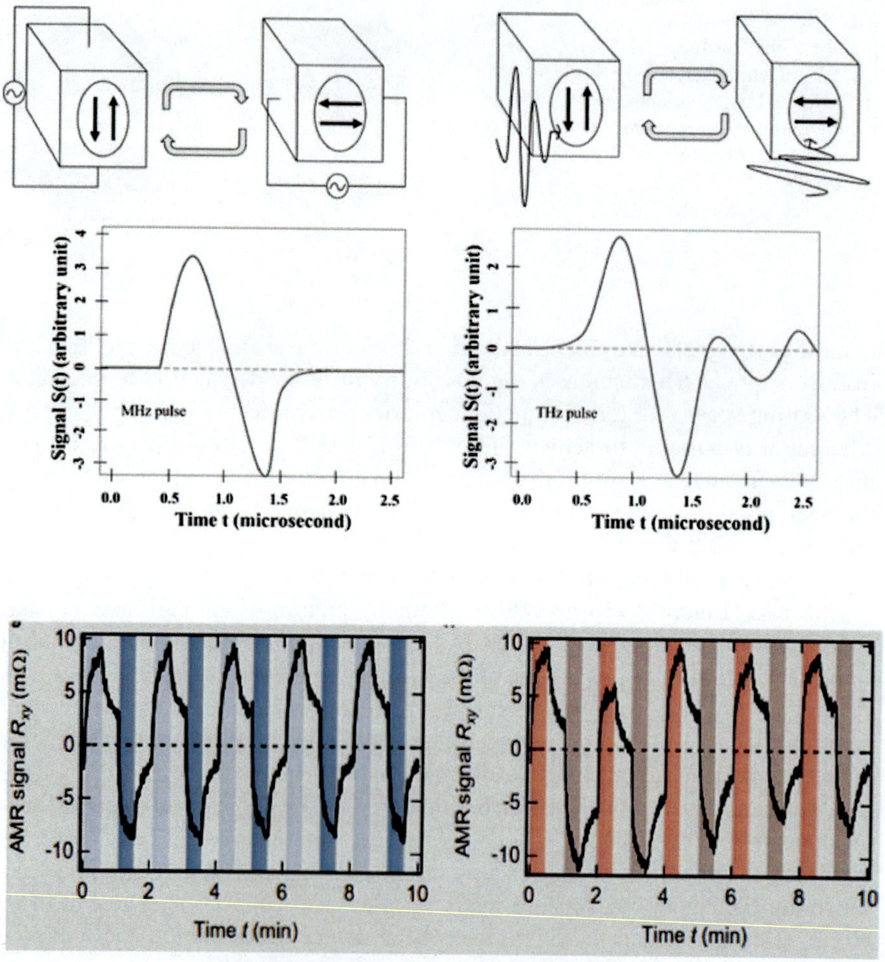

Fig. 8.14 MHz and THz writing scheme. Both are similar in nature. [Adapted and redrawn from Olejnik et al., Science Advances (2018) and T. Seifert, PhD thesis. Refs. Seifert (2017) and Olejník (2018)]

of bilayers or multilayers ferromagnetic (FM) and non-magnetic (NM) metals are competent enough to offer a broadband and continuous gap-free spectrum. It has been reported that a 5.8-nm-thick W/$Co_{20}Fe_{60}B_{20}$/Pt trilayer generates ultrashort THz pulses fully covering the 1-to-30-THz range. Significantly, the THz emission from the spintronic emitters is largely independent of the pump wavelength. Since they are produced from the same source, the focus diameter, energy, and duration of the pump pulses are reasonably independent of the wavelength in the range of 1000–1300 nm. A comparison has been made between spintronic terahertz emitters and other commonly used THz emitters in the market. The basic principles associated in this process are spin-dependent Seebeck effect (SDSE), Spin Seebeck Effect (SSE) and inverse spin Hall effect (ISHE). Spin current is generated through SSDE/SSE, where as spin current is detected through the ISHE, which transforms a spin current into a transverse charge current in materials with strong spin–orbit interaction. The unanswered fundamental problem in spintronic is the electrical writing speed limit, which is in the GHz range. Terahertz writing speed is a realistic prospect in CuMnAs and can open the door of an ultrafast complete write–read cycle in antiferromagnetic memories. The ultrafast writing of antiferromagnetic bit cells can possibly be explored without separate THz speed processors.

8.8 Exercises

Q.1. What is TeraHertz radiation? Give some examples of scientific phenomena observed in picosecond scale (THz frequency range).

Q.2. Explain the importance of THz radiation in science and technology. Why do we need THz spintronics?

Q.3. Mention the observations that lead to the implementation of spin-dependent phenomena in THz scale.

Q.4. What are different types of THz emitters? Explain the basic principle of operation of metallic spintronic THz emitter.

Q.5. Explain the criteria for the selection of material towards implementation of THz emitter. Compare the performances between metallic magnet (Fe) and non-metallic magnet (YIG) in THz emitter.

Q.6. Discuss the role of complex magnetic compounds in terahertz emission.

Q.7. What do you mean by THz writing scheme? How can it be implemented in spintronics?

References

L. Bocklage, Coherent THz transient spin currents by spin pumping. Phys. Rev. Lett. **118**, 257202 (2017)

K.J. Chau, A.Y. Elezzabi, Photonic anisotropic magnetoresistance in dense Co particle ensembles. Phys. Rev. Lett. **96**, 033903 (2006)

F. Freimuth, S. Blügel, Y. Mokrousov, Direct and inverse spin-orbit torques. Phys. Rev. B. **92**, 064415 (2015)

S.D. Ganichev, E.L. Ivchenko, V.V. Bel'Kov et al., Spin-galvanic effect. Nature **417**, 153–156 (2002)

R.I. Herapath, S.M. Hornett, T.S. Seifert et al., Impact of pump wavelength on terahertz emission of a cavity-enhanced spintronic trilayer. Appl. Phys. Lett. **114**, 041107 (2019)

T. Kampfrath, Probing and controlling spin dynamics with THz pulses (2018). https://magnetism.eu/esm/2018/slides/kampfrath-slides.pdf

T. Kampfrath, M. Battiato, P. Maldonado et al., Terahertz spin current pulses controlled by magnetic heterostructures. Nat. Nanotechnol. **8**, 256–260 (2013)

A.V. Kimel, A. Kirilyuk, P.A. Usachev et al., Ultrafast non-thermal control of magnetization by instantaneous photomagnetic pulses. Nature **435**, 655–657 (2005)

Y.S. Lee, Principles of terahertz science. Springer Publication. ISBN 978-0-387-09539-4 (2009)

S. Lendinez , Y. Li, W. Wu, et al., Terahertz emission from magnetic thin film and patterned heterostructures. In: *Proceedings of the SPIE 11090, Spintronics XII, 1109013 (16 Sept 2019)*. https://doi.org/10.1117/12.2526194

G. Li, R. Mikhaylovskiy, K. Grishunin et al., Laser induced THz emission from femtosecond photocurrents in Co/ZnO/Pt and Co/Cu/Pt multilayers. J. Phys. D. Appl. Phys. **51**, 134001 (2017)

U. Nandi, M.S. Abdelaziz, S. Jaiswal et al., Antenna-coupled spintronic terahertz emitters driven by a 1550 nm femtosecond laser XE "Femtosecond LASER" oscillator. Appl. Phys. Lett. **115**, 022405 (2019)

P. Němec, M. Fiebig, T. Kampfrath et al., Antiferromagnetic opto-spintronics. Nat. Phys. **14**, 229–241 (2018)

K. Olejník, T. Seifert, Z. Kašpar, et al., Terahertz electrical writing speed in an antiferromagnetic memory. Sci. Adv. **4**(3), eaar3566 (2018). https://doi.org/10.1126/sciadv.aar3566.

E. Saitoh, M. Ueda, H. Miyajima, G. Tatara, Conversion of spin current into charge current at room temperature: inverse spin-hall effect. Appl. Phys. Lett. **88**(18), 182509 (2006)

T.S. Seifert, Spintronics with terahertz radiation: probing and driving spins at highest frequencies. Ph.D. Dissertation. Veritus Iustitia, Freie Universitat Berlin (2017)

T.S. Seifert, S. Jaiswal, U. Martens, et al., Efficient metallic spintronic emitters of ultrabroadband terahertz radiation. Nat. Photon. **10**, 483–488 (2016)

T. Seifert, S. Jaiswal, U. Martens et al., Efficient metallic spintronic emitters of ultrabroadband terahertz radiation. Nat. Photonics **10**(7), 483–488 (2016)

T. Seifert, S. Jaiswal, M. Sajadi, et al., Ultrabroadband single-cycle terahertz pulses with peak fields of 300 kV cm^{-1} from a metallic spintronic emitter. Appl. Phys. Lett. **110**, 252402 (2017)

T. Seifert, U. Martens, S. Günther et al., Terahertz spin currents and inverse spin hall effect in thin-film heterostructures containing complex magnetic compounds XE "Complex magnetic compounds" . SPIN **07**(03), 1740010 (2017)

Spintronics for Next Generation Innovative, *Edited by Katsuaki Sato, Chapter 7* (Wiley, Ltd, 2015).

G. Torosyan, S. Keller, L. Scheuer, R. Beigang, ETh. Papaioannou, Optimized spintronic terahertz emitters based on epitaxial grown Fe/Pt layer structures. Sci. Rep. **8**(1), 1311 (2018)

J. Walowski, M. M€unzenberg, Perspective: ultrafast magnetism and THz spintronics. J. Appl. Phys. **120**, 140901 (2016)

J. Walowski, M. Münzenberg, Perspective: ultrafast magnetism and THz spintronics. J. Appl. Phys. **120**, 4250–4717 (2016)

L.K. Werake, B.A. Ruzicka, H. Zhao, Observation of intrinsic inverse spin hall effect. Phys. Rev. Lett. **106**, 107205 (2011)

Y. Wu, M. Elyasi, X. Qiu, et al., High-performance THz emitters based on ferromagnetic/nonmagnetic heterostructures. Adv. Mater. **29**, (2016)

D. Yang, J. Liang, C. Zhou, et al., Powerful and tunable THz emitters based on the Fe/Pt magnetic heterostructure. Adv. Opt. Mater. **4**, 1944–1949 (2016)

Chapter 9
Semiconductor Spintronics

9.1 Introduction

Faster, better, cheaper, anywhere and everywhere are the needs of today's device-happy and data-centric world. To meet these demands, we need novel technologies towards storing and processing of information. Spin-based electronics (spintronics) has come out with interesting solution for information storing device. In the present scenario, though the storage devices are based on magnetic metals and insulators, yet information-processing and communication devices rely on semiconductor devices. Since the discovery of the giant magnetoresistance (GMR) effect, the spin transport has been studied in several metals and semiconductors including copper, aluminium, silicon, germanium, gallium-arsenide and grapheme. Conventionally, motions of charges are controlled in information-processing devices to achieve switching mechanism, while magnetic domains are reoriented in the magnetic storage devices for information storage and retrieval. Semiconductor and ferromagnetic materials are essential for the present-day electronics industries. Semiconductor spintronics may offer a promising path for the development of new devices where all three operations like logic, communications and storage can be combined together on a single chip to produce a multifunctional device that could replace several components. It has already been observed that the magnetic random access memory (MRAM) based on the tunnelling magneto-resistance effect has taken a major role in information storage industry. In MRAM device, the resistance depends on the spin orientations between two ferromagnetic electrodes. But, it is not too easy to fabricate an active device, such as a logic circuit from only ferromagnetic metals. However, it can be easily made with semiconductor materials since channel properties are controllable with electric field. This chapter discusses the different types of promising materials that can be explored to design semiconductor spintronics devices. Spin-based semiconductor devices like spin LED, spin transistor, spin laser and spin FET have also been discussed.

P. Dey and J. N. Roy, *Spintronics*,
https://doi.org/10.1007/978-981-16-0069-2_9

9.2 Issues to Be Considered

Integrating semiconductor and magnetism is essential since it may offer novel opportunities for utilizing spin degrees of freedom in semiconductor devices. The charge of electrons and holes are exploited in the present-day electronic and photonic devices for information processing or light emission (De Cesari et al. 2017; Frougier 2014; Hirohataa et al. 2020; Bortolotti 2020; Hirohata and Takanashi 2014; Van Roy et al. 2004; S´anchez et al. 2007). The comparatively novel field of semiconductor spintronics looks for utilizing the spin of charge carriers to make transistors, light emitting diodes and lasers. The usefulness of such devices will depend on the availability of materials having capability to manipulate or control of spin injection, transport and detection well-matched with accessible semiconductor materials. Two important aspects are to be taken into account for the selection of materials towards the realization of semiconductor spintronic devices. The first aspect is the retention of ferromagnetism at and above room temperature, i.e., greater than 300 K. The second aspect is the availability of already existing technology base for the desired materials in another use.

Development of practical semiconductor spintronic devices relies on the fulfilment of some vital and important requirements. They are as follows

(i) Efficient electrical injection of spin-polarized carriers,
(ii) High transmission efficiency of carriers within the host semiconductor or conducting oxide,
(iii) The capacity of detection or collection of the spin-polarized carriers,
(iv) Controlling or manipulation of the spin transport by external agency like as biasing of a gate contact on a transistor structure.

Extensive research on semiconducting materials shows that the bulk and thin film semiconductors exhibit long spin lifetime and spin diffusion length. This is due to their easy tuning of doping profile. The spin lifetime varies from tens of picoseconds to several nanoseconds, whereas the spin diffusion length is usually hundreds of nanometers. But several readings show that the spin lifetime and spin diffusion length are larger in semiconductor nanowires compared to their bulk/thin film counterparts. Theoretical studies reveal that the spin relaxation can be considerably concealed in quasi-one-dimensional (1-D) nanostructures. This opens up a significant attention to learn the electrical spin injection and transport in nanostructures.

Demonstration of several spintronic devices depends on successful spin injection into semiconductors. Achieving efficient electrical spin injection is much easier in ordinary metals and can be done through metallic spin valve structures. However, this process is much more complex in case of semiconductors and is governed by numerous factors:

(1) Conductivity mismatch: spin injection efficiency is very small due to the huge difference in conductivity between ordinary ferromagnetic metals (FM) and semiconductors (SC).

(2) Decrease of spin lifetime: To lessen the conductivity mismatch, the doping concentration of the semiconductor is increased. This intern would reduce the spin lifetime due to the aggravated spin relaxation from impurity scattering.
(3) The localized states at the FM/SC interface and the surface roughness could considerably make matters worse and put at risk the spin injection process.

9.3 Materials, Structures and Spin Injection

There are several methods for achieving formation, injection and detection of spin in semiconductors (De Cesari et al. 2017).

First method: It utilizes heterostructure with magnetic components where spin polarization comes out in usual way in magnetic materials. The principles of injection as well as detection are simple and are sketched in Fig. 9.1.

As the device is heterojunctions, the essential parameter is efficient spin-dependent transport across the interface between different materials with high degree of carrier spin polarization and the interfacial transparency. One of the ways to attain spin-polarized carrier injection can be done by injecting the carrier from ferromagnetic metals into a semiconductor. To achieve fruitful result, the ferromagnets must satisfy some strict criteria. They must be: (i) grown epitaxially on the semiconductors; (ii) thermodynamically stable with no interfacial reactions and (iii) morphologically stable on the semiconductors. In this respect, the permutation of MnAs metal and GaAs semiconductor may be a good choice. Because,

- This hybrid structures can use GaAs and Si for epitaxial growth.
- They also have stable heterointerfaces, as arsenic atoms are present in both MnAs and GaAs.
- Moreover, the structure can be grown with on hand III–V MBE technology.
- Growth of MnAs/III–V/MnAs trilayers is also feasible, which plays an important role in magnetic tunnel junctions.

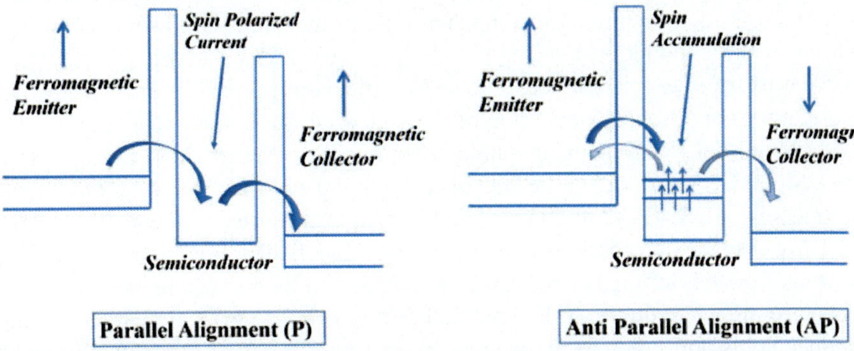

Fig. 9.1 Sketches illustrating spin injection (up) and spin detection (down) (Figures adapted and redrawn from De Cesari et al. 2017.)

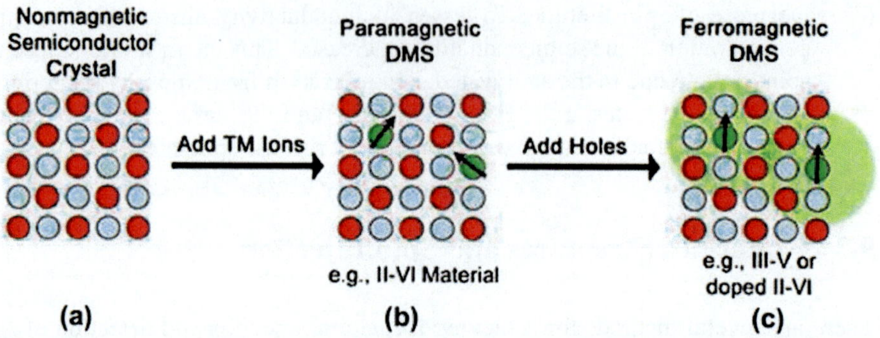

Fig. 9.2 Three types of semiconductors (Figure taken from https://www.researchgate.net/figure/Three-types-of-semiconductors-a-nonmagnetic-semiconductor-which-contains-no_fig28_319256244.)

- A gate electrode may also be used on the semiconductor for additional control of spin and charge transport.

Second method: It exploits magnetic semiconductors as a spin emitter, which can have fully, spin polarized carriers at the Fermi level. The prospect of altering the physical properties with an external electrical field is an important characteristic of semiconducting materials (https://www.researchgate.net/figure/Three-types-of-semiconductors-a-nonmagnetic-semiconductor-which-contains-no_fig28_319256244; https://physicsworld.com/a/the-spintronics-challenge/; Lanje 2014).

There are three types of semiconductors: (i) non-magnetic semiconductor where magnetic ions are absent; (ii) diluted magnetic semiconductor (DMS), semiconductors that, when doped with impurity atoms, display ferromagnetism and (iii) DMS with ferromagnetic order mediated by charge carriers (see Fig. 9.2).

Diluted magnetic semiconductors (DMS) are the materials of the utmost attention in semiconductor spintronics industry. The term DMS refers to the fact that some portion of the diamagnetic atoms in a non-magnetic semiconductor is substituted by transition metal atoms. Such extremely diluted materials are paramagnetic in nature. DMS behaves like a semiconductor when no external field is present. Giant spin splitting of the conduction band and the valence band is observed when DMS is subjected to some external electric field. Complete spin polarization is achievable in laboratory conditions. The source of the spin splitting is the exchange coupling sp–d between delocalized carriers and core spins. However, attempt to inject spins from DMS to GaAs is much more feasible and may reach near 100%. Among several materials, the III–V, II–VI and IV–VI DMS have drawn great interest. Particularly, the III-Mn-As DMS has been grown and confirmed to be ferromagnetic. Building of spin transistors exploiting the properties of DMS is more promising.

The prominent example of this material family is (Ga, Mn)As in which Mn ions introduce spins and holes to the valence band. Other types of magnetically doped p-type compounds are also available, in which holes originate from point defects, like (Pb, Sn, Mn)Te, or from shallow acceptor impurities, e.g., (Cd, Mn)Te/(Cd,

Mg)Te:N and (Zn, Mn)Te:N, instead of Mn ions. Ground-breaking spintronic functionalities have been exhibited and theoretically presented for (Ga, Mn)As and related systems. Report shows that magnitudes of Curie temperature T_C have reached 200 K in (Ga, Mn)As. The ferromagnetism of (Ga, Mn)As can be controlled by external electric field, in particular, the critical temperature can be changed with application an electrical field. It is also an excellent source of polarized electrons, which permits injection of spins to normal semiconductors.

Now the question is why Mn is so important? Let us have a look on electronic configurations of Ga, As and Mn.

$$\text{Ga: [Ar] } 3d^{10}4s^{2}4p^{1}$$

$$\text{As: [Ar] } 3d^{10}4s^{2}4p^{3}$$

$$\text{Mn: [Ar] } 3d^{5}4s^{2}4p^{0}$$

- Mn has half filled 3d shell, which leads to S = 5/2 magnetic moment
- Mn has missing 4p electron and hence can act as acceptor (p-type conductor)

Mn has a peculiar property characterized by deep emplacement of the impurity levels in the valence and conduction bands. As a consequence of that the ferromagnetic order in Mn-based DMS is mediated by carriers present in relatively wide valence bands, while the d levels of other transition metals reside in the band gap of III–V and II–VI compounds. In this scenario, spin–spin interactions are dominated. Mn compounds are divalent in II–VI and characterize by $S = 5/2$ and $g = 2.0$. The spin-dependent hybridization between anion p and Mn d states leads to the superexchange among the Mn moments, and MnAs or MnTe compounds. This makes them antiferomagnetic. On the other hand, antiferromagnetic superexchange can be overcome by ferromagnetic interactions mediated by band holes in DMS. As an outcome, DMS can turn into ferromagnetic. For III–V compounds, Mn behaves as an effective mass acceptor (d5 + h) in the case of antimonides and arsenides.

DMS based on wide band gap semiconductor like gallium nitride has drawn an immense attention as semiconductor spintronics materials as GaN is indisputably one of the most competent materials for application in electronic and optoelectronic devices. The remarkable achievement of DMS based on GaN doped with Mn is its curie temperature, which is above 300 K. Other materials that show room temperature ferromagnetism are GaMnO, (Cd, Mn)GeP_2, (Zn, Mn)GeP_2, $ZnSnAs_2$, (Zn, Co)O, Co(Ti, Sn)O and Eu chalcogenides (see Fig. 9.3).

As far as applications are concerned, the chalcopyrite semiconducting materials are very promising. As an example, $ZnGeP_2$ shows strange non-linear optical properties, which can be exploited to design optical oscillators and frequency converters. $ZnSnAs_2$ anticipates far-IR generation and frequency conversions. Lattice matching of wide band gap chalcopyrites, such as $ZnGeN_2$ and $ZnSiN_2$ with GaN and SiC, respectively, and the attainment of ferromagnetism in these materials would enable for direct integration of magnetic sensors and switches with blue/green/UV lasers and light-emitting diodes, fabricated in the GaN and SiC.

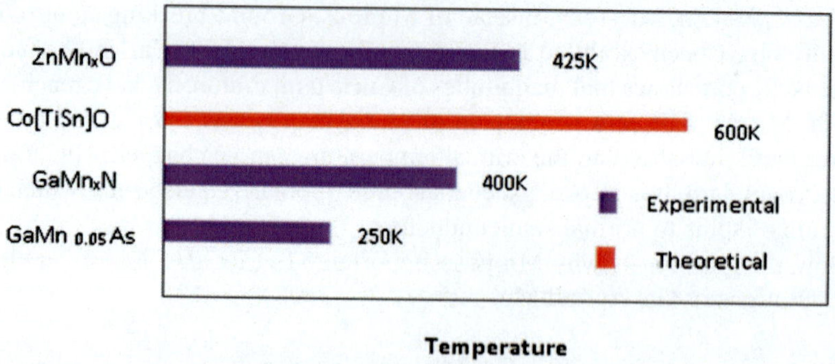

Fig. 9.3 Curie temperature of some ferromagnetic materials

The rapid development of spintronics shows that spintronic elements can be fabricated with different types of materials such as ferromagnetic metals or ferromagnetic semiconductors. Very recently, concentrated magnetic semiconductors such as Eu chalcogenides have demonstrated outstanding novel topological properties by exchange splitting of interfacial bands via a ferromagnetic proximity effect. At the same time, $HgCr_2Se_4$ (TC = 110 K) appears to be a Weyl semimetal. Furthermore, fabricated antiferromagnetic EuTe layers make up a test-bench for the promising field of antiferromagnetic spintronics. The ferromagnetic non-magnetic EuS-PbS heterostructure is emerging as promising materials for potential applications in the field of optoelectronics and photonics. EuS is a model non-metallic Heisenberg ferromagnet whereas PbS is the narrow energy gap IV–VI semiconductor compound. The electronic structure of EuS-PbS multilayers can find its use as multiple quantum well. The excellent usefulness and robustness have already been exhibited by EuS as ferromagnetic spin filter material in heterostructures with metals.

Third method: It is based on digital structures and also known as spin filtering tunnelling structures. GaSb/Mn digital alloys are successful in the realization of spintronic devices. They can also be grown by molecular beam epitaxy. Applied electric bias or photoexcitation can change the carrier densities and ferromagnetism is observed up to 400 K. A handsome number of different tunnelling structures based on spin–orbit interaction have been reported to create spin-polarized current.

9.4 Spin-Polarized Semiconductor Devices

Spintronic devices are predicted to work more rapidly than conventional charged based devices and estimated to generate less heat than conventional microelectronic components. One of the definitive objectives is to fabricate a spin-based transistor that would substitute usual transistors in integrated logic circuits and memory devices. On the other hand, spintronics also gives the birth of exclusively innovative types of device, such as *Spin-polarized light-emitting diodes (spin LED)* that generate left or right circularly polarized light for use in encrypted communication, spin-polarized laser, spin interference device, etc. (https://nptel.ac.in/courses/115/103/115103039/; Zuti et al. 2020; Pearton et al. 2003, 2005; Nitta 2004; Tang and Wang 2015; Kantser 2006; Lee et al. 2014). Spin light emitting diodes (spin-LEDs) and spin-polarized lasers (spin-laser) are an important class of semiconductor spintronic devices. In these devices, spin-polarized electrons (or holes) are injected into an active region where they recombine radiatively with unpolarized holes (or electrons) to emit preferentially right or left-circularly polarized light.

Various approaches have been made to manage of the electron spin degree of freedom that works through one of the following interactions:

- *Magnetic interaction,*
- *Spin–orbit interaction (SOI)*
- *Exchange interaction*
- *Hyperfine interaction with nuclear spins.*

SOI is believed to be the most important channel for spin controlling due to relatively large magnitude and potential of using the electric fields, optical radiation and others for modification of spin states. The schemes of useful control and handling of the spin system through SOI mechanisms in semiconductor spintronics can be placed into three categories:

- Optical
- Electrical
- Magnetic.

9.4.1 Three Terminal Spintronic Devices

Basic block diagram of three terminal devices consisting of source, gate and drain is depicted in Fig. 9.4.

Major three terminal devices can be grouped into different categories such as spin FET, spin LED, spin RTD and coulomb blockade, etc. Some of their characteristics are given in Table 9.1.

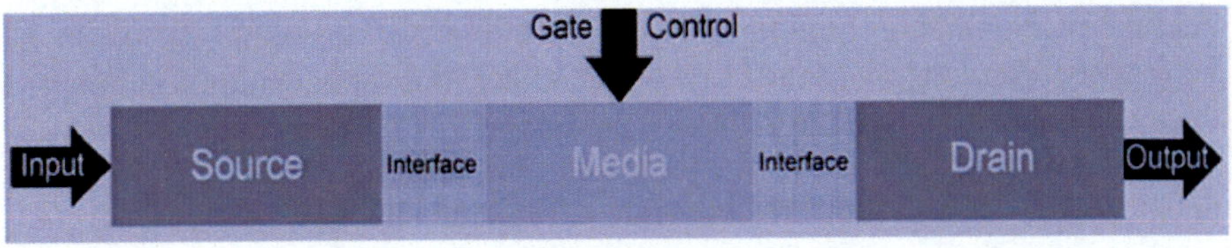

Fig. 9.4 Schematic representation of three terminal devices

Table 9.1 Some features of three terminal spintronic devices

Type	Spin FET	Spin LED	Spin RTD	Coulomb blockade
Source	FM	DMS	Double tunnel barriers	FM
Gate	Bias voltage	Bias voltage	Bias voltage	Bias voltage
Drain	FM	Quantum well	Quantum well	FM
Input	Spin-polarized electrons/holes	Spin-polarized electrons/holes	Spin-polarized electrons/holes	Spin-polarized electrons
Output	Electrical signals	Circularly polarized EL	Circularly polarized EL	Electrical signals

FM: Ferromagnet, *DMS*: Diluted Magnetic Semiconductor, *EL*: Electro Luminescence

9.4.2 Spin-Polarized Field-Effect Transistors (Spin FET)

Potential applications of ferromagnetic semiconductors include designing new and improved devices, such as spin transistors. 'Spin transistors are expected to be used as the basic element of low-power consumption, non-volatile and reconfigurable logic circuits. In 1990, Datta and Das reintroduced a spin FET, which needs well-organized spin injection into a semiconductor. First, electron spins are injected from a source. Then spins are modulated by gate bias and finally are detected at a drain. Working principle and its advantages over ordinary (charge based) have been discussed in detail in Chap. 10.

9.4.3 Spin Light Emitting Diode (Spin LED)

A Spin LED is a spin-optoelectronic device that converts the spin information contained in a population of spin-polarized carriers into circularly polarized light. In spin-polarized light sources, polarized electron is injected from a magnetic layer into a non-magnetic semiconductor structure through drift and diffusion mechanisms. They recombine radiatively with unpolarized holes injected from a non-magnetic contact in the active medium of the structure. If the carriers spin lifetime is greater than the recombination time in the active medium, then the spin orientation does not wholly relaxed by the time of recombination. Thus the radiation resulting from the recombination of the spin-polarized carriers will be partially circularly polarized according to the optical quantum selection rules (In a conventional LED, in contrast, unpolarized electrons and holes combine to produce unpolarized light). Figure 9.5 gives a schematic drawing of a spin light emitting diode (LED).

Fig. 9.5 Schematic drawing of a spin LED

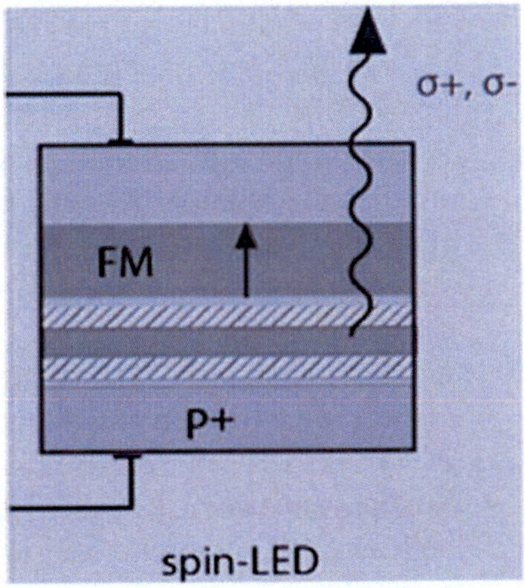

9.4.3.1 Design of Spin-Polarized Light Sources

General considerations: The initial parameter that is to be taken into account is the distance between the spin injector and the active medium of the device. This distance should be such that the travel time between the ferromagnetic spin polarizer and the QWs becomes very small. It serves two purposes. In one hand, it enables to maximize the spin collection and, on the other hand, it minimizes the spin depolarization through relaxation mechanisms. Nevertheless, a negotiation has to be established between minimizing the spin transport length and limiting the interdiffusion of magnetic impurities into the active region.

Emitting geometry considerations: The second important condition to think is the emitting geometry of the device. Based on emitting geometry, Light emitting diodes can be classified into different categories. Figure 9.6 depicts the schematics of various types of LED such as edge emitting, surface-emitting LED and oblique Hanle geometry LED. We shall now discuss briefly about each LEDs below.

- **Edge-emitting LED**: The circular polarization of the light reflects only the electron spin component parallel to the propagation direction of the light. Prima facie

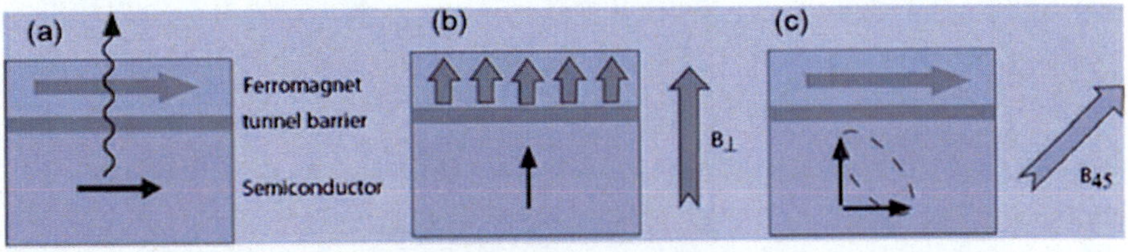

Fig. 9.6 Schematic representation of different LED structures **a** edge emitting, **b** surface emitting, **c** oblique Hanle geometry (Figures adapted and redrawn from https://nptel.ac.in/courses/115/103/115103039/.)

this configuration gives the impression that it would be the best suited for applications as spin manipulation can be done with the application of modest magnetic. In fact, the device can operate in a remanent state and switching can simply be performed with a weak magnetic field. But the significant drawback of this architecture is the orthogonal orientation between the confinement axis and the spin polarization of the injected carriers. It greatly hampers the spin conversion efficiency. In addition to that, the light has to travel a larger distances to escape the structure. This may cause strong photons reabsorption and make the measurement complex. In spite of all these limitations, the spin injection performances are competitive with the values obtained in surface-emitting geometry in AlGaAs/GaAsQW Spin-LEDs.

- **Surface-emitting LEDs**: The consistent configuration both for quantum wells and bulk active regions is to use surface-emitting LEDs. This configuration provides a small escape distance for the photons coming out from the radiative recombination, which significantly reduces photon reabsorption and recycling. However, a fairly large external magnetic field (~1–2 T) is required to overcome the shape anisotropy of the ferromagnetic thin layer. This large field may result the parasitic effects due to Zeeman splitting in the semiconductor and hence considered as a main drawback of this configuration.
- **Hanle geometry**: The oblique Hanle geometry as shown in Fig. 9.7 uses spin manipulation inside the semiconductor to obtain a circular spin component. A small field applied at an angle (ideally 45°, $B_{45} = 0.1$ T to 0.5 T) with the surface causes the injected electron spins to precess around **B** with the Larmor frequency $\Omega = g^*\mu_B/(\hbar \boldsymbol{B})$. Here g^* is the effective g factor ($g^* = -0.44$ for electrons in GaAs); $\hbar$ is Planck's constant; and μ_B is the Bohr magneton. At small fields ($\Omega\ T_S \ll 1$), the spins are hardly disturbed due to the small average precession

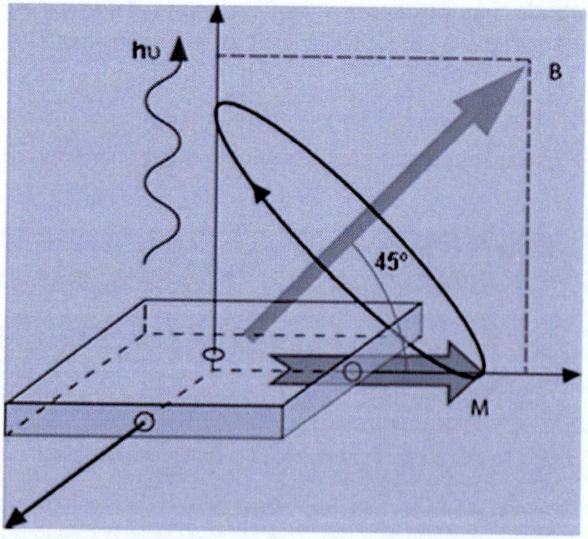

Fig. 9.7 Schematic drawing of oblique Hanle measurement geometry (Figures adapted from https://nptel.ac.in/courses/115/103/115103039/.)

angle. However, at larger applied fields ($\Omega\, T_S \gg 1$), the spins go through many precession cycles before they recombine, resulting effectively in the projection of the injected spins on the oblique field.

The advantage of the oblique Hanle technique is that it is possible to extract the loss in polarization during the interval between injection and recombination and hence to determine the spin polarization of the electrons.

Detection Geometry Considerations:

Three measurement geometries are typically employed for the characterization of spin-polarized light sources: Faraday, Voigt and oblique Hanle. The best configuration for a particular experiment is determined by the details of the heterostructure and device design.

- **Faraday geometry**: This geometry is the frequently used geometry as the optical selection rules are uncomplicated in this arrangement permitting a direct readout of the spin injection efficiency (Fig. 9.8a).

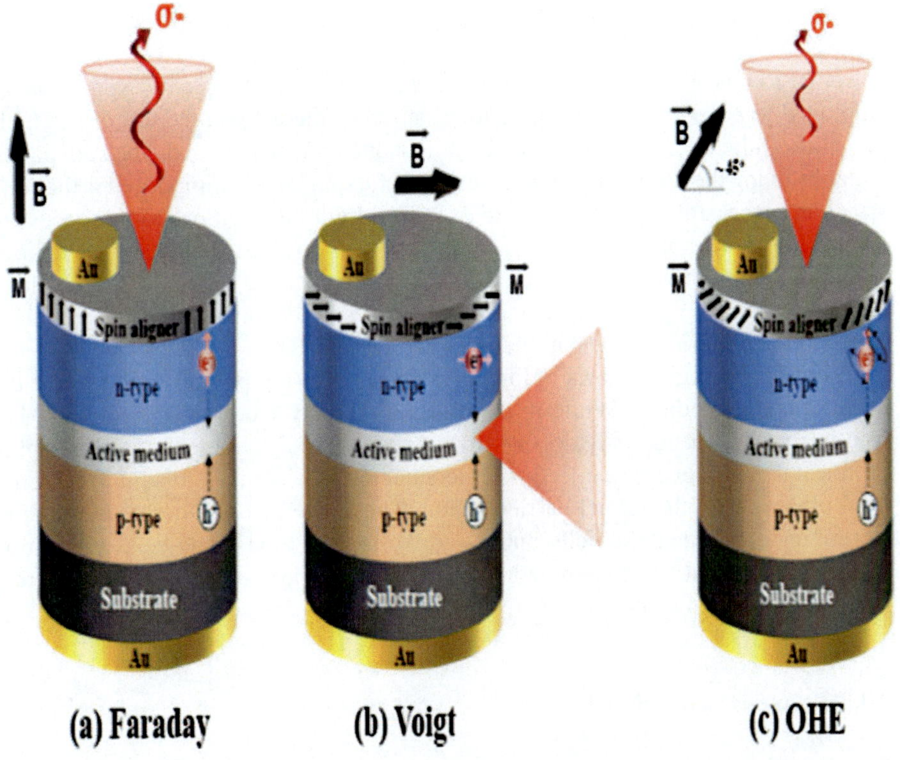

Fig. 9.8 Schematic representation of the three detection geometries for spin LEDS: **a** Faraday geometry **b** Voigt geometry: injection of **c** oblique Hanle effect (Figure taken from Ph.D. thesis of Julien Frougier. https://tel.archives-ouvertes.fr/tel-01127040; Frougier 2014.)

- **Voigt geometry**: This geometry is applied to characterize the seldom used edge emitting spin LEDs (Fig. 9.8b). The selection rules applicable to this configuration with QWs active medium are no longer suitable because the injected carriers are spin polarized in the direction perpendicular to the quantification axis. In most cases, the performance shown by Voigt geometry devices is below the results obtained with surface-emitting devices. The main benefit is the possibility of operation at magnetic remanence as most thin film ferromagnets easy axis is in-plane.
- **Oblique Hanle geometry**: This detection method involves the application of an oblique magnetic field B making an angle of roughly 45 degree with respect to the horizontal axis. B induces a precession of the carrier's spin with the Larmor frequency and assigns a perpendicular component to the spin vector detectable through the emitted light degree of circular polarization. This configuration is a clever way to effectively detect spin injection from an in-plane ferromagnetic contact (Fig. 9.8c). The application of a small oblique B manipulates spins sufficiently enough during transport.

9.4.3.2 GaAs-Based Spin LED

A spin LED is a LED configuration caped with a ferromagnetic spin injector employed to polarize the spin prior to injection in the active medium of the LED structure. Quite a lot of sophisticated semiconductor structures have been explored after observing the fascinating phenomena that the spin information stored in the solid state can successfully be converted to polarized light information. Most successful one is a n–i–p heterostructure that includes heavily doped p-type and n-type regions separated by a lightly doped 'near' intrinsic semiconductor region. It is feasible to establish a confined potential such as QWs or QDs in the intrinsic region (Fig. 9.9). The outline of fabrication technique is given in Fig. 9.10.

The p-doped region shows a doping gradient from the p+ substrate toward the intrinsic active medium. While in the n-region, the layer in contact with the MTJ spin injector is considerably doped with respect to the active medium. The doping profile at the interface spin injector/semiconductor is to be tuned to adjust the Fermi level pinning near the tunnel barrier region. The active medium consists of one or several undoped quantum wells. InGaAs QWs may be a wise choice as absorption of the emitted light can be avoided. Utmost attentions are to be given for selecting the structural parameters.

9.4.3.3 Ge-Based Spin LED

To realize the room temperature, spin LED, investigations have been made for spin injection in Ge. This type of spin-LED relies on a Fe/GeO_2 contact. However, a strong magnetic field (around 4T) is required. The FM tunnel contacts expose spin accumulation as given by three and four terminal device geometries (Fig. 9.11a, b).

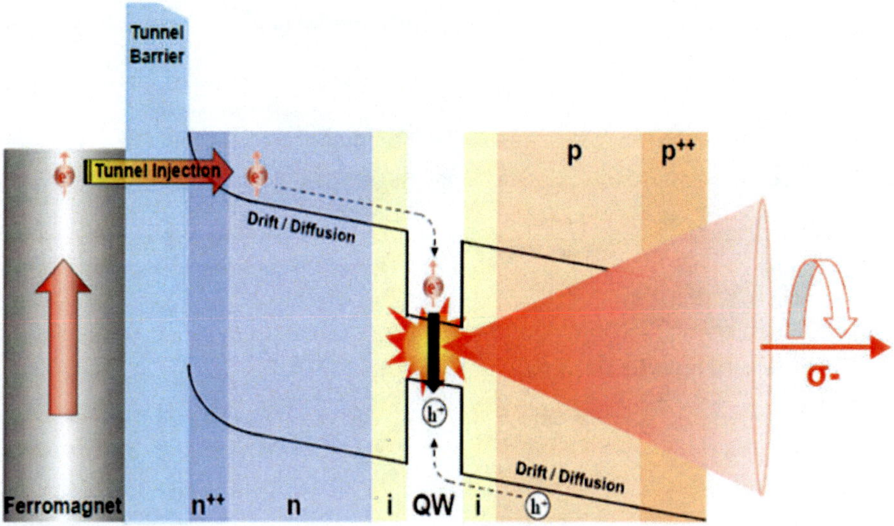

Fig. 9.9 Schematic band structure of a n–i–p Spin-LED (Figure taken from Ph.D. thesis of Julien Frougier. https://tel.archives-ouvertes.fr/tel-01127040; Frougier 2014.)

The exercise of ferromagnetic contacts for electrical spin injection can give rise a spin-photodiode device working at room temperature. This result makes up a significant step towards the achievement of a Fe/MgO/Ge-based spin-light emitting diode (Fig. 9.11c).

9.4.4 *Spin-Polarized Resonant Tunnelling Diodes (Spin RTD)*

A *resonant tunnelling diode* (RTD) is a commonly studied quantum mechanical semiconductor device where the electrons *tunnel* through two barriers separated by a well in flowing source to drain (shown in Fig. 9.12). Resonant tunnelling occurs when the incident bunch of electrons matches its energy with the quantum level formed in the well, thereby working as an energy filter to a good extent. Diode bias controls the electrons flow. This matches the energy levels of the electrons in the source to the quantized level in the well so that electrons can tunnel through the barriers. The energy level in the well is quantized due to the small well dimension. When the energy levels are equal, a *resonance takes place and* electrons are allowed to flow through the barriers. For no bias, source and well energy levels are not matched and hence, no conduction of electron takes place. Small bias causes matched energy levels (resonance) and allows conduction. Further increase in biasing results in energy level mismatch and hence yields no conduction.

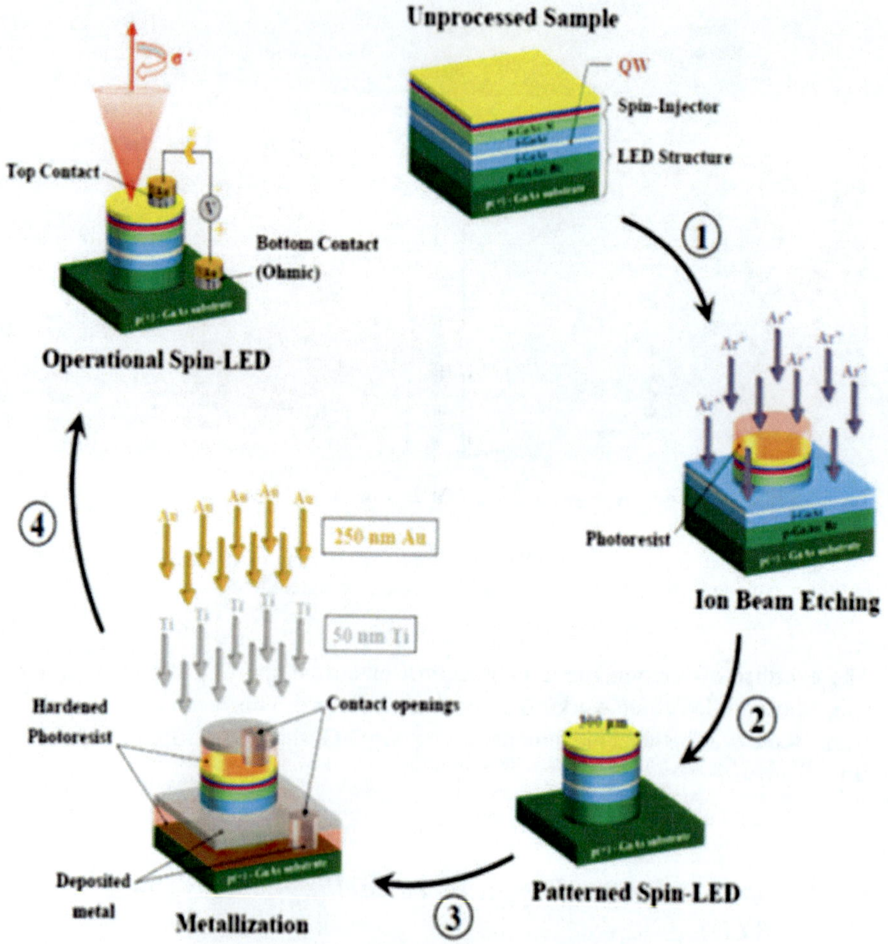

Fig. 9.10 Outline of the fabrication procedure of spin LED (Figure taken from Ph.D. thesis of Julien Frougier. https://tel.archives-ouvertes.fr/tel-01127040; Frougier 2014.)

Doping of quantum well with Mn ions gives DMS a paramagnetic character. When the magnetic field is absent, the well sub-bands are spin degenerate. However, the application of small magnetic field yields a giant Zeeman splitting (of the order of a few meV) due to the exchange interaction between the localized magnetic moments (the Mn++ ions) and the conduction band electrons. As a result of that, the bottom of the well energy sub-bands splits and two conduction channels (one per spin) are able to take part in the transport, see Fig. 9.12. Using bias voltage to select a specific transmission resonance leads to the generation of spin polarized current with predominantly spins up or down.

The spin filter effect in the paramagnetic RTD is restricted to below the room temperature and needs the strong external magnetic field. These restrictions cause that more interest is directed towards the exploitation of the ferromagnetic III–V

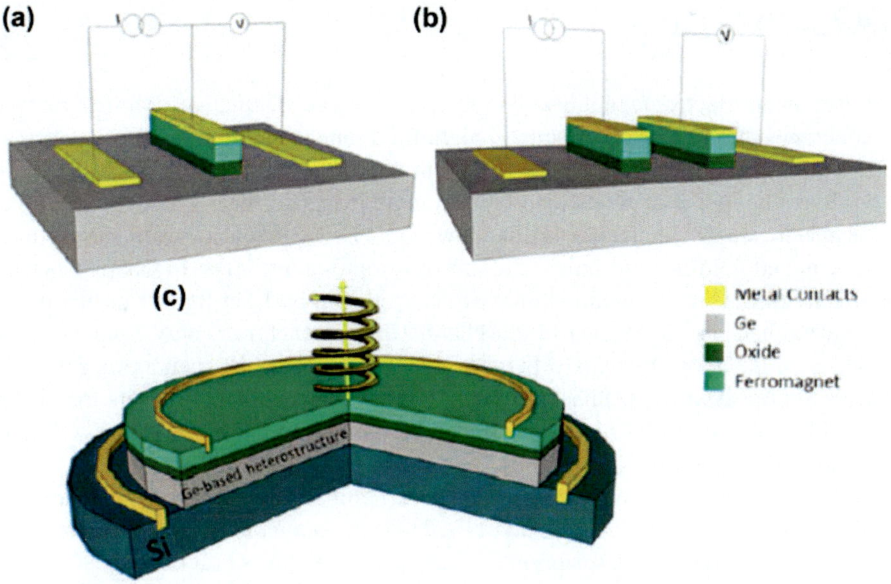

Fig. 9.11 Schematic representations of **a** three terminal and **b** four terminal devices **c** Ge-based spin LED (Figure taken from De Cesari et al. 2017.)

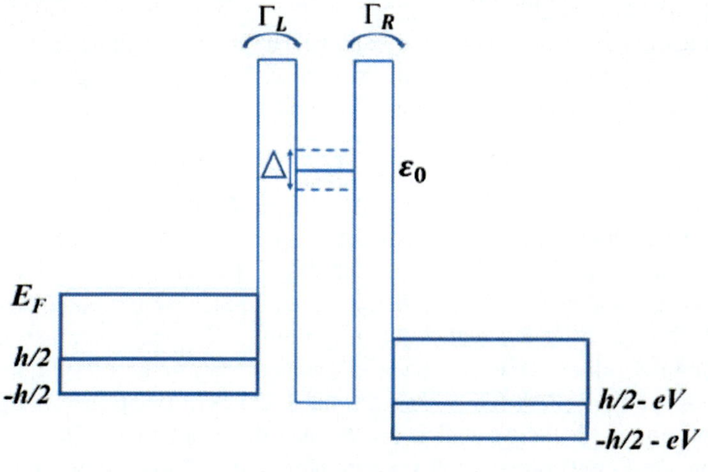

Fig. 9.12 Schematic representation of the energy profile of a double barrier tunnel diode with a single energy level in the quantum well (Figure adapted and redrawn fro S´anchez et al. 2007.)

semiconductors, particularly with high Curie temperature can eliminate these limitations. Materials of renewed interests are GaMnAs or GaMnN. Ferromagnetic RTD based on GaMnAs demonstrates spin splitting without external magnetic field but it still its function limited to very low temperature. However, GaMnN has shown evidence of the ferromagnetic properties above room temperature.

9.4.5 Spin Laser

Laser is an electromagnetic radiation having some distinctive features like high coherency, high monochromaticity, high directionality and high intensity. It is one of the outstanding inventions of twentieth century and has brought the revolution in diversified field of science and technology. It is said that 'wherever you go and whatever you do: the Laser will follow you'. Laser has also given the birth of a specialized field in spintronics and known as opto-spintronics. In Chap. 7, we have already discussed the manipulation of electron spin by laser light. But that was the conventional laser light that uses the charges of current carriers only. Spintronics also opens up the door of exclusively innovative types of device, such as *spin-polarized light source* (*Spin LED and Spin Laser*). Interestingly, the spin lasers may offer a corridor to several realistic room temperature spintronic devices, which will not be limited to magnetoresistance only. In this section, we will discuss the construction and working principle of spin laser and brief comparison between conventional and spin laser (Zuti et al. 2020; Lee et al. 2014). As semiconductor lasers are bipolar devices, simultaneous description of electrons and holes is crucial.

9.4.5.1 Construction

Both the conventional lasers and spin lasers have three basic elements: (i) the active (gain) region, (ii) the resonant cavity and (iii) the pumping mechanism. The major difference of spin lasers is the spin imbalance in the active region, which yields critical changes in their action. This spin imbalance is accountable for emission of circularly polarized light. This is the consequence of the conservation of the total angular momentum during electron hole recombination.

Semiconductor-based Vertical-Cavity-Surface-Emitting-Lasers (VCSELs) and Vertical-External-Cavity-Surface-Emitting-Lasers (VECSELs) are considered to be the ideal choice for the realization of spin-polarized laser sources (Fig. 9.13). The resonator (cavity) is made up of two semiconductor Distributed Bragg Reflectors (DBRs). Each DBR is constructed by multiple layers of alternating materials exhibiting dissimilar refractive index. The gain region is based on quantum wells (QWs) or quantum dots (QDs) and a total thickness of only a few micrometres. For applications, the active region can be electrically pumped. The most common emission wavelengths of VCSELs are in the range of 750–980 nm (GaAs/AlGaAs QWs). Larger wavelengths of 1.3, 1.55 μm (for application in telecom industry) or even beyond 2 μm (suitable for gas sensing technology) can be obtained with dilute nitrides (GaIn-NAs/GaAs QWs) and from devices based on indium phosphide (InAlGaAsP/InPQWs). The nature of pumping mechanism for monolithic VCSEL, VECSEL can be optical, electrical or mixed pumping.

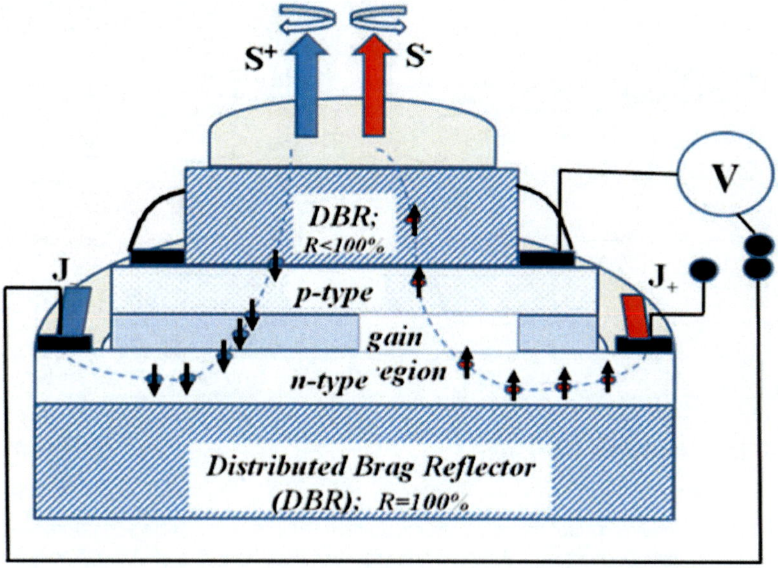

Fig. 9.13 Spin-laser scheme (Adapted and redrawn from Igor Zuti et al. 2020.)

9.4.5.2 Working Principle and Water Bucket Model

Spin lasers can be expressed as a special case of conventional lasers. Unpolarized spin injection can transform the spin lasers to conventional laser operation. The polarization properties of the gain medium and optical cavity determine the polarization character of semiconductor lasers. In Quantum Well V(E)CSEL, the emitted light is circularly polarized. This can be explained with the help of optical quantum selection rules:

(i) spin-up electrons recombine with spin-up HH (Heavy Hole (Valence Band)) and consequently emits a σ^- polarized photon or
(ii) spin-down electrons recombine with spin-down HH and accordingly emits a σ^+ polarized photon.

Thus, spin-polarized electrons couple selectively to one of the two lasing modes and subsequently, emits either left-circular or right-circular polarized light.

An insightful representation has been done by Zutic et al. to explain the basic differences in working principle between conventional and spin polarized lasers. This is popularly known as spin bucket model (Fig. 9.14).

In this model

- Adding of water to the bucket represents the injection of carriers in the laser.
- Coming out of water from the bucket represents emission of light.
- The small holes in the bucket correspond to loss of carriers by spontaneous recombination.
- The large opening near the top describes the lasing threshold.

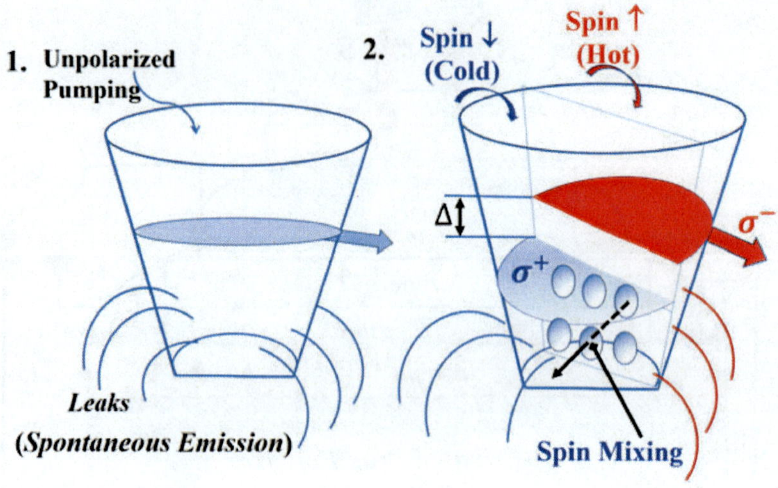

Fig. 9.14 Spin bucket model for conventional and spin-polarized lasers. **1** represents conventional laser and **2** represents spin laser (Adapted and redrawn from Igor Zuti et al. 2020.)

Conventional lasers: As long as the injection or pumping is low (J is below threshold level), the laser operates in a spontaneous emission like LED and radiation is incoherent. The water (unpolarized injection) fills up the bucket till the threshold is achieved. Stimulated emission is found to predominate over spontaneous emission when the injection current J crosses the injection threshold J_T. Water starts stream out (stimulated emission) of the large slit (see Fig. 9.14a). Hence, we can say that a conventional laser works in two operating modes (ON and OFF).

Spin lasers: Two halves of the bucket represent two separate spin populations (symbolized by hot and cold water). They are independently filled. Now, in this case, in addition to the On and Off regimes, one can infer a system where only hot water will stream out. This corresponds to the spin-filtering regime between two discrete lasing thresholds. The openings in their partition allow mixing of hot and cold water to model the spin relaxation. With an unequal injection of hot and cold water, the injection spin polarization can be defined as, $P_J = (J_+ + J_-)/J$, where J represents the total injection $J = J_+ + J_-$. The difference in the hot and cold water levels gives rise to the three operating zones and $J_{T1} \neq J_{T2}$ $(J_{T1} < J_T < J_{T2})$. When the value of J is low (hot and cold water levels below the large slit), up-spin and down-spin carriers are in the off (LED) mode. When the value of J is high, the hot water reaches the large slit and it flows out (see Fig. 9.14b), though the quantity of cold water flowing out is very small. It shows that the majority spin is lasing, whereas the minority spin is at a halt in the LED regime. Two significant findings can be noted: (a) A spin-laser will lase at a smaller J than a corresponding conventional laser (b) Even a small $P_J \ll 1$ can lead to highly circularly polarized light. Spin amplification is also predicted in the interval $J_{T1} < J_T < J_{T2}$. Some comparisons between conventional lasers and spin lasers are given in Table 9.2.

Table 9.2 Comparison between conventional lasers and spin lasers

	Conventional laser	Spin laser
Injected carriers	Spin unpolarized	Spin polarized
Emitted light	Unpolarized	Circularly polarized
Lasing thresholds	One	Two (one for majority spin carriers and another for minority carriers)
Gain spectra	When pumping is done with an equal number of up-spin and down-spin, there is overlapping of gain spectra for σ^+ and σ^- polarized modes 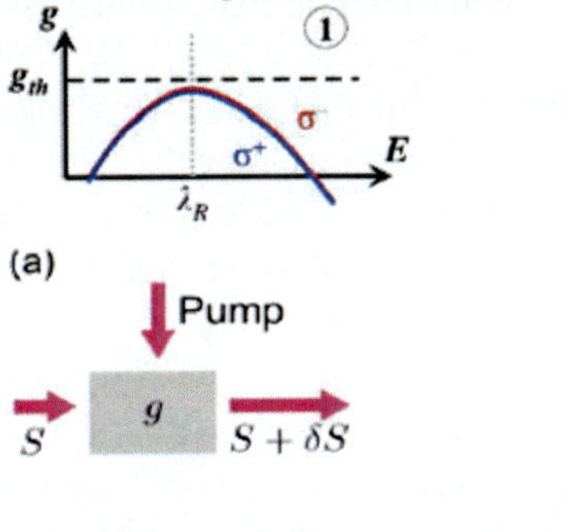	When the pumping is done with a spin-polarized current, a gain anisotropy favouring the majority carriers instantaneously appears 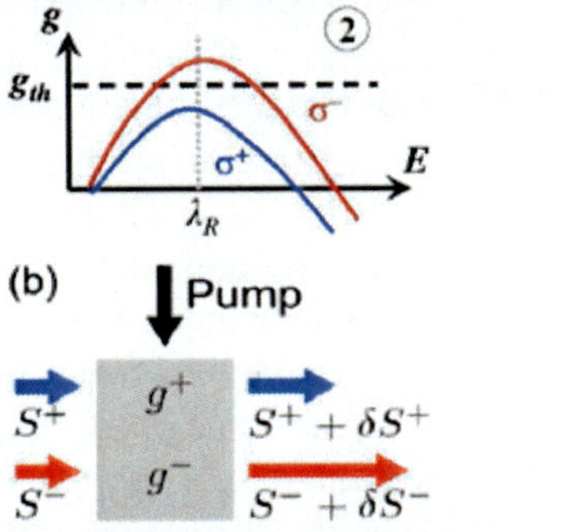
	(b) With pumping/injection, a photon density S increases by δS when it moves through the gain medium	(b) The increase of photon density depends on the positive (+)/negative (−) helicity of the light. Here, optical gain is denoted by g

9.5 Conclusion

In spite of the rapid advances in metal-based spintronics devices (such as GMR devices), a major focus for researchers has been to find new ways to generate and utilize spin-polarized currents in semiconductors. This is significant because integration of conventional semiconductor technology with semiconductor-based spintronics devices can easily be implemented. Further, spins in semiconductors can be more easily manipulated and controlled. (Ga, Mn)As and (In, Mn)As had taken the major focus of attention where samples were carefully grown single phase by molecular beam epitaxy (MBE). A remarkable research has yield some fruitful results in terms of very long spin lifetimes and coherence times in GaAs and the capability to attain spin transfer through a heterointerface, either of semiconductor–semiconductor or metal–semiconductor. Combination of electronics, photonics and magnetic have made available novel spin-based multifunctional devices such as spin-FETs (field-effect transistors), spin-LEDs (light-emitting diodes), spin-RTDs (resonant tunnelling devices), spin lasers, etc. Operational principles of these devices have been discussed in this chapter.

9.6 Exercises

Q.1. Mention some vital and important requirements needs to be fulfilled for the development of practical semiconductor spintronic devices. Also explain why the achievement of efficient electrical spin injection is complex in semiconductor?

Q.2. Explain several methods for achieving formation, injection and detection of spin in semiconductors.

Q.3. The combination of MnAs metal and GaAs semiconductor may be a good choice for semiconductor heterojunction spintronic devices: Comment on it.

Q.4. What are diluted magnetic semiconductors (DMS)? Why are they so important in semiconductor spintronics industry?

Q.5. Write down some peculiar property of Mn that makes them so essential in semiconductor spintronics. What are the uses of wide band gap semiconductor?

Q.6. What do you mean by three terminal semiconductor spintronics device? Name some of the devices and compare their characteristic features.

Q.7. *Designing of spin-polarized light sources needs some general considerations*: Give views.

Q.8. What is spin LED? Summarize the different types of spin LED. Briefly discuss different detection geometries for spin LEDS with the help of schematic representation. Compare the performances of GaAs-based spin LED and Ge-based spin LED.

Q.9. What is RTD and how does it work? What types of material are used to fabricate RTD?

Q.10. In which respect spin laser differs from conventional laser? Explain the construction and working principle of spin laser with spin bucket model.

References

P. Bortolotti, Opportunities and challenges for spintronics in the microelectronics industry. Nat. Electron. **3**, 446–459 (2020)

S. De Cesari, E. Vitiello, A. Giorgioni, F. Pezzoli, Progress towards spin-based light emission in group IV semiconductors. Electronics **6**, 19 (2017)

J. Frougier, *Toward Spin-LED and Spin-VECSEL Operations at Magnetic Remanence* (Université Paris Sud, Paris XI, 2014). Other[cond-mat.other]. English. NNT: 2014PA112175, https://tel.archives-ouvertes.fr/tel-01127040

A. Hirohata, K. Takanashi, Future perspectives for spintronic devices. J. Phys. D Appl. Phys. **47**, 193001 (2014)

A. Hirohataa, K. Yamadab et al., Review on spintronics: principles and device applications. J. Magn. Magn. Mater. **509** (2020)

https://physicsworld.com/a/the-spintronics-challenge/

https://nptel.ac.in/courses/115/103/115103039/

https://www.researchgate.net/figure/Three-types-of-semiconductors-a-nonmagnetic-semiconductor-which-contains-no_fig28_319256244
V.G. Kantser, Materials and structures for semiconductor spintronics. J. Optoelectron. Adv. Mater. **8**, 425 (2006)
A.S. Lanje, *Functional Nanomaterial Synthesis and Charaterization Electrical, Magnetic and Optical* (Lambert Academic Publishing, 2014). ISBN: 978-3-659-11316-1
J. Lee, S. Bearden, E. Wasner, I. Žutić, Spin-lasers: from threshold reduction to large-signal analysis. Appl. Phys. Lett. **105**, 042411 (2014)
J. Nitta, Semiconductor spintronics. NTT Tech. Rev. **6**, 30 (2004)
S.J. Pearton, C.R. Abernathy, D.P. Norton et al., Advances in wide bandgap materials for semiconductor spintronics. Mater. Sci. Eng. R **40**, 137–168 (2003)
S.J. Pearton, D.P. Norton, R. Frazier et al., Spintronics device concepts. IEE Proc. Circuits Dev. Syst. **152**, 312 (2005)
D. S´anchez, C. Gould, G. Schmidt, L.W. Molenkamp, Spin-polarized transport in II–VI magnetic resonant tunneling devices. IEEE Trans. Electron. Dev. **54**, 1 (2007)
J. Tang, K.L. Wang, Electrical spin injection and transport in semiconductor nanowires: challenges, progress, and perspectives. Nanoscale **7**, 4325 (2015)
W. Van Roy et al., Phys. Stat. Sol. B **241**, 1470 (2004)
I. Zuti, X. Gaofeng, M. Lindemann et al., Spin-lasers: spintronics beyond magnetoresistance. Solid State Commun. **316–317**, 113949 (2020)

Chapter 10
Spintronics Applications

10.1 Overview

Spintronic devices exploit the spin, as well as the charge, of electrons and could bring new capabilities to the microelectronics industry (Hirohataa and Yamadab 2020; Potter 1974; Berger 1988; Dieny 1991; Lederman 1999; Berg 1996; Sakakima 2000; Veloso 2000; Parkin 1999; Granley et al. 1996; Ziese and Thornton 2001). The induction of mass production of magnetic memory in 2018 makes the spintronics technology accepted for the mainstream technology in microelectronics industry. Industrialization of devices based on giant magnetoresistance, tunnel magnetoresistance and spin transfer torque bears perspectives of applications in the fields as diverse as ultra-low power electronics, Internet of Things (IoT), RF communication, energy harvesting, artificial intelligence, cryo-electronics, quantum engineering and many more. This chapter illustrates modern advances in spintronic devices that have the potential to impact the key areas of information technology and microelectronics. Spintronic devices are promising candidates for future low-energy electronics that take advantage of the non-volatility of nanoscale magnets. In these devices, the spin polarization is controlled either by magnetic layers used as spin polarizers or analysers or via spin–orbit coupling. For example, magnetic random access memory (MRAM) is emerging as a universal integrated on-chip memory. Magnetic devices and systems exploiting a spin quantum number, a hard disk drive (HDD), had global market revenue of approximately $11bn in 2018. The other booming field in spintronics application is magnetic field sensors, which had market revenue of ~$19bn in 2018. However, the field of semiconductor devices had a much larger market of ~$469bn in 2018.

The Nobel Prize winning discovery of GMR in 32 years ago initiated the field and remarkable quantity of scientific and technological know-how has been built up in the mean time. GMR-based spin valves and magnetic tunnel junctions (MTJs) quickly found large-scale commercial applications in diversified field of science and technology. The innovation of spin transfer torques (STT), tunnelling magnetoresistance (TMR) and magnetic tunnel junctions (MTJs) have made it possible to build scalable

P. Dey and J. N. Roy, *Spintronics*,
https://doi.org/10.1007/978-981-16-0069-2_10

non-volatile magnetic random access memories (MRAMs). Recently, STT-MRAMs have been commercialized as a replacement for SRAMs in embedded cache memories due to their small energy consumption, rapid switching and better endurance. These have been discussed in subsequent sections.

In the past 10–15 years, the world market has observed a significant development in spintronics-based magnetic sensor technology. GMR, AMR and ordinary Hall effect-based sensor technologies are steadily being substituted by TMR sensors owing to their higher output and signal to noise ratio (SNR), good thermal stability, compatibility with CMOS integration, reduced cost. Main industrial markets today include:

- The automotive sector [(angular, speed, current, position/proximity sensors), ABS (antiblocking systems), drive by wire, engine management and ESP (electronic stabilization program)];
- The full electrification of vehicles and other transportation systems;
- Broader industrial environment industry 4.0 with current and power sensors;
- Linear and angular encoders;
- Scanners, and consumer electronics/smartphones (3D magnetometers/digital compasses);
- Brand new applications in the IoT,
- Biomedical devices (Spintronics biochip for recognition of proteins, or DNA; detection of cells/bacteria).

At the same time as, from the semiconductor community, dilute magnetic semiconductors (DMS) have been the subjects of a large number of studies. Spin field-effect transistor (spin FET), three-terminal device imagined by Datta and Das, was presented, which is a foundational concept of a large number of discoveries. Future ICT needs Gbps data transfer rates via optical fibres. A necessary part of such data transfer is the high-speed operation of the diode laser, which depends on the efficiency of the optical isolator. An optical isolator consisting of a DMS is the first practical use of spintronics based on large magneto-optical effects. Beyond Boolean logic, spintronics also put forward various potential paths to unconventional computing schemes. Spintronics is promising for very low power neuromorphic computation. Spintronics can play a pivotal role in next-generation quantum computation schemes also.

Length in Magnetic scale

Figure 10.1 represents the classic length scales of magnetic devices and systems. Transfer of spin polarization takes place through conduction electrons in case of a conductor. However, in case of insulator, it is done by spin-wave propagation across local magnetic moments. In bulk magnets, the demagnetizing field from the edges breaks up the magnetization into domains of different sizes. The *domain* size depends on the materials and the domains may range between tens of nanometers and micrometres or even millimetres. Domain walls separate these domains whose width results from a trade-off between exchange energy and anisotropy energy. The width of the domain wall can vary from tens of nanometers to microns. The exchange

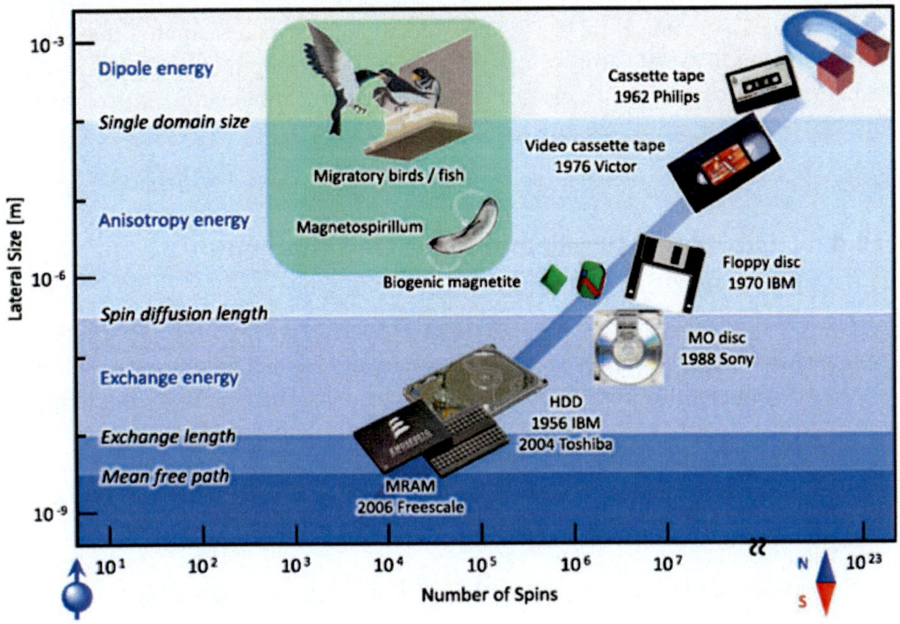

Fig. 10.1 Typical magnetic length scales and development of magnetic storage devices (Taken from Hirohataa and Yamadab 2020)

interactions take a leading role if the system size is greatly reduced and results in a single-domain state. Magnetic distortions may be present at the exchange length whose typical value may be of the order of a few nanometers in materials such as Co or NiFe. From electronic transport point of view, there are two main length scales: (i) elastic mean free path and (ii) spin diffusion length. Elastic mean free path is spin-dependent in a magnetic material and can vary from a fraction of nanometre to several tens of nanometres. Whereas the spin diffusion length (which is basically the distance over which an electron can keep the memory of its spin) depends on the strength of spin–orbit interaction. For strong SOI, this length can range from a nanometer to a few hundred of nanometers. It may be even more in non-magnetic materials with weak spin–orbit interaction. It is exciting to link these length scales with characteristic length scales encountered in the advancement of magnetic storage and non-volatile memory technology. Magnetic media from tapes to HDD are prepared with granular ferromagnetic materials. Here, the grains are single domain and effectively not coupled. The digital information (0 or 1) is written in the form of magnetic domains (the bits) magnetized in one direction or the opposite one along the easy axis of anisotropy. These domains occupy cluster of tens or hundreds of grains. Grain size reduction is an important aspect for increase in areal storage density. In audio tapes, the grain size is in micron order whereas the size of the grain reduces to few nanometers in state-of-the-art hard disk drives. In magnetic random access memory, the usual thickness of the storage layer of each magnetic tunnel junction (MTJ) lies between 1.4 and 2 nm and the MTJ is patterned in the form of a cylinder of diameter in the range 20–50 nm. If the device considered length scale is smaller than the spin diffusion length, the electron spin is preserved. If the length scale is much smaller than the mean free path of the electrons then the electron momentum is also preserved during transport (ballistic regime).

10.2 Historical Advancement and Development of Spintronic Devices

Spintronics started its journey after the discovery of giant magnetoresistance in 1988. Within 10 years of this discovery, the industry enjoyed the benefit by implementing into hard disk drives, the most common storage media. Such fast development is really remarkable in any scientific field. Spintronics research is now moving into second-generation spin dynamics and beyond. In this section, we mention the historical advances in spintronics device applications together with the background physics (see Table 10.1).

Some important breakthroughs in the field of spintronics development are given below:

- In the year 1988, the phenomenon of Giant Magneto Resistive Effect has been discovered by Albert Fert in France and Peter Gruenberg in Germany.
- In the year 1990, an entirely new concept comes into the scenario to attain desired spin manipulation in semiconductor device, as published by Datta and Das and named Spin Field-Effect Transistor (spin FET).
- In 1996, *ulltrafast* control of the electron spin with the help of ultrashort light pulses emerged as one of the hottest topics of the fundamental magnetism and generated significant interest in the area of information technology. This has given the birth of optospintronics and femtomagnetism.
- In the year 1997—first GMR hard disk head was introduced by IBM.
- The year 2002, a novel device has been realized utilizing the spin polarization of an electron to decide the switching of the device.
- In the year 2004, the University of Utah in the USA fabricated the first organic 'spin valve'.
- In the year 2006, room temperature spin Hall effect was detected.
- In the year 2009, spin battery and magnetic super atom have been exhibited.
- In the year 2011, development of room temperature spintronics transistors has taken place. Discovery of a spin-valve-like magneto-resistance of an antiferromagnet-based tunnel junction was also observed.
- In 2013, the room temperature tunnelling anisotropic magnetoresistances have been realized in antiferromagnetic material.
- The year 2014 witnessed the emergence of a room temperature antiferromagnetic memory resistor.
- The year 2016 observed the reversible electrical switching in antiferromagnets. In 2016, Everspin started shipping 256 Mb ST-MRAM samples to customer.

Table 10.1 Historical developments in spintronics phenomena and devices

Electrical spin generation	1957 RKKY	1975 Juliere	1988 GMR spin valve	• 1995 RT-TMR 1996 STT theory • 1999 spin injection • 1999 STT experiment	• 2000 conductance mismatch • 2001 giant TMR theory •2003 spin oscillator • 2004 giant TMR • 2004 LLG equation	2016 neuromorphic operation
Spin-orbit effects	• 1958 Skyrmion theory • 1960 DMI theory • 1968 SOT theory	1971 spin-Hall theory			• 2004 domain motion by a current • 2004 spin-Hall experiment • 2006 inverse spin-Hall • 2009 Skyrmions	
Electric field application			1989 FM DMS	1990 spin-FET concept	2000 voltage-controlled FM	
Electromagnetic wave application				• 1995 photoexcitation • 1998 spin-STM	• 2002 spin pumping • 2002 FMR	2010 Magnonics
Spin-band splitting				• 1993 spin injection • 1999 spin-LED		
Influence of thermal gradient					2008 spin Seabeck effect	2017 spin Nernst effect
Geometrical phase	1958 AB effect		• 1981 AAS effect • 1984 Berry phase	• 1992 persistent current theory • 1999 ballistic MR		
Mechanical rotation	1915 Barnett effect					• 2011 spin mechatronics theory • 2016 hydrodynamic spin current • 2018 MOKE detection
Materials	1903 Heusler alloy discovery		• 1983 half-metallic Heusler alloy • 1988 DMS		2005 topological insulator	
Products	***1968 HDD***	***1972 MRAM concept***		***• 1997 GMR HDD • 1995 GMR sensors***	***• 2002 MRAM • 2008 TMR-HDD***	***• 2011 Racetrack memory prototype • 2016 TMR sensors • 2019 STT-MRAM***
	1970	**1980**	**1990** 1G	**2000** 2G	**2010** 3G	**2020**

Data taken from Atsufumi Hirohata et al. (Hirohataa and Yamadab 2020)

- 2017 saw the experimental demonstration of single-pulse all-optical switching in Pt/Co/Gd stacks using linearly polarized laser pulses. Spin Nernst effect was also observed in this year.

Spintronic devices

Spintronics devices can broadly be classified into two types (i) Mott-type and (ii) Dirac type. This classification is done on the basis of electron/hole spins as well as spin waves and spin/orbit moments. Figure 10.2 shows an outline of spintronic devices. Fundamental phenomenon involved in Mott-type devices is giant magnetoresistance and tunnel magnetoresistance, while the essential phenomena associated with Dirac-type devices are SOT. These devices can also be categorized into three generations. The first generation is based on electrical spin generation and spin

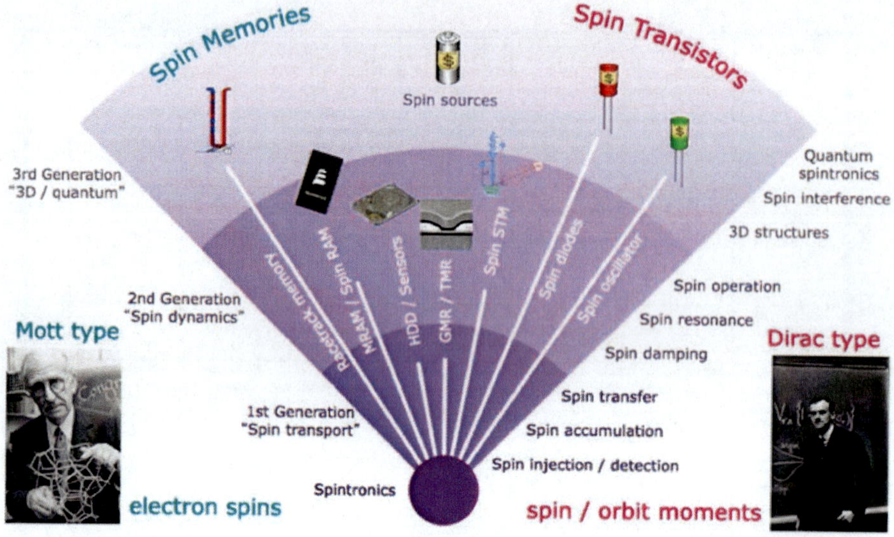

Fig. 10.2 List of spintronics devices (Taken from Hirohataa and Yamadab 2020)

transport. The second generation utilizes spin–orbit effects, electric field and electromagnetic wave applications. The most recent development has been entered into the third generation employing three-dimensional structures and quantum engineering, including the applications for quantum computation. However, in order for spintronic devices to meet the ever-increasing demands of the industry, innovation in terms of materials, processes and circuits are required.

10.3 Read Head in Magnetic Data Storage

10.3.1 Application of GMR Effect

- The very basic **application of GMR effect** is the advent of magnetic field sensors, which are largely employed as read head sensor in hard disk drives to read data, biosensors, micro electromechanical systems (MEMS) and other devices.
- **GMR multilayer structures** can also be utilized as cells that are capable to store one bit of information in magnetoresistive random access memory (MRAM) devices.
- Automotive sensors, solid-state compasses, non-volatile magnetic memories are some other diversified applications of GMR-based devices.

10.3.2 Role of GMR in Magnetic Data Storage

The largest technological application of GMR has been realized in the data storage industry. Based on GMR technology, IBM was the pioneer to produce the *market hard disks*. In recent time, the working of most of the disk drives is exploiting this technology. There is an aspiring demand and hence, a steady rise of requirement for magnetic data *storage* stimulated from various applications, such as novel and new generation of PC operating systems, increasing applications of internet systems, advent of diversified modern days consumer electronics. This leads to the development of '*Hard disk based magnetic data storage device*'. Such development in turn motivates the scientific and industrial community to look for ***new solutions***, ***suitable media***, ***heads***, ***tribology*** and ***system electronics***.

10.3.3 Internal Structure of a Magnetic Hard Disk

The hard disk drive is basically the integration of many key technologies and the associated components, which include storage medium, read/write transducer, channel coding/decoding, servo control, head/disk interface, tribology, electromechanical and electromagnetic systems. Let us discuss the physics and material aspects of the storage medium and read/write heads, related to the magnetic recording (Fig. 10.3).

Hard disk drive consists of thin-film structures, which is composed of multiple layers of thin films. Such structure is fabricated on NiP-coated aluminium alloy, glass substrate, etc. In earlier days, a single magnetic layer is only employed as the recording layer in the hard disk media. Polycrystalline alloy of Co, Cr and Pt was normally used for that purpose, along with additional elements like Ta or B in order to improve the magnetic properties. In recent time, magnetic hard disk media involves multiple magnetic layers as the recording layer. In an attempt to improve

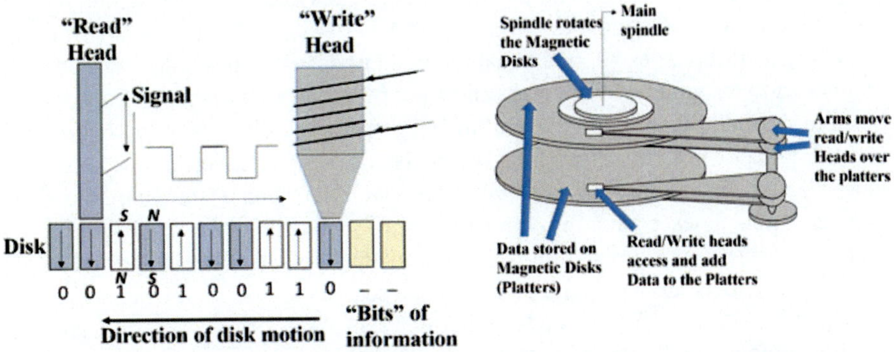

Fig. 10.3 Schematic diagram of read and write head sensor for magnetic storage devices and media (Adapted and redrawn from **IGCSE ICT—Magnetic Storage Devices and Media** https://www.ictlounge.com/html/magnetic_storage_media.htm)

upon the thermal stability of the bits, encoding certain information and recorded on the hard disk, antiferromagnetically coupled (AFC) media has been realized. For the purpose of recording or writing the information onto the magnetic hard disk (Fig. 10.3), a *'writer'* should be such that it can produce sufficiently high magnetic field. Such strong magnetic field would be capable to switch the magnetization of a localized nanosized area of the media to one of two fixed directions. Among those two directions, one of them represents digit '1' and the other digit '0'. Hence, each of those nanosized localized areas of the hard disk, where information is being stored, can be considered as one bit (Fig. 10.3). Noteworthy, each bit consists of a number of partially exchange-coupled magnetic grains.

In early disk drive, while extracting and reproducing the saved information from the disk, the same write head was employed as the read head. This implies that the read head detected the signal while passing through the recorded media by measuring the voltage change, induced across the coils because of the variation of magnetic flux. Areal density in commercial drives has been found to increase steadily at a rate of nearly 100% per year and one of the most significant technologies for large capacity hard disks was the development of magnetic heads (Hirohataa and Yamadab 2020). Vital technology that is essential to increase the memory capacity, while keeping the physical dimension of the hard disk intact, should be capable to write and read of recorded information at *ultra-high* densities. Magnetoresistance is a physical phenomenon, adopted and applied by magnetic read heads while reading information. It is the evolution of the magnetoresistance element that in turn causes the advancement of magnetic heads and thereby breaking the data storage barrier for hard disks.

10.3.4 Operation of Magnetoresistance Read Heads

Figure 10.4 exhibits schematic representation of a shielded magnetoresistance sensor used as read head in magnetic hard disk applications. Each bit is recorded in the hard disk in the form of tiny magnet.

Now, binary bits, i.e., '0' and '1' are stored either in two possible orientations of magnetizations, namely, left to right and right to left, of tiny magnets. For example, one may assign '0' to the 'left to right' magnetic polarity, whereas '1' to that of 'right to left' magnetic polarity. *Flying* magnetoresistance sensor, hovering just above such nano-sized magnetic bits where information is stored, i.e., data storage media, detect magnetic field lines emanating from those tiny magnets. As a result, they could identify the direction of polarization of those tiny magnets and thereby the bit encoded with them. This is the way magnetoresistance sensors act as 'read head sensor'. As shown in Fig. 10.4, such read head magnetoresistance sensors are generally shielded, in order to increase linear resolution and improve high frequency response (Potter 1974).

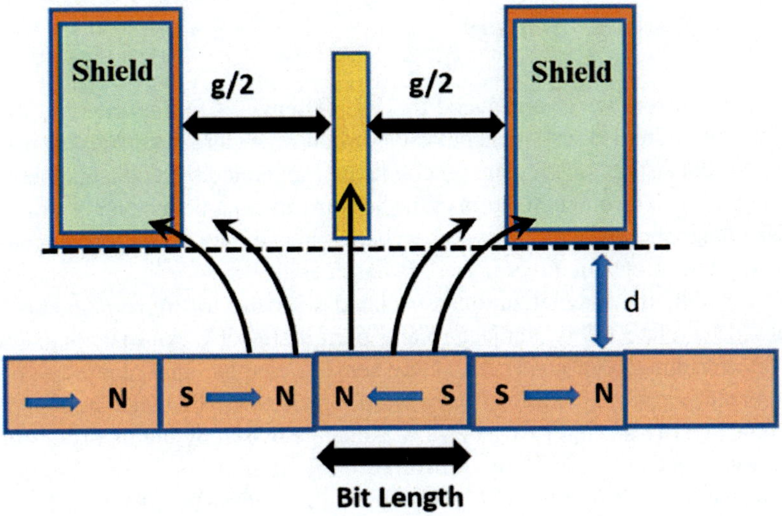

Fig. 10.4 Schematic representation of a magnetic recording media, where magnetic bits are written, and shielded magnetic read head

10.3.5 Evolution of Magnetoresistance Read Head Sensor

Initial magnetoresistance read heads were based on anisotropic magnetoresistance (AMR) effect. Introduction of AMR read head has been done by IBM in 1990 that in turn led to a rapid increase in areal density of about 100% per year. As already briefly discussed in Chap. 3, in case of AMR, the sensor resistance is proportional to the square of the cosine of the angle between the magnetization and the sensing current. This sensor would be effective for hard disk drive having areal densities up to 8 bits/μm^2. With the advent of nanotechnology and related technologies, areal densities of hard disk drive have exceeded above 8 bits/μm^2. Consequently, difficulties were evolved in maintaining the desired bit error rate with AMR sensors.

In 1997, GMR read heads started to replace AMR heads. Spin-valve GMR read heads, exhibiting magnetoresistance in the range of 6–8% at room temperature, have been employed for this purpose (Berger 1988). Following the same line, there was increasing demand for even higher areal densities of disk drive (150 bits/μm^2, bit size 40 nm, written track width = 0.15 μm). This requires read-out head, exhibiting magnetoresistance more than 10% and other types of read heads with low resistance spin tunnel junction.

10.3.6 Spin Valve Sensors

Spin valve sensors were introduced in 1991 (Dieny 1991). Figure 10.5 shows the schematic of a simple four-layer spin valve structure. As already discussed in Chap. 2, spin valve consists of two ferromagnetic layers, separated by a non-magnetic spacer layer (e.g., Cu). The magnetization of the bottom ferromagnetic layer is pinned by the antiferromagnetic layer by virtue of exchange bias effect, while the magnetization of the top ferromagnetic layer is free (Binasch et al. 1989).

As a result, the magnetization associated with the top ferromagnetic layer is supposed to rotate freely, subject to any perturbation by external magnetic field. The free ferromagnetic layer acts as the sensing element and generally comprises of Co or $Co_{90}Fe_{10}$ or a $Ni_{80}Fe_{20}$/Co or $Ni_{80}Fe_{20}/Co_{90}Fe_{10}$ bilayer. The pinned ferromagnetic layer (e.g., Co or $Co_{90}Fe_{10}$) is exchange biased by the incorporation of an antiferromagnetic (e.g., $Mn_{76}Ir_{24}$ or $Mn_{50}Pt_{50}$) (Lederman 1999) or a synthetic antiferromagnetic layer (e.g., NiO/Co/Ru/Co, $Mn_{76}Ir_{24}/Co_{90}Fe_{10}/Ru/Co_{90}Fe_{10}$) (Berg 1996). In case of first generation of top-pinned and bottom-pinned spin valves, where pinned layer resides above and below the spacer layer (e.g., Cu), respectively, typical magnetoresistance values have been found to range from 6 to 10%.

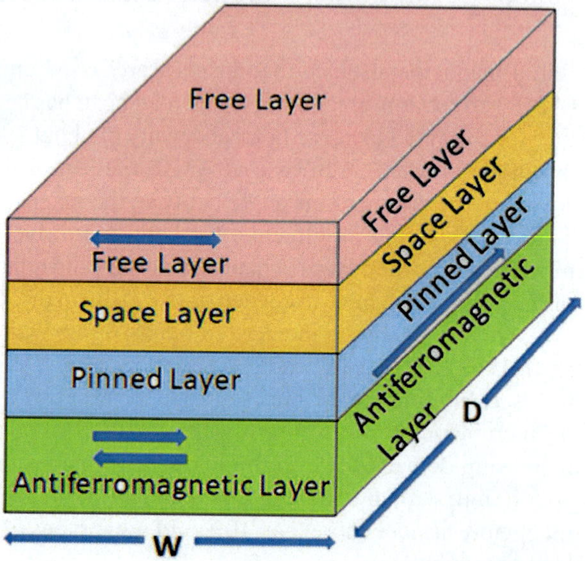

Fig. 10.5 Schematic view of a simplified four-layer spin valve structure

10.3.7 Some Important Features of Spin-Valve GMR Read Head Applications

In order to achieve optimized read head applications, two important deciding factors are: ***(a)*** large magnetoresistance values and ***(b)*** effective exchange field, created at the pinned layer/exchange layer interface. Obviously to ensure the room temperature application of the device, the blocking temperature, i.e., the temperature where the exchange field vanishes, should exceed 300 °C. This, in turn, prevents the accidental depinning of the pinned layer during the fabrication of read head. Furthermore, the exchange energy should be large enough (>0.2 mJ/m^2) so that it predominates over the demagnetizing fields at read head level. Now, we will discuss those two important factors, which play a key role to yield proper and improved read head applications, as mentioned above, consecutively.

(a) **Large Magnetoresistance values**: Enhancement in magnetoresistance values can be achieved in two ways:

(1) ***A Dual Symmetric Spin Valve***: In this kind of device, two spin valves, one of them is bottom pinned another top pinned, share a common free ferromagnetic layer. Such ***dual symmetric spin valve*** offers magnetoresistance signals that surpass even 20%. However, spin valve of this kind has not been employed in hard disk read head because of its larger thickness.

(2) ***A Specular spin valve***: In this kind of device, fine nano oxide layers (NOL) are deposited on both sides of Co-Fe/Cu/Co-Fe standard spin valve structure (Sakakima 2000). Figure 10.6 represents the schematic diagram of specular spin valves (Veloso 2000), where the incorporation of NOL layers enhances magnetoresistance ratio from 6 to 14%.

(b) **Effective Exchange field**: There are two approaches, which must be followed to increase exchange energy and blocking temperature of exchange fields. ***First*** approach is the utilization of a synthetic antiferromagnet (SAF) that

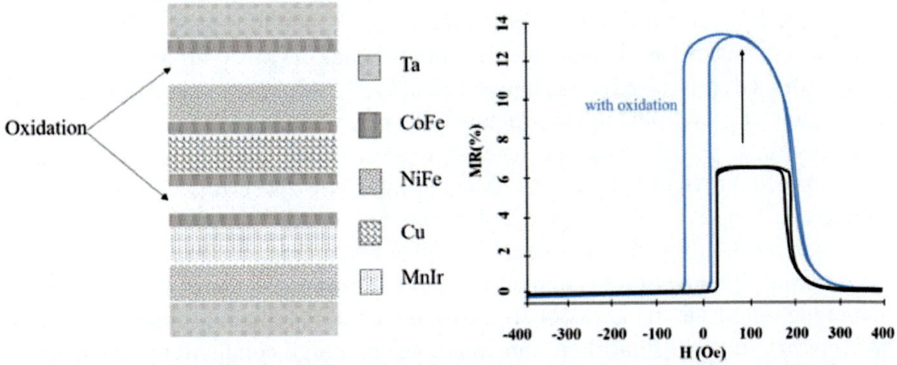

Fig. 10.6 Schematic representation of a specular spin valve structure and the variation of magnetoresistance (MR) as a function of applied field (H)

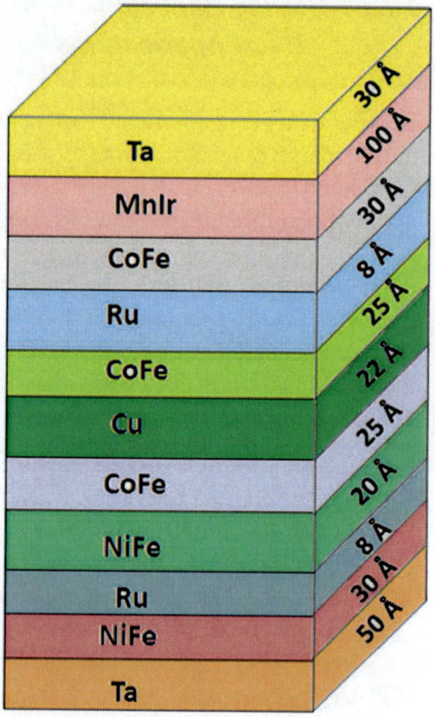

Fig. 10.7 Schematic representation of a spin valve structure with both synthetic-free and pinned ferromagnetic layers

is composed of two ferromagnetic layers, separated by a thin non-magnetic layer. For example, Co or Co–Fe layers, having similar thickness, are strongly coupled antiferromagnetically through 0.5–0.7 nm of Ru (Berg 1996).

Figure 10.7 exhibits the spin valve structure having both synthetic-free and pinned layers. These synthetic antiferromagnetic structures have certain advantages. Effective magnetic moment of the pinned ferromagnetic layer is low. As a result, its effect on the demagnetizing field as well as on the coupling fields acting on the magnetization of the free ferromagnetic layer is much feebler than in typical spin valves. The second approach is based on the increase of exchange energies (>0.3 to 0.4 mJ/m^2) at bottom pinned spin valves ($Mn_{76}Ir_{24}$ and $Mn_{50}Pt_{50}$), which is achieved through proper controlling of growth and microstructure tailoring. With this attempt, exchange fields in excess than 80 kA/m can be reached in spin valves having blocking temperatures exceeding room temperature. Merged synthetic antiferromagnetic structures, employing $Mn_{50}Pt_{50}$ as reference antiferromagnet, have been found to provide the best thermal stability and the largest exchange energies.

Additionally, the improvement of spin valve sensor could also be obtained through the reduction of the thickness of the ferromagnetic free layer. In this direction, one attempt is to utilize 'spin filter' spin valve where a high conductivity layer, such as Cu, is placed under the $Ni_{80}Fe_{20}$ free layer. Another attempt is to employ a synthetic free layer in the spin valve structure. Reduction of the magnetic thickness of the ferromagnetic free layer leads to the enhancement of the output signal of the read-head.

NOTE: CPP and CIP Configuration

It is understandable that electric current is passing in two different ways through magnetic super lattice structure. In one hand, current in plane (CIP) geometry signifies current flowing along the layers where electrodes are locating on one edge of the device structure. In other hand, current perpendicular to plane (CPP) configuration implies current passing perpendicular to the layers where the electrodes are positioning on different sides of the super lattice structure. The results obtained in case of CPP geometry exhibits more than twice higher value of GMR. However, in practice, CPP geometry is more complicated to realize than CIP configuration. CPP and CIP are basically two different forms of spin-valve sensors (Dieny et al. 1991).

Although in present days, spin-valve/GMR-based spintronics devices or sensors CIP configuration is dominant, CPP configuration is expected to play a major role in future terabit recording systems. CPP spin valve can be effectively replaced by the magnetic tunnel junction (MTJ), in which current also passing perpendicular to the plane of the layer. However, there is major difference in the construction between a CPP spin valve and MTJ. As it is well known, MTJ consists of two ferromagnetic layers separated by an insulating spacer layer, instead of a metallic element. As a result, electrical conduction in MTJ takes place through quantum–mechanical tunnelling. However, the large junction resistance of MTJ, compared to that of CIP or CPP sensors, may possibly affect its performance as a read sensor. This is due to the enhancement in thermal noise and the reduction in bandwidth, arising out of the reduction in junction size to a certain value.

10.4 Magnetic Random Access Memories (MRAM)

10.4.1 Application Based on Spin-Tunnel Junctions

Magnetoresistance values of Magnetic Tunnel Junctions, i.e., MTJ, ranging from 20 to 40% at room temperature, have initiated a new domain for technological applications (Parkin 1999; Koga et al. 2002; Ravi et al. 2014). In effect, MTJs, governed by spin-dependent tunnelling phenomenon, can be implemented both as read head sensors in hard disk drives for very high-density magnetic data storage and as non-volatile memory cells in *Magnetic Random Access Memories*, i.e., MRAMs. Barrier fabrication plays a critical role in the fabrication and operation of MTJ. In case of read head applications, junction resistance should be less than 10 $\Omega\ \mu m^2$. Concerning the junction electrodes, all improvements made on spin valve sensor technology are in general applicable here.

10.4.2 Background

The motivation behind the invention of spin valves and MTJs is the hard disk drive applications. However, later they have been proved to be effective and beneficial in the applications of memory devices too. Computer memories endowed with the following attributes are having high demand. The attributes are as follows: **(1)** *non-volatility*, **(2)** *high density*, **(3)** *short cycle time*, **(4)** *low power consumption*, **(5)** *low cost*, **(6)** *high reliability* and **(7)** *infinite lifetime*. Precisely speaking, in case of hard disk drive applications, the spin valves and MTJs are utilized as magnetic field sensors. Such sensors sense the magnetic fields from the information bits, which are nothing but nanosized magnet, recorded on the hard disks. In case of memory applications, when those spin valves and MTJs are employed as cell elements, the sensors themselves act to store the information. The basic structure of the memory cell is still the same as that of the sensor. Noteworthy, in case of memory cell, the saturation region of magnetization of spin valves and MTJs are utilized, whereas in case of read head sensors, the linear region of magnetization is used.

10.4.3 Why MRAM Should Be Used?

DRAM has the advantage of being cheap but is comparatively slow and data are lost when power is off. SRAM, on the other hand is faster than DRAM. But it can cost up to four times as much as DRAM and data are lost when power is turned off. FLASH memory saves data when the power is off, but the process is too slow and consumes a lot of power. MRAM potentially combines the density of DRAM and the high speed of SRAM and the non-volatility of FLASH memory or hard disk, and all this is done using very less power. MRAM can resist high ionizing radiation, can operate in extreme temperature conditions and thus is very suitable for aerospace applications when combined with suitable CMOS technology. The most prominent and vital characteristics of MTJs are their large magnetoresistance values, high resistance and low operating voltage. Owing to these attributes, MRAM devices are quite prospective for implementation in laptop computers so that it does not need to boot up and in cell phones promising improved battery time and enhanced capabilities.

10.4.4 About MTJ-Based Magnetic Random Access Memories (MRAM)

Recently, there is a strong interest in the development of non-volatile memory devices based on magnetic materials, i.e., Magneto-resistive Random Access Memories (MRAMs). The key attributes are (i) non-volatility; (ii) low operating voltage; (iii) limitless endurance of reading and writing capacity; (iv) high-speed read and write operation; (v) radiation hardness and (vi) very large ($>10^{15}$) read–write cycle capability (Granley et al. 1996). Combination of these attributes in MRAM provides performance, cost-effectiveness and several exclusive characteristics for diversified applications. MRAM memory technology merges MR device, with MTJs as storage element and standard Si-based microelectronics to yield the functionality. Some basic characteristics of MTJs are as follows:

(a) As already discussed, operation of MTJ is governed by quantum mechanical tunnelling of spin-polarized electrons through a very thin insulating spacer layer.
(b) Resistance of the MTJ depends on the relative magnetization directions of two ferromagnetic layers separated by an insulating spacer layer.
(c) MRAM cells are constructed in such a way so that it possesses two stable magnetic states corresponding to high or low resistance values. More important point is that those high or low resistance values should be retained without application of any external power.
(d) High value of MR, obtained in MTJs, promises implementation in case of large read signals for MRAM cells.
(e) Moreover, higher resistance of the cell and enhanced MR value causes large output signal of cell voltage and optimal matching of impedance with peripheral sensing circuitry.

These advanced attributes allow MTJ-based MRAM to be commercially competitive. This approach is quite different from usually obtainable commercial memories, such as (dynamic) DRAM and Flash memory. Their working principle is generally based on stored charge. It has been found that MRAMs could be as fast as Dynamic Random Access Memories (DRAMs) and almost as tiny as Static Random Access Memories (SRAM) in the dimension of the cell.

10.4.5 Basic Cell Operation

In case of MTJ-based MRAM device, a single MTJ defines the memory cell. MTJ cells are arranged in array as shown in Fig. 10.8a. In order to read a single bit, a bias voltage is applied across that particular bit line where the target bit is located. In the following manner, the memory state of the bit is decided from the amount of current passing through the bit. It is well understood that a MTJ bit consists of a

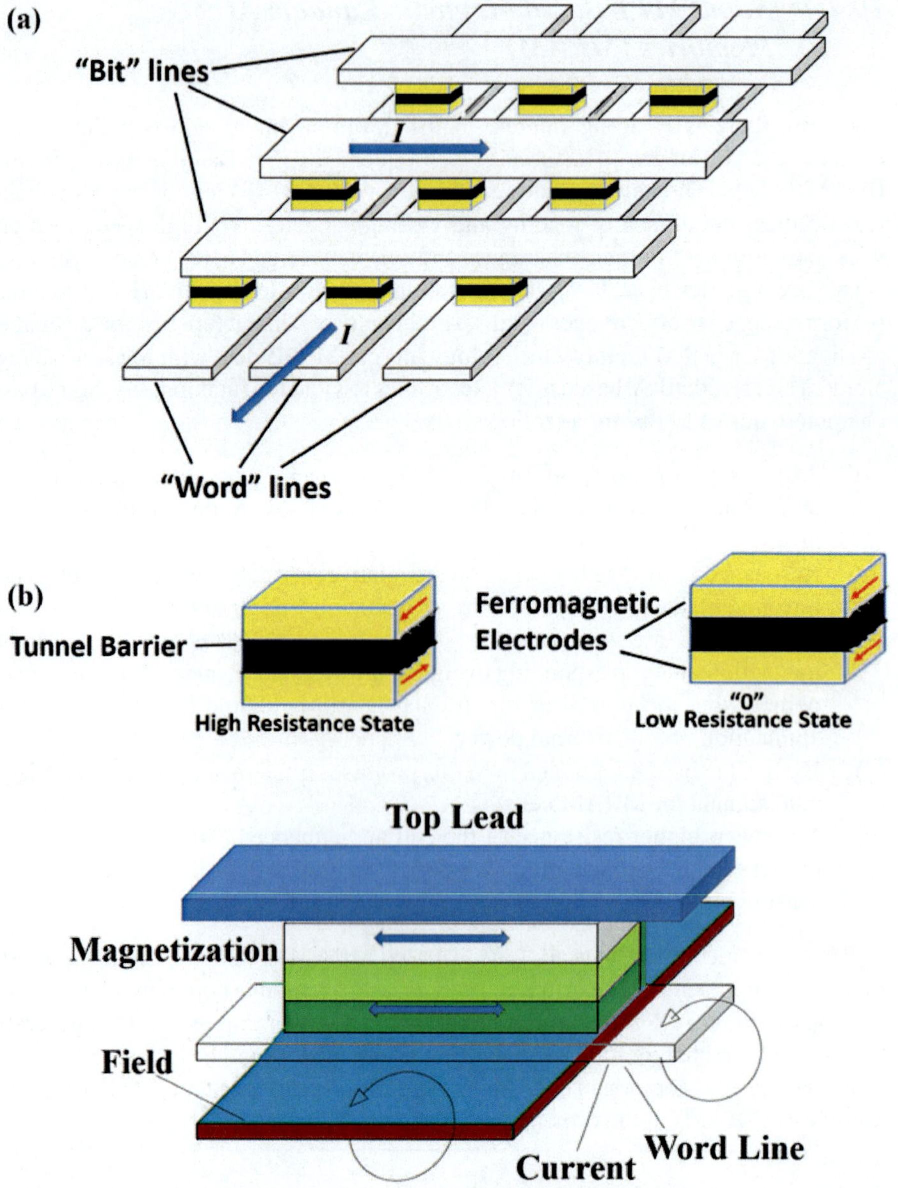

Fig. 10.8 **a** Schematic layout of the MRAM matrix and **b** each cell consisting of a magnetic tunnel junction (Figure (a) adapted and redrawn from Granley et al. 1996)

free magnetic layer, a tunnelling barrier and a fixed magnetic layer (Fig. 10.8b). The fixed layer magnetization is pinned against the rotation under the application of a magnetic field through exchange coupling to additional antiferromagnetic layer. It is the magnetization orientation of the free ferromagnetic layer that is utilized for the storing of information. Depending on the parallel or antiparallel orientation of the magnetization between the free and fixed ferromagnetic layer, the resistance of the memory bit becomes either low or high. This approach is somewhat similar to the GMR spin valve memory cell. The condition to be achieved here is that the free layer magnetization reverses its direction for a write operation, whereas the magnetization of another ferromagnetic layer is fixed. Crucial point that needs to be understood is that the fixed ferromagnetic layer must *rigidly* hold its magnetization direction even when exposed to applied magnetic fields that switch the free ferromagnetic layer. Such condition could be achieved either through pinning by the incorporation of an adjacent antiferromagnetic layer or simply the ferromagnetic layer may be composed of a high-coercivity material.

Write Operation of MRAM—In an attempt to execute '*Write Operation*', a small electrical current is driven in the write lines that in turn create a magnetic field. Such magnetic field causes flipping of magnetic moments in the storage layer of the MTJ, thereby accomplish writing of binary data in two possible orientations of those magnetic moments. This, in turn, causes variation in the resistance of those MTJ cells. Generally, data are read through the tunnelling current or resistance of the MTJ. Indeed, a low power solution is the requirement for these frequently occurring 'write operation'. The process involved with the write operation is divided into two parts:

Bit-cell value detection: In order to identify an already stored bit-cell, its resistance state must be detected by driving a test current through it. As discussed earlier, such operation is accomplished employing a *memory read operation* and is referred to as ***Internal Read Operation***.

Actual write operation: Once the bit-cell value is identified, it is then compared with the value to be written. If the existing value is different from the value to be written, then the actual write operation is done. Otherwise, there is no question of write operation. This intriguing conditional implementation of write operation is accomplished employing a special write activation circuit.

Read Operation of MRAM—At the onset, we must highlight the fact that in case of MRAM, information is stored in the relative orientation of the magnetization of the two ferromagnetic layers. Earlier, storing phenomenon lies in the charging/discharging of capacitors. This information storing, based on magnetization orientation, in turn, promises for scaling down towards tiny structures, at least, in terms of output signal. Initially, the difference between reading of '1' and '0' was about 1 mV in practical devices. The application was confined to mainly for military applications, small capacity non-volatile memories and 16 Kbit integrated MRAM chips. In order to select the cell for reading is to introduce an isolation transistor in each cell.

Figure 10.9 exhibits the typical response of resistance of a MTJ bit as a function of applied magnetic field. Let us suppose a magnetic field is applied along the length

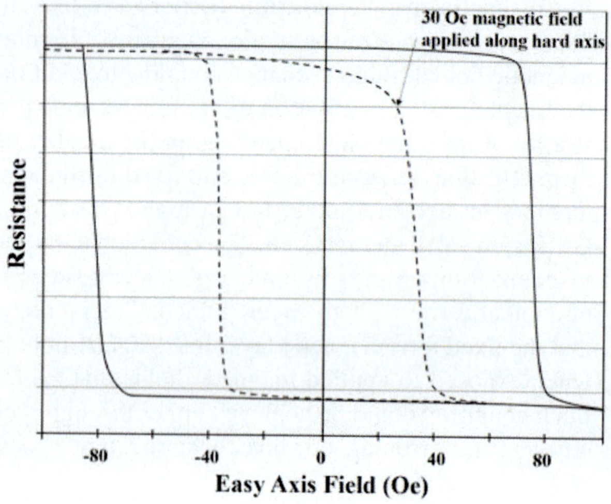

Fig. 10.9 Schematic demonstration of resistance versus magnetic field curves for a MTJ bit with the application of hard-axis field (dashed) and without application of hard-axis field (solid)

of the bit, which is known as the easy-axis of the bit. The magnitude of this applied magnetic field is greater than the switching field of the magnetization. Such applied field forces the magnetization orientation of the free ferromagnetic layer to align along its direction. For instance, as shown in Fig. 10.9, the switching field of the bit is about 80 Oe. Now, magnetic field applied in a direction transverse to the length of the bit, which is also known as the bit hard-axis, would not be able to switch the magnetization of the free ferromagnetic layer along its own direction. Rather, in this case, magnetization orientation of the free ferromagnetic layer makes an angle with the applied magnetic field. Such canting of magnetization makes it possible for a simultaneously applied magnetic field of lower value (for instance in this case less than 80 Oe) along easy-axis direction to switch the magnetization direction of the free ferromagnetic layer. This phenomenon is graphically represented as a 50% reduction in the switching threshold under the application of a 35 Oe hard-axis field (Fig. 10.9). As it comes out the resistance *versus* magnetic field (applied along easy axis) response of the MTJ bit is hysteretic. According to this graphical representation, when no magnetic field is applied, bit will remain in its last-selected state and thereby attain its non-volatile character.

The bits are arranged in array as shown in Fig. 10.10. Such arrangement of array exploits the switching properties of the magnetization of free ferromagnetic layer to write any given bit within that array without unsettling other bits. Such ***Selective bit Programming*** can be achieved by driving currents through a current line above the bit another through a perpendicular digit line just below the bit. Figure 10.10a exhibits how a hard-axis magnetic field is created by the current passing in the line beneath the bits of an array. Such hard-axis magnetic field triggers the magnetization of the free ferromagnetic layers of all the MTJ bits in an array to undergo canting in an arbitrary direction. Such tilted bits are now said to be in '*half-selected*' state. In this scenario, following the dashed hysteresis curve in Fig. 10.9 when current flows through the line above the MTJ bits in an array (Fig. 10.10b), the easy-axis magnetic

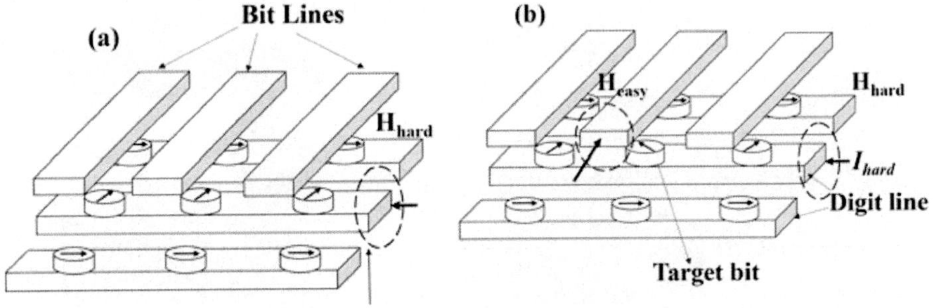

Fig. 10.10 Schematic demonstration of selective write process for an array of bits using bit and digit lines to generate hard and easy-axis for fields for switching. The free layer and direction of magnetization of the free layer of each bit are shown. In **a**, field from the hard-axis line causes the magnetizations to cant, but not switch. In **b**, field arising from current in the easy-axis line combining with magnetic field from the hard-axis line causes magnetic bit at the intersection to switch

field is so generated that it causes only the half-selected bit beneath the line to switch its direction along the applied field. Interestingly, in this process, the bits that are not half-selected remain unaffected.

10.4.6 Applications of M-RAM

MRAM has a potential in all memory applications in these devices, such as Digital Cameras, Cellular Phones, MP3, HDTV, Laptops, etc. In case of first MRAM devices, toggle memory switching was utilized where magnetic field is employed to change the electron spin orientation. Such toggle MRAM was quite easier to develop. However, such device was not easy to scale up. In case of second-generation MRAM devices, different architectures employing spin-polarized current for switching the electrons spin have been utilized. Very recently, there is advent of MRAM device based on spin transfer torque (STT) effect. Such newly developed STT-MRAM devices are faster, more energy efficient and easier to scale-up compared to its earlier versions.

10.5 Spin Transfer Torque (STT)—MRAM

10.5.1 Introduction

As a prelude, we may say that Spin Transfer Torque (STT)-MRAM is a highly developed kind of MRAM devices. This advanced device offers higher densities, low power consumption and reduced cost compared to regular MRAM devices. The prime advantage of STT-MRAM over regular MRAM device lies in its capacity to

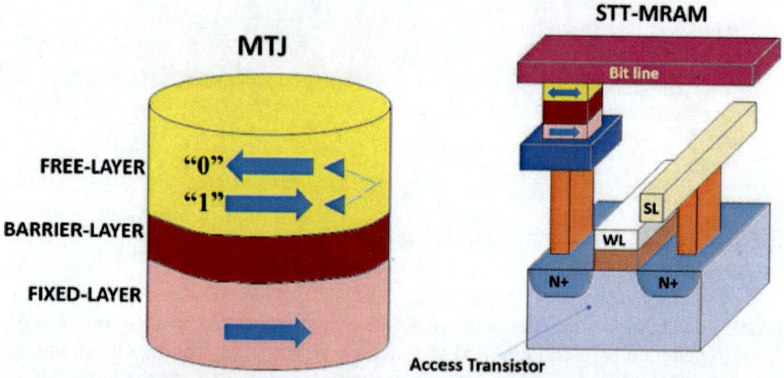

Fig. 10.11 Schematic demonstration of MTJ MRAM and STT MRAM

scale down STT-MRAM chips in order to attain even higher densities and that at lower cost compared to regular MRAM device.

10.5.2 What Is STT-MRAM?

In case of a spin transfer torque (STT)—MRAM device, electron spins are flipped employing spin-polarized current, following the technique as described in Chap. 5. With some structural engineering, this effect is realizable in a magnetic tunnel junction (MTJ) or in a spin-valve. In case of fabrication of STT-MRAM devices, STT-MTJ is employed (Fig. 10.11). According to simplified description of the operation of such device, we may recall that spin-polarized current is produced by passing current through a thin ferromagnetic layer. This current is then directed.

into an even thinner ferromagnetic layer where exchange of angular momentum takes place between the magnetic moment of spin polarized current and the magnetic moment associated with that thinner ferromagnetic layer. Consequently, magnetization orientation of thinner ferromagnetic layer changes in the direction of the net spin magnetic moment of the spin-polarized current.

10.5.3 What Is Perpendicular STT-MRAM?

In general, typical STT-MRAM structure employs an in-plane MTJ structure, generally referred to as *iMTJ*. There are some STT-MRAM devices using a more optimized device structure, where the magnetic moments are oriented perpendicular to the surface of the silicon substrate. Such MTJ structure is named as perpendicular MTJ, i.e., *pMTJ*.

Advantages of *pMTJ*-based STT-MRAM are that it is even more scalable in devices compared to *iMTJ*-based STT-MRAM. Hence, *pMTJ*-based STT-MRAM offers better compactness in devices, size miniaturization and also more cost-competitive than *iMTJ*-based STT-MRAM. Recently, *pMTJ*-based STT-MRAM has emerged out to be more promising technology for replacing DRAM and other memory technologies.

10.5.4 Application of STT-MRAM

STT-MRAM has enormous potential to revolutionize information technology by its superb density, higher speed, cost-effectiveness and low power consumption, and thereby has become a leading storage technology. STT-MRAM leads to the development of high-performance memory device that can challenge DRAM and SRAM. Moreover, this device can be scaled well below down to 10 nm and may challenge low cost flash memory. There are several companies, such as IBM, Samsung, Everspin, Avalanche Technologies, Spin Transfer Technologies and Crocus, those are developing STT-MRAM chips. It is noteworthy that Everspin has already announced commercialization of 256 Mb STT-MRAM in April 2016. This novel chip demonstrates very high interface speeds comparable to DRAM and its volume production is also expected 'soon'. Furthermore, in August 2016, Everspin has initiated development of *pMTJ*-based STT-MRAM chips. Initially, those chips were also 256 Mb in size, however, there are some special features associated with *pMTJ* versions that include improved performance, higher endurance, lower power consumption and better scalability compared to that of *iMTJ*-based STT-MRAM products. Nowadays, Everspin has started production of 256 Mb *pMTJ* STT-MRAM chips that lead to the development of a scaled-down 1 Gb version. A comparison on the performances of MRAM versus other random access memories have been presented in Table 10.2.

10.5.5 The Latest STT-MRAM

Numem is a renowned high-performance STT-MRAM developer and has been selected to supply low-density STT-MRAM, i.e., ***NumemNuRAM MRAM-based Memory*** to a NASA AI core project. According to the demands of Numem, its memory enables a two to three times smaller memory area and 20–50 times lower standby power compared to SRAM.

Table 10.2 MRAM versus other random access memories

	SRAM	DRAM	NOR Flash	MTJ-MRAM	STT-MRAM
Access time	<1 ns	260 ps	25 ns	35 ns	<10 ns
Endurance (write cycles)	Infinite	Infinite	10^5	Infinite	Infinite
Retention	0 (volatile)	0 (volatile)	>10 years	>20 years	>20 years
Cell size (F^2)	100	8	6	10	<4

Adapted from Heidecker (2012)

10.6 Spintronics Sensors

Using either the spin in tandem with the charge or alone, spintronics has some advantages over conventional semiconductor electronics including higher integration density, non-volatility, decreased power dissipation and faster processing speed (Dieny et al. 1991; Koga et al. 2002; Ravi et al. 2014; Wlodarczyk et al. 2011; Baibich et al. 1988; Zutic et al. 2004; Rashba 2002; Grunberg et al. 1992). Spintronics sensor, acting as solid-state magnetic sensor over the last few years, has attracted huge attention and research effort owing to their diversified attributes, such as excellent sensitivity, low power consumption, compactness, wide bandwidth, room temperature operation, magnificent ability of measuring and COMS compatibility. Spintronics sensor is capable to provide data associated with the magnetic field. Such data enable us to extract some magnetic field-related parameters of several real-world objects. Furthermore, it possesses several unique benefits that could act as significant basis for IOT (Internet of Things), which predicts the connectivity of billions of things surrounding us to the Internet as well as experts for the information usage.

Basic principles of Spintronic sensors

Operation of spintronic sensors is based on dependence of its resistance on both the magnitude and direction of the applied magnetic field, i.e., the so-called magnetoresistance effect. Spintronic sensors can be categorized into **Anisotropic Magnetoresistance** (AMR), **Giant Magnetoresistance** (GMR) and **tunnelling Magnetoresistance** (TMR) sensor depending on their distinct underlying mechanisms (Fig. 10.12).

In recent time, motivated by the novel applications, traditional methods of magnetic field sensing are often being challenged and substituted by emergent technologies. In this direction, attention has been focussed on solid-state magnetic sensors, magnetodiodes, magnetotransistors, Hall effect-based devices and magnetoresistors. Noteworthy, Hall effect sensors are quite well accepted in industry, whereas magnetoresistive sensors are continuously attracting industrial interest owing to its technological potentials. Although, both Hall effect and magnetoresistive sensors are having compatibility with current CMOS fabrication processes, magnetoresistive sensors offer some intrinsic advantages. Generally, magnetoresistive sensors are more sensitive than Hall effect-based ones at room temperature.

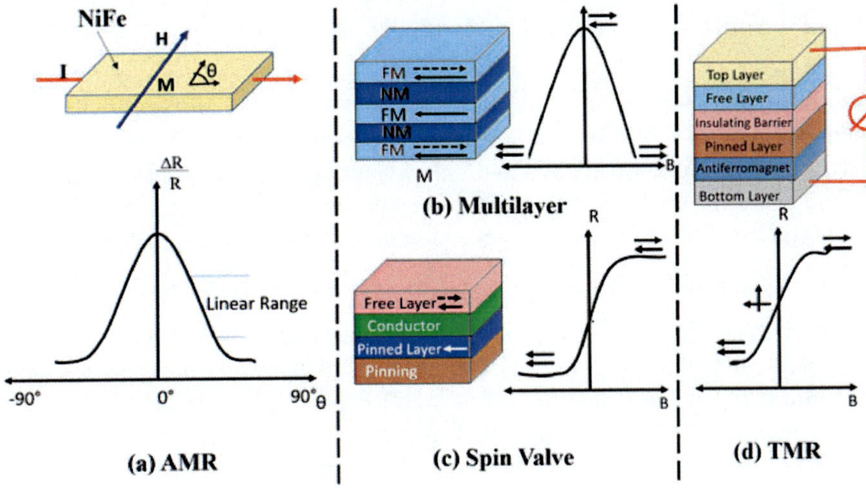

Fig. 10.12 Different kinds of spintronics sensors. Structure and response of **a** AMR sensor; **b** multilayer sensor; **c** spin valve sensor; **d** TMR sensor (Figure adapted and redrawn from Rolfe et al. 2011)

Moreover, novel magnetoresistance-based devices offer higher scale of integration. Additionally, in-plane fields can be measured with magnetoresistive sensor, which is particularly advantageous for instance, electric current sensors.

10.6.1 Anisotropic Magnetoresistance Sensor

Anisotropic magnetoresistance (AMR) effect in ferromagnetic metals and alloys originates from the anisotropic scattering of conduction electrons mediated by spin–orbit interactions. Owing to the fact that the magnetization direction of the AMR material can be altered by the influence of both magnitude and direction of an external magnetic field, there is an interconnection between its resistance and the external magnetic field.

In a broad sense, we may state that the information of magnetic field can be sensed by using AMR sensors. The direction sensing ability of the AMR sensor can be employed to decide the rotational conditions of any ferromagnetic object. Consequently, AMR sensor could act as contactless magnetometers, angle or rotation sensor.

10.6.2 Some GMR-Based Sensors Applications

After the introduction of the first commercial GMR sensors in 1995, there has been a rapid evolution of GMR sensors technology. Eventually, this has explored a wide and promising range of applications. Nowadays, GMR-based sensors are being applied in diverse fields, such as engineering, physics, biology, space, etc. Following discussions are focussed on the direct and indirect GMR-based applications.

Electrical Current Sensing

In conventional approach, electrical current is measured by using shunt resistances, coils and solid-state sensors. On the other hand, solid-state magnetic sensor can sense magnetic field produced due to current flow (Fig. 10.13). This is a quite generic approach and can be employed to both printed circuit board or integrated circuit in order to measure current whether flowing through a wire (Fig. 10.13a) or by a conductive strap (Fig. 10.13b). Measurement of AC/DC current can be done with small, cheap and contact-less systems.

Incorporation of GMR sensors for this purpose has been motivated by their excellent sensitivities (Dieny et al. 2020; Hirohata and Takanashi 2014; Heidecker 2012; Datta 2018; Datta and Das 1990; Chuang et al. 2015; Liu et al. 2016). In this direction, application of GMR-based electrical current sensors can be considered as the final implementation of an ammeter. The operational principle is shown in Fig. 10.14a. Evidently in this configuration, the flow of current takes place from left to right above R1 and R3, and from-right-to-left above R2 and R4. As a result, when current flow occurs from A to B, then depending on its sign, the values of resistances R1 and R3 increase/decrease and that of resistances R2 and R4 decrease/increase. Therefore, a full Wheatstone behaviour can be obtained (Fig. 10.14b). The input is given to the sensor through terminals *a* and *b* and the output is extracted between terminals *c* and *d*. In order to make this sensor insensitive to external magnetic field, a particular

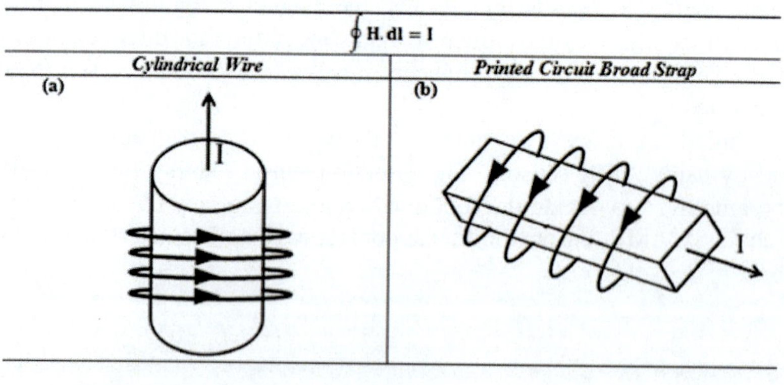

Fig. 10.13 Magnetic field generated in case of current flowing through **a** a wire or by **b** a conductive strap (Figure adapted and redrawn from Berger 1988)

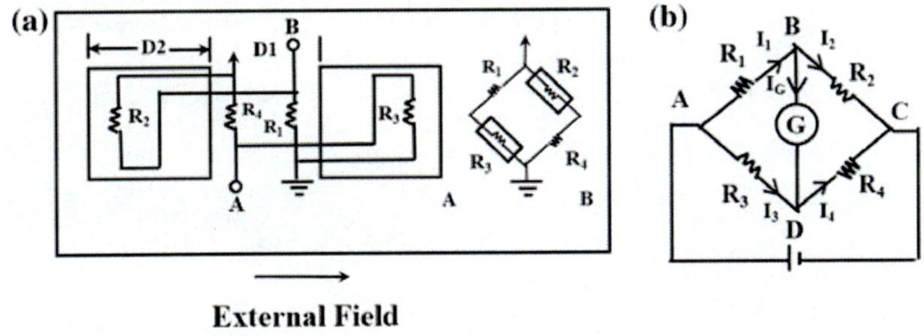

Fig. 10.14 GMR circuit technique. **a** GMR resistors configured as Wheatstone bridge sensor; **b** Wheatstone bridge circuit

arrangement of the magnetoresistors is done. This, in turn, *theoretically* minimizes the possibility of interferences.

Measurement of Differential Currents

GMR sensors can be effectively used in current wires and transmission lines (Fig. 10.15). GMR sensors can also measure the differential currents. In this case, the sensor is placed between two Helmholtz coils through which, the current to be compared, is flowing. It is well known that when currents carried by the two coils are identical, the magnetic field generated at the middle point of the coils is zero. Accordingly, the output voltage of the sensor is also zero.

Engineering Applications

In an attempt to sense magnetic field employing GMR-based technology, sensing of the field is executed through the measurement of perturbations produced in the Earth's magnetic field by the ferrous body specifically under consideration. For instance, GMR sensors could possibly detect and assess the small perturbations of magnetic field triggered by the iron body of the car *locally* over that of the Earth's magnetic field (Fig. 10.16). In case of a *GMR gradient type sensors*, it is the magnitude of the variation of the magnetic field that decides the output signal. No compensation is required for the additionally applied external magnetic field. Thus, when a car is running close to a sensor, a voltage response is detected from its differential output. In a combined approach, incorporation of an additional sensor, placed to a well-defined distance could also measure the speed of the car. This principle can also be comfortably adopted for measuring vibrations in industrial machines. The subtle variation of magnetic field generated by the vibration of the ferromagnetic components in industrial systems, over that of the Earth's magnetic field, can be efficiently converted into variations of resistances by utilizing GMR-based magnetic field gradient devices (Fig. 10.16).

Furthermore, there are also demands of angle and circular positional detectors in the industry for automotive applications, rotational machinery etc. These kinds of sensors are generally designed as contact-less systems in which GMR-based magnetic sensor senses the relative angular position of a rotating magnet.

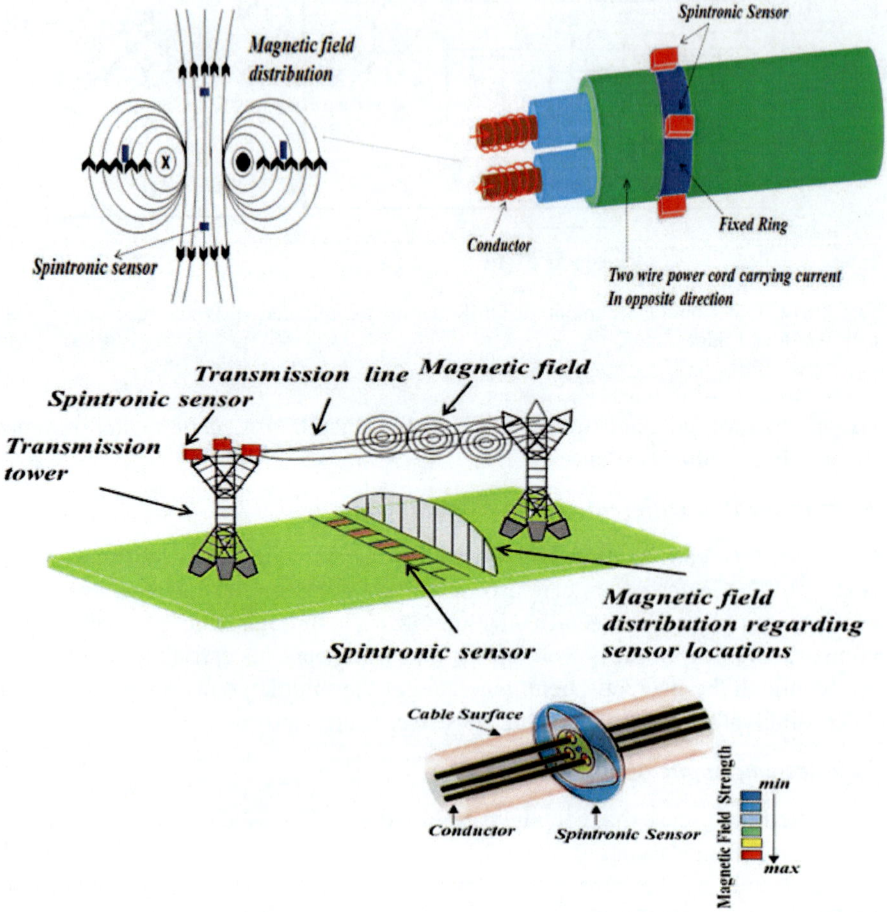

Fig. 10.15 GMR sensors used in current wires and transmission lines (Figure adapted and redrawn from Liu et al. 2016)

Biological Applications

Among widely studied magnetoresistance-based biosensors, GMR-based sensors have attracted a great deal of attraction (Fig. 10.17) (Wlodarczyk et al. 2011). Two different kinds of layer structures have been introduced in biosensing: ***(i)*** multilayer system and ***(ii)*** spin-valve system. Initially, the surface of the sensing area is functionalized with the probe molecules. Such probe molecules can specifically bind to the target analyte molecules. In recent researches, the target analytes are mainly DNAs.

or proteins. In case of detection of DNA sequence, the complementary DNA chains are employed as probes, whereas in order to detect protein molecules, antibodies are utilized for binding the target protein. After the application of the sample solution on the sensor surface, magnetic labels are being attached to the surface of the sensor. Under the excitation with an applied magnetic field, a fringe field is produced by

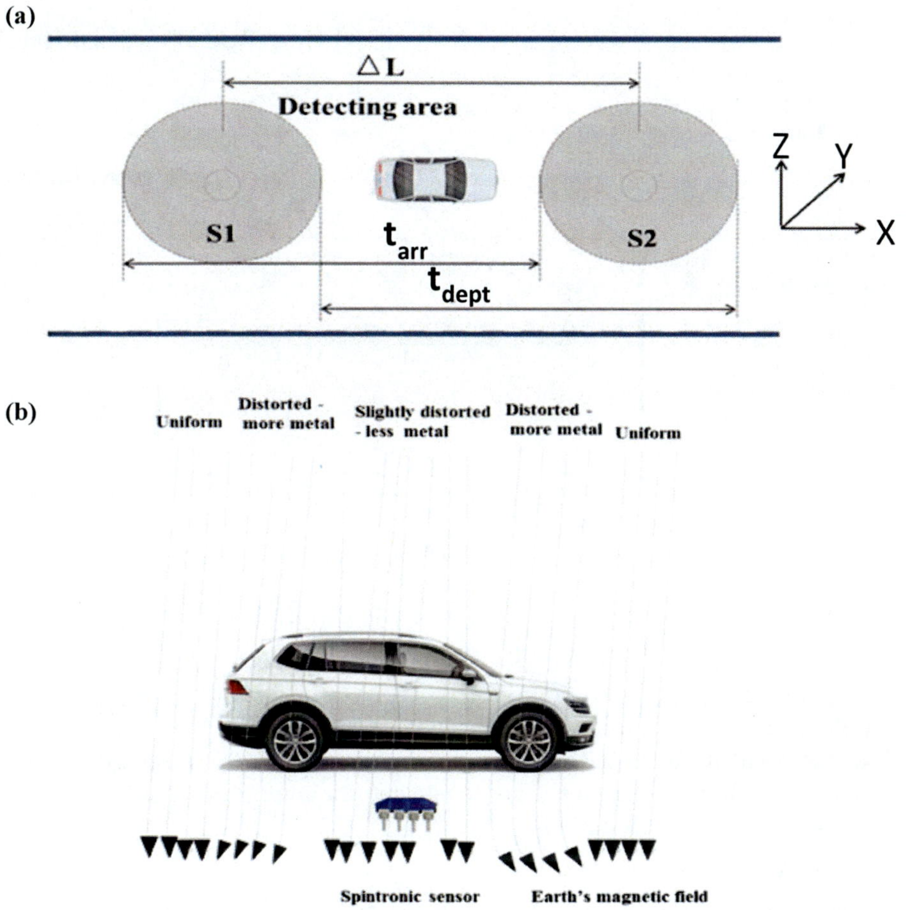

Fig. 10.16 GMR sensors used in **a** detection of object; **b** vehicle detection (Figure adapted and redrawn from Liu et al. 2016)

those magnetic labels. Such fringe field can be detected by the bottom sensors. Quantitative suggestions for the level of concentration of the target analyte in the sample solution can indeed be obtained from those detected magnetic signals. GMR-based biosensing systems have been widely implemented for immunochemical assay. First, GMR-based immunochemical assay was done in 2003 by utilizing spin valve sensors in order to detect a single MNP via biotin–streptavidin interaction. GMR-based biosensing system has been successfully experimented for the detection of prostate cancer, ovarian cancer, lung cancer, etc. Several other types of analytes, such as viruses, bacteria and toxic ions can also be detected. Recently, multiplexed detection is also feasible using portable GMR biosensing systems having capacity to detect up to 12 kinds of tumour biomarkers.

Such GMR sensor-based Point of Care (POC) diagnostic devices, having ability to perform multiplexed detection, are quite promising for diagnostic applications.

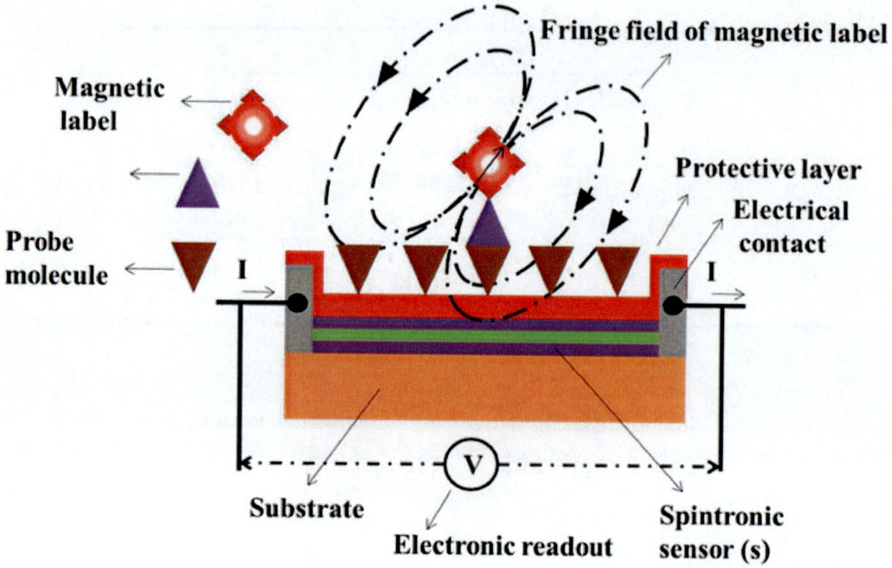

Fig. 10.17 GMR sensor used in biodetection process (Figure adapted and redrawn from Liu et al. 2016)

Furthermore, with the recent development of nano technologies, the technical issues associated with the increased number of sensing elements can be addressed with ease. A lot of effort has also been dedicated to integrate the amplification and denaturation processes with the GMR biosensors.

Spintronic Couplers

Employing the sputtering technology, thin film *on-chip* coil could possibly be fabricated. Current in this coil, in combination and cooperation with an on-chip GMR magnetic sensor, can connect an electrical signal across an insulating layer, thereby acquiring galvanic isolation. Similar to sensor, these components can also be merged with other semiconductor functions in order to manufacture a very high-speed digital isolator.

Construction of a GMR Coupler

In a spintronics coupler, four GMR resistors construct a Wheatstone bridge (Fig. 10.18). Isolation of several thousand volts from the input coil can be achieved by introducing a thin polymer dielectric barrier layer. A magnetic field, which is proportional to the input current, is produced below the winding of the coil. This magnetic field results in flipping of the electrons spin in the GMR resistors and thereby varying their resistance. A magnetic shield is normally employed to protect the sensor from external stray magnetic fields. The output of a GMR bridge is given to the input of a comparator, which runs either a CMOS output stage or an open drain transistor output. GMR resistors have been observed to be sensitive to the magnetic

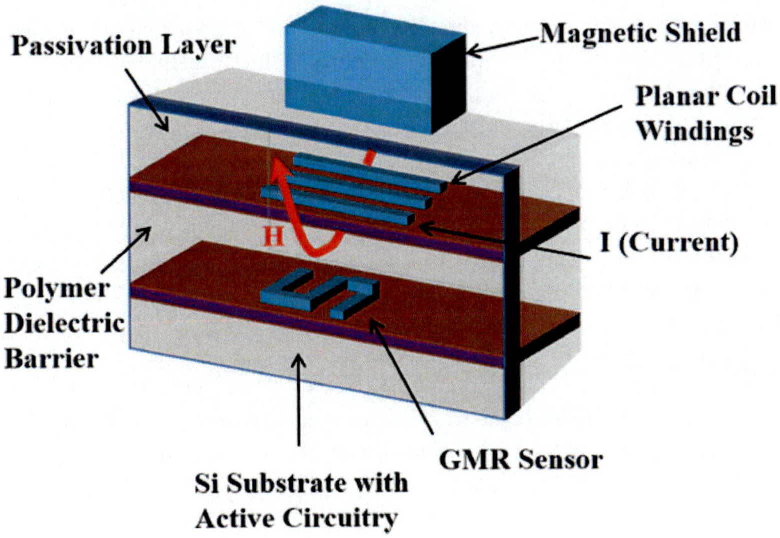

Fig. 10.18 Schematic demonstration of spintronics coupler

field, applied in the substrate plane. This offers a more compact integration scheme than that would be obtained with a Hall sensor, measuring fields perpendicular to the substrate. Furthermore, high sensitivity of GMR results in shorter propagation delay of spintronic couplers and more stability than a simple MEMs-based pick-up coil.

Advantages of Spintronics Coupler

Spintronics coupling has certain advantages in comparison to that of conventional opto-couplers. The benefits are ***high speed***, ***small footprint, superb noise immunity*** and ***unlimited life***.

High Speed—Speed of spintronics couplers are at least double than even the fastest opto-couplers. Accordingly its rise, fall and propagation times are faster than opto-couplers. Such faster time scale, associated with Spintronics couplers, in turn also reduces power consumption in the device. This is feasible by time minimization in active regions.

High Channel Density—Another intriguing quality of spintronics coupler is its size miniaturization. It can be understood that opto-couplers and transformers are inherently bulky, whereas spintronics couplers are monolithic in true sense and hence can be fabricated in less than 1 mm^2 of die area per channel. *In technical sense, less board real estate means lesser costs and more scope for other functions.* Moreover, smaller die size in case of spintronic couplers offers less cost than opto-couplers exhibiting high performance.

Excellent Noise Immunity—Spintronic couplers have up to 25 kV/μs transient immunity, compared to 10 kV/μs for opto-couplers. This attribute of spintronics coupler is important in harsh industrial and process control environments.

Unlimited Life—Opto-couplers have limited life time, which degrade over time. On the contrary, spintronics couplers have unlimited life time that last for an indefinite period without any degradation of its performance over time.

Similar to opto-couplers, spintronics couplers have a current-mode input since the electronic signal is applied directly to the coil. Such devices can be driven by any input voltages up to 400 V, applied across a current-limiting resistor.

On the other hand, dissimilar to opto-couplers that typically need at least two volts for the light-emitting diode to turn on, in case of spintronics couplers, the coil-drive input requires only just millivolts. This, in turn, leads to typically less requirement of drive current.

Some Other Applications

- ***Conditioning Circuits***: Presently, research activities are also being carried out concerning the electronics required for biasing and conditioning of GMR-based sensors. In case of GMR sensors, both applied magnetic field and appropriate circuitry can decide the accurate bias point of the sensor.
- ***Space Applications***: Spanish National Institute of Aerospace Technology (INTA) is presently carrying out research in the direction of introducing a miniaturized three axis GMR sensor to the altitude controlling system.

10.6.3 Some Magnetic Tunnel Junction-Based TMR Sensor Applications

- Sensors based on magnetic tunnel junctions, i.e., tunnel-magnetoresistance (TMR) sensors can be used in cars' steering columns to improve the accuracy of automated parking capabilities. As such, these advances move the automotive industry closer to required instantaneous processing and actions for fully autonomous driving.
- Various electronic companies have developed TMR sensors. There is ongoing mass production of TMR sensors since 2014. It is indeed mentioned worthy that TMR sensors achieve an impressively high output. For instance, about 20 times higher than conventional anisotropic magnetoresistance (AMR) sensors, and nearly 6 times higher than giant magnetoresistance (GMR) sensors.
- Additional advantages of TMR sensors include high accuracy, high stability with less temperature drift and less aging deterioration. These sensors are well employed as angle sensors, position sensors and rotary sensors.
- A very simplified description of the TMR sensor acting as angle sensor is presented here: When a magnet is rotated on top of a TMR sensor, the orientation of magnetization of its free ferromagnetic layer follows the direction of this rotating magnetic field. Since the resistance value of the TMR sensor is proportional to the relative angle of the magnetization directions of its fixed and the free ferromagnetic layer,

the corresponding resistance of the sensor fluctuates continuously. The fluctuation of resistance of this TMR sensor is in accordance with the rotation of the top magnet and in this way, the TMR sensor can be employed as an angle sensor.

- In this context, we should mention that in comparison to GMR-based sensors, MTJ sensors offer a higher MR ratio and sensitivity. However, the top electrodes of MTJ sensors are generally thicker than GMR sensors. This, in turn, makes the sensing field of the free ferromagnetic layer smaller and thereby limiting its application. Another challenge lies on the high intrinsic noise of MTJ sensors for achieving *ultra*-high sensitivities.

TMR-Based Biosensors

MTJ-based sensors have also found their application as ***Biosensors***. The MgO-based MTJ sensor is bio-functionalized with *Alpha Fetoprotein* (AFP) capture antibodies where Metabolic Nutrition Programs (MNPs) are functionalized with AFP detection antibodies. A sandwich-assay configuration is generally adopted for the detection of AFP. A complete binding of target AFP to capture antibodies is achieved. Finally, the excess MNPs are eliminated by cleansing the Phosphate Buffered Saline (PBST) solution.

TMR-Based Sensors for Genotyping

A proof-of-concept detection of DNA, employing MTJ-based sensors was exhibited in 2005. In that case, MTJs with Al_2O_3 barrier layer were utilized to detect single-stranded DNA attached with $MnFe_2O_4$ nanoparticles. However, the first quantitative detection of DNA was accomplished in MgO-based MTJ sensors.

10.7 Spin FET

Spin-based logic is presently being investigated as one of the beyond-CMOS computing technologies to supplement complementary metal oxide semiconductor (CMOS) field effect transistors (FET) in microprocessors. Strong-willed research in spintronic has resulted in commercialization of MRAM and STRAM and had an immense influence on day-to-day life. Now the question comes up whether the spintronic devices are competent enough for similar success in logic. One of the strong contenders may be the spin field-effect transistor (spin FET), which is basically a three-terminal device that utilizes ferromagnetic contacts in the source and drain as spin injector and detector.

Spin FET, visualized by Datta and Das in 1990, opens a gateway to spin information processing (Datta 2018; Datta and Das 1990). Although the researches on semiconductor spintronics have made coherent manipulation of electron spins in semiconductors feasible yet the realization of a functional spin field-effect transistor for information processing has not been achieved. The numbers of primary challenges are low spin-injection efficiency due to resistance mismatch, spin relaxation, and the spread of spin precession angles. Supriyo Datta said that they were inspired by optical phenomena to put forward the concept of spin transistor. Electronic device was proposed similar to optical devices by exploiting the similarities between the spin of an electron and the polarization of a photon.

10.7.1 Working of Spin FET

The basic idea of the spin FET, as proposed by Datta and Das, is to control the spin orientation by applying a gate voltage. Schematic of a spin FET is shown in Fig. 10.19. A spin FET consists of ferromagnetic electrodes and a semiconductor 2DEG (two-dimensional electron gas) channel with a gate electrode, It is basically similar in structure to a conventional FET but with contacts that are magnetic. One magnetic contact is known as source (S). It injects electrons whose spins are aligned with its magnetization. Whereas the other magnetic contact is known as drain (D) and lets through electrons with properly aligned spin. If S injects up-spin electrons and the D detects down-spin electrons, then the transmission will be a minimum. Now the transmission can be increased if the spin be rotated as it passes through the channel linking source to drain.

The subsequent challenge is to rotate the spin. It can be done by two ways (i) applying magnetic field and (ii) using electric field, provided the 2DEG channel

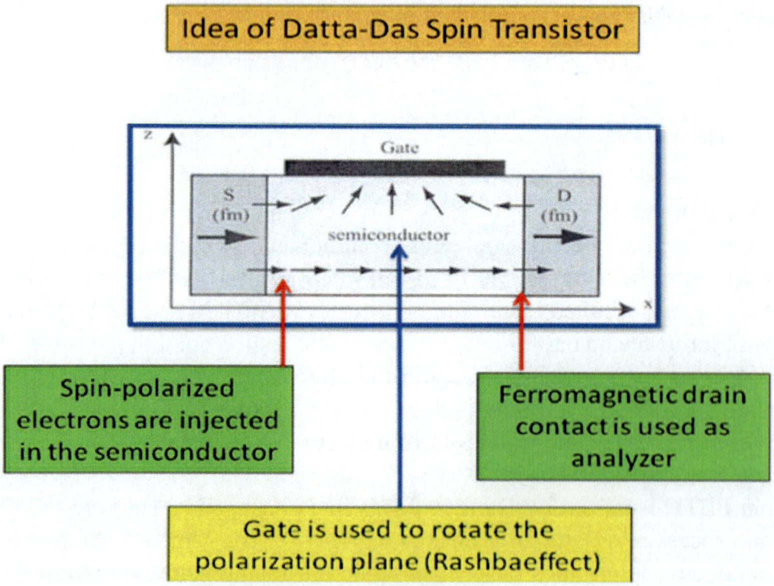

Fig. 10.19 Schematic diagram of a spin FET

exhibits spin–orbit interaction. The origin of the spin–orbit interaction is the relativistic effect, and an electron feels the effective magnetic field B_{eff} perpendicular to its motion in the electric field E and propagation velocity direction v, such that $B_{eff} \propto (v \times E)$. A gate voltage can alter the spin–orbit interaction and is used to tune the spin precession angle. The spin–orbit interaction is modified by the gate voltage (the effective magnetic field is changed). Therefore, the spin precession angle can be controlled by the gate voltage. In a spin FET, current flow is modified by the spin precession angle.

The injection and detection of polarized spins with semiconducting channels is complicated. Finally, Hyun Cheol Koo and colleagues working with Johnson demonstrated one version of the device in 2009 and then another version in 2015 where a spin FET with the help of stray magnetic field induced by ferromagnetic patterns is recommended. At source, the spin polarized electrons are generated by the Zeeman splitting. The selected spins are sent out by the spin filtering effect at drain. This device is frequently called a spin transistor, may be due to the fact that the gate voltage changes its resistance from a high value (OFF) to a low value (ON), like a real transistor. But a typical FET has an OFF/ON resistance ratio of about 10^5. By contrast, the spin transistor can at best reduce the current a little bit because the strong randomization of spins can be taken place by scattering processes in solids depending on the temperature and the channel length.

Alternative designs need to be implemented to overcome this constraint. Since 1990, researchers are looking for a device to replace the FET. But it seems more practical to find ways to complement the FET. For example, small magnets are unstable, making the resistance of an MTJ fluctuate with time. It has been argued that such a stochastic MTJ could be combined with FETs to build a probabilistic bit or a p-bit: a three-terminal device whose output alternates between ‘0’ and ‘1’ when the input is weak, but locks into ‘0’ for large negative input and to ‘1’ for large positive input.

10.7.2 Advantages of Spin FET

Though the spin FET has not been realized yet but is regarded as one of the most advanced applications of spintronics in the future. A spin FET would have several advantages over a conventional FET: It could be used not only as a logic gate but also as a *non-volatile* memory component. Less energy would be necessary to flip an electron’s spin, and the spin could be flipped much faster than electrons can be pushed away from their channel by a gate voltage. It may also be possible to change the magnetization direction of the electrodes by applying a magnetic field.

10.8 Racetrack Memory

10.8.1 What Is Racetrack Memory?

Racetrack memory (RM) is future generation memory and expected to replace the existing memory types, including hard disk drives (HDDs) and flash (see Fig. 10.20). RM is a comparatively new technology that exploits moving magnetic domain walls or skyrmions* along structures like magnetic nanowires. The information bit can be coded in magnetic regions separated by domain walls (DWs), which can be moved by means of the spin transfer torque (STT). Conventional computer hard drives use a motor to rotate discs on which magnetic bits are stored in a thin film. The racetrack memory uses electric currents to move the magnetic domain walls or skyrmions in the nanowire without displacing any atoms at all. It falls under the category of non-volatile solid-state memory. IBM predicts that it will be able to store data, which is hundred times more than that of technology commercially available today. 'The technology could have the potential to enable hand-held devices to hold a few thousand movies, run for weeks at a time on a single battery and be practically unbreakable', IBM claims.

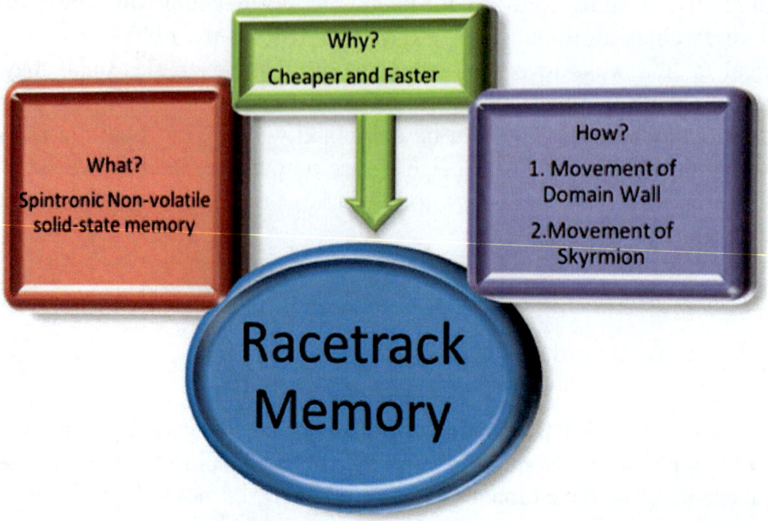

Fig. 10.20 Racetrack memory: what, why and how?

10.8.2 Advantages and Disadvantages

The major benefit of racetrack memory is its cost-effectiveness. Drives using racetrack memory will also be considerably cheaper than flash memory. This is because the racetrack memory can bundle massive amount of data onto single nanowires. Reading and writing of data are much faster than that of any existing technology. It is predicted that it will be able to read or write a bit of information in ten nanoseconds or less. By contrast, a hard disk would take around three million nanoseconds to do the same.

Racetrack memory is still considered an experimental technology. Tackling the movement of domain wall in nanoscale is controlled by quantum mechanical phenomena. One of the major challenges encountered is with domain wall motion. They are capable of moving or stopping at random, with no definite way of prevention. Another frequent challenge is the significant reduction of read/write times even due to the smallest imperfection in the wire.

10.8.3 How Does it Work?

Racetrack memory relies on spintronics. Here, data are stored as a sequence of magnetic domains along a nanowire. There are two promising approaches to encode bits: (i) by nanoscale domain walls (DW) or (ii) by skyrmions. The main challenges here are to make smaller bits and then efficiently move these at high speeds.

RM With DW Motion: Spin transfer torque (STT) concept is applied to displace domain wall (DW) by flowing an electrical current through DW. Domain wall displacement as discussed in Chap. 6 can be utilized to fabricate a racetrack memory as a high-density shift register memory as shown in Fig. 10.21. It consists of an array of magnetic nanowires, measuring around 50 nm in diameter, set horizontally or vertically on a silicon chip. These nanowire structures are known as so-called **racetrack** coined by IBM researcher Stuart Parkin in 2007. The nanowires are like skyscrapers with each floor of each skyscraper containing a single bit of data. A transistor at the bottom of the wire shoots spin-polarized electric currents up and down the wires, which moves the data up and down. Like other electronic memory, the racetrack memory employs reading and writing heads located near the storage medium. However, it also significantly differs from the existing electronic memory. It does not write just to a single side of the storage medium. Instead, it writes on the whole of 'domain wall' of the nanowires. Information is written by a domain injection pad and read by a spin tunnel junction at the end of the wire. This technique employs multiple magnetic domain walls per racetrack. Bit length is defined by the spacing between successive magnetic domain walls and is controlled by pinning sites. Practically, domain walls need to be pinned at precise positions following the wall motion is induced by a current. This enables the device to store huge data on a single wire and makes an example of 3D data storage.

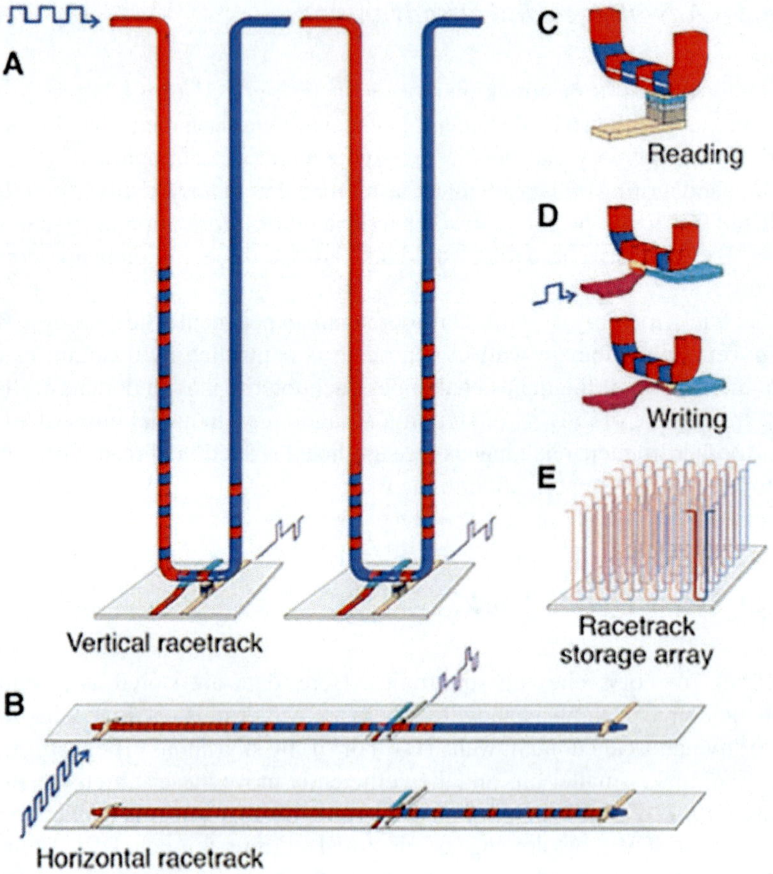

Fig. 10.21 Schematic diagram of a racetrack memory concept in (A) vertical and (B) horizontal configurations. (C) Reading and (D) writing operation can be carried out electrically. (E) 3D-array concept (Taken from Hirohataa and Yamadab 2020)

In order to realize the racetrack memory, we need to fabricate micro- or nanometre wide FM wire with very smooth edge to make sure uniform DW displacement. To widen the racetrack concept to 3D, it would be essential to fabricate these wires with 'U' shape as shown in Fig. 10.21. A FM micro-wire has been made with DW having current density between 107 and 108 A/cm^2 for the wall motion. Of late, DW motion with speed 750 m/s has been reported in synthetic antiferromagnetic wires with perpendicular anisotropy.

RM with Magnetic skyrmion: Owing to their distinctive vortex-like spin texture, skyrmions show many interesting features and are helpful for their efficient manipulation (Tomasello, et al. 2014). They could be wonderful as storage bits in racetrack memories and logic devices because they can be made much smaller than the magnetic domains used in modern disk drives. Flipping (e.g., switching device's

Table 10.3 Comparison between skyrmions and domain walls

	Skyrmion	Domain wall
Size (nm)	10–50	20–50
Speed (m/s)	~100	~500–750
Critical current (A/Cm^2)	10^2	10^7

memory state from 1 to 0) of all the spins in the conventional domains needs substantial power where as skyrmions need fewer spin flips to switch. Another advantage is that the final spin state is not easily disrupted either, which makes these skyrmion structures steadier than their usual counterpart.

Magnetic skyrmions can be driven by a spin transfer torque mechanism at a very low current density. This is a technologically exciting property and has made skyrmion a promising candidate for racetrack memory. The skyrmion motion has been demonstrated in a multilayer of Ta/CoFeB/Ta-O at the speed of ~10 m/s at room temperature. Further increase in the speed >100 m/s can then be achieved in Pt/CoFeB/MgO. Using such current-induced skyrmion motion, skyrmion logic has been proposed and demonstrated. Using a 'Y'-shaped wire, both AND and OR operations have been verified. The size of the skyrmions can also be controlled by applied magnetic field. Both the methods (i.e., DW and Skyrmion) have their own advantages and disadvantages. A comparative chart is presented in Table 10.3.

A skyrmion RM can be obtained in four different situations (a), (b), (c) and (d) as shown in Fig. 10.22. This can be done by combining the two types of skyrmion Bloch

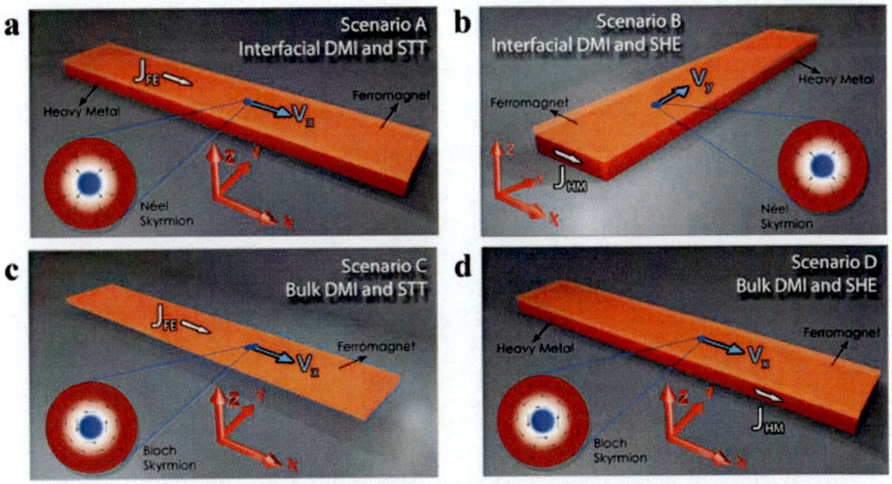

Fig. 10.22 Four different scenarios for the design of a skyrmion racetrack memory. **a** Néel skyrmion motion driven by the STT; **b** Néel skyrmion motion driven by the SHE; **c** Bloch skyrmion motion driven by the STT; **d** Bloch skyrmion motion driven by the SHE. The arrows are related to the in-plane components of the magnetization. The current flows along the x-direction. The skyrmion moves along the x-direction in the scenarios **a**, **c** and **d** along the y-direction in the scenario **b** (Taken from Tomasello et al. 2014)

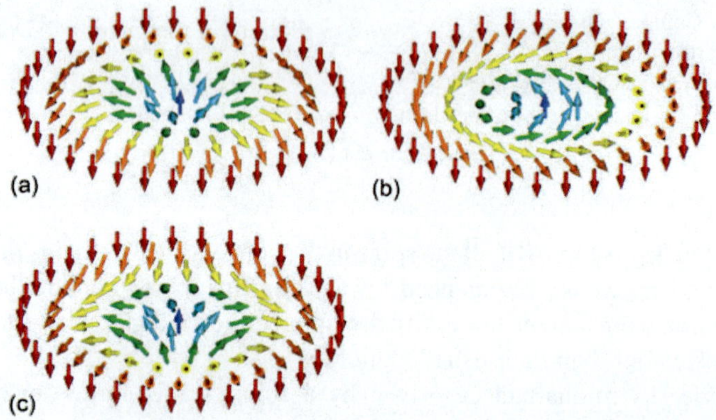

Fig. 10.23 Skyrmions in a 2D FM with uniaxial magnetic anisotropy **a** Bloch type by 2π around an axis perpendicular to the diameter and **b** Neel type by 2π around the diameter and **c** anti-skyrmion

(azimuthal magnetization in the skyrmion boundary) or Neel (radial magnetization in the skyrmion boundary) and its motion driven by the STT or the SHE.

Notes on Magnetic Skyrmions

Skyrmions are quasiparticle magnetic spin configurations with a whirling vortex-like structure that can be stabilized by Dzyaloshinskii–Moriya interactions (DMIs) in chiral bulk magnets such as MnSi, FeGe etc. The Dzyaloshinskii–Moriya interaction (DMI) is described by

$$H_{DMI} = D(si \times sj)$$

where D is the DMI vector based on the crystalline structure. The cross product in the above equation bears a chirality. Rotations of spins can be observed in any one of the two directions: (i) either along the radius or (ii) along the circumference to form a vortex configuration. Dzyaloshinskii–Moriya vector determines the difference in the rotation. A magnetic skyrmion is a quasi-particle theoretically predicted by Skyrme. The skyrmions can be classified into three categories: Bloch, Neel and anti-skyrmions as shown in Fig. 10.23. In Bloch type skyrmion, spins are rotated continuously across the skyrmion radius from perpendicular-to-plane to in-plane and back to perpendicular-to-plane, the in-plane component of the magnetization being along the radius. In Neel type skyrmion, spins are rotated uniformly but the in-plane component of the magnetization is tangential to the radius. The anti-skyrmion is a combination of these two types with in-plane spin rotation along two directions. Magnetic skyrmions are solid states topologically protected defects. However, stable smectic liquid-crystalline structure of skyrmions has also recently been observed.

It is believed that racetrack memory should, 1 day, work in a similar way to punched tape technology back when computers were still in their infancy. During those days, 'hole' was used to represent 1 and 'no hole' was used to represent 0. In future, a 'skyrmion' would represent 1 and 'no skyrmion' would represent 0. Voltage should permit the skyrmions to travel over to a read head, one after another.

10.9 Quantum Computing

Computer is the most essential device in our modern-day life. With the advent of research in semiconductor technology and information processing, as the time elapsed, the size of our computer became smaller and smaller and its computational power increased manifold. But our binary computer is not fast enough to calculate more difficult problems where higher computing power is needed. For example, finding an item from an unsorted big database or factoring big integers needs high computational power, sometimes so high that practically limits our binary computers to solve the problems. So, we need some special type of computer which should be very fast compared to classical binary computer. To meet the need, a new type of computers based on the principle of quantum mechanics are being studied and developed globally. These computers are called quantum computer.

Quantum computer is acting on the principle of quantum mechanics. Our current binary computers manipulate and store information as binary bits (0 and 1). In the physical level, binary bits correspond to binary states (i.e., high or low voltage) of logic devices. On the other hand, quantum computers process information as quantum bits or Qubits. Unlike a binary bit, which corresponds to a binary electronic state of logic device and has only value either 0 or 1, a quantum bit is superposition of binary states, which can hold both the values 0 and 1 at the same time. For example, electrons with up or down spin orientations represent the binary values 0 or 1. Superposition of electronic states leads us to the realization that the electron is spinning a little bit up and little bit down at the same time, i.e., the electron could represent both the bit values 0 and 1, simultaneously. This is quantum bit or Qubit.

10.9.1 Bloch Sphere Representation

We may use Bloch sphere to realize the possible quantum states associated with a single Qubit, as shown in Fig. 10.24. In this sphere, the position of classical bits

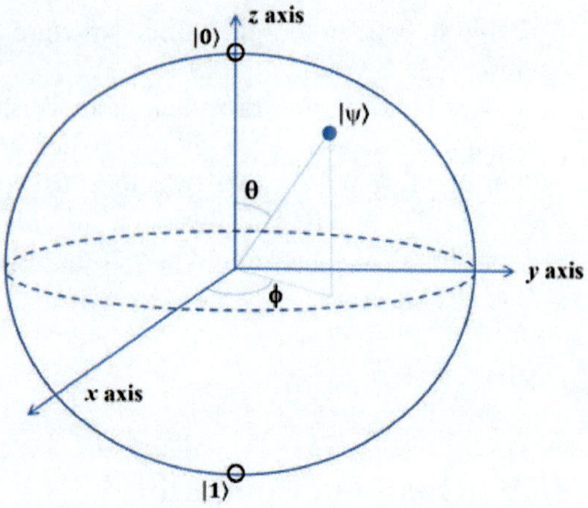

Fig. 10.24 The Bloch sphere representing geometrically a two-level quantum system

could be at the conventionally perceived 'North Pole' or the 'South Pole', i.e., at the positions where $|0\rangle$ and $|1\rangle$ are located, respectively. It should be noted that this particular choice of the polar axis is arbitrary and classical bits have no access on the rest of the surface of the Bloch sphere. On the contrary, a pure Qubit state can be represented by any point anywhere on the Bloch sphere, i.e., a Qubit can have quantum states at any point on this two-dimensional surface. In order to simply understand the concept, let us consider a Qubit represented by a point $|\psi\rangle$ (associated function), as shown in Fig. 10.24. It can be understood from Fig. 10.24 that the probability amplitudes for the superposition state of this Qubit can be written as,$|\psi\rangle = |0\rangle \cos\frac{\theta}{2} + |1\rangle\, e^{i\phi}\sin\frac{\theta}{2}$. It is evident that the surface of the Bloch sphere is a two-dimensional space. This, in fact, represents the ***state space*** of the pure Qubit states. Such ***state space*** is associated with two local degrees of freedom, which are basically two angles, θ and ϕ, as shown in Fig. 10.24.

In conventional electronics, charges are simply moved around a circuit and the spin of electrons is ignored. This means that eight bits are required to represent numbers between 0 and 255 and only one number can only give given at any particular moment in time. Spintronics uses Qubits, so the 'spin-up' and 'spin-down' superposition states of 0 or 1 allow 8 Qubits to represent every number between 0 and 255 simultaneously. Power consumption is also dramatically reduced.

Qubits in reality

- Electron spin (up or down)
- Photon polarization (horizontal or vertical)
- Spin of an atomic nucleus
- Current in a superconducting loop
- Quantum bit represents a unit of quantum information. Different physical objects can be employed as Qubits such as atoms, photons or electrons. Electron spins can represent '0' or '1' or simultaneously as a superposition of both '0' and '1'. Superposition is based on the quantum mechanical idea that an object can

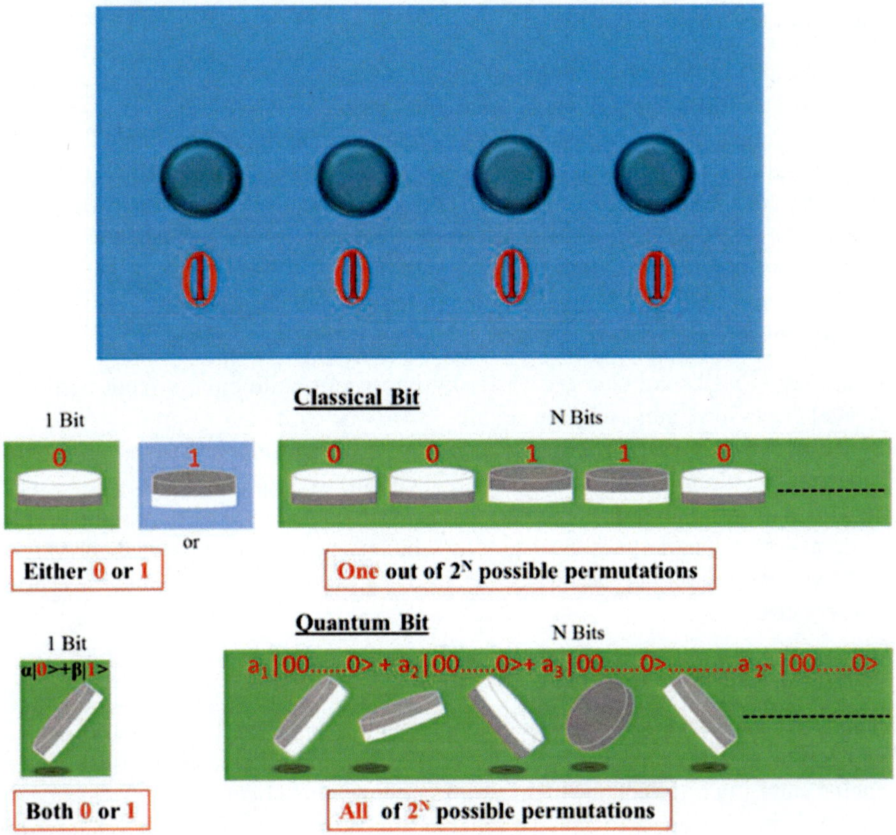

Fig. 10.25 Representation of classical and quantum bits

simultaneously possess two or more values for the same quantity or state at any given time (Fig. 10.25).

10.9.2 Quantum Properties and Computing

The quantum properties that are used in quantum computations are **superposition, entanglement** and **interference**. Superposition is the combination of multiple independent quantum states. Due to the superposition of states, a Qubit can be 0 and 1 state simultaneously. Entanglement is a quantum property by which multiple quantum states can be correlated with one another. Interference of quantum states happens due to phase of wave function of states. In computing, superposition of quantum states in Qubits allows parallel processing, unlike a binary computer, where parallel processing of information is not possible. Entanglement and interference are used in computing. Entangled Qubits can be used in error fixing.

Advantages

- Could process massive amount of complex data.
- Ability to solve scientific and commercial problem.
- Process data in a much faster speed than the normal binary computer.
- (10 Qubits = 1000 calculations and 30 Qubits = 1 billion calculation)
- These are used to protect secure web pages, encrypted email and many other types of data.
- Computing power is exponential.

Limitations

- Quantum states are changed easily by interaction with environment and noises and errors have crept into the signal.
- Error correction is harder as an arbitrary unknown quantum state (of a Qubit) cannot be directly copied (no-cloning theorem) to check for errors.
- Quantum computing is not energy efficient till now due to the requirement of low temperature (sometimes, near to 0 K) which uses a lot of energy for cooling.
- Quantum computing has the power to crack current cryptographic encryptions very easily.

Spintronics and Quantum computing

Though there are many ways to achieve Quantum computing like using polarization of photons or using nuclear spins. One of the most critical problems researchers are facing currently is the change of quantum sates in Qubits (decohorence) and creeping of noises in the signal. To achieve reliable results, we need to preserve the sates of Qubits for sufficiently long time to finish the computations. Fortunately, spintronics, especially Organics Spintronics can solve the problem caused by decohorence (Tang and Slyke 1987; Heeger 2000; Tang et al. 1989; Kumar et al. 2006; Burroughes et al. 1990; Bennati et al. 1996; Jiang et al. 2008; Rolfe et al. 2011; Nguyen et al. 2012; Naber et al. 2007; Francis et al. 2004; Liu et al. 1985). In organic carbon-based small molecules, spins of the electrons interact less with the environment and tend to be preserved for longer amount of time compared to other materials, up to a few microseconds! Though, in organic molecules, a high level of electrical noise due to the hopping of electrons is a serious limitation to study quantum computing, we may avoid the hopping using spin pumping technique to manipulate spin quantum states.

10.10 Conclusions

In conclusions, we have presented an overview of the application of spintronics, its historical advancement and development of spintronic devices. We have presented a detail discussion on the Read Head technology in Magnetic Data Storage. Here, we have highlighted the application of GMR effect in magnetic read head, the role of GMR effect in Magnetic Data storage, internal structure of a magnetic hard disk, operation of Magnetoresistance Read Heads and finally the evolution of magnetoresistance read head sensor with the implementation of GMR sensor. We have also carried out discussion on spin valve sensors and some important features of spin-valve GMR read head applications. A detailed discussion on Magnetic Random Access Memories (MRAM) has been presented here. In this direction, we have addressed application based on spin-tunnel junctions, its background, and *'Why MRAM should be used?'*. We have elaborated the discussion on the basic cell operation of MTJ based—Magnetic Random Access Memories (MRAM). Furthermore, Spin Transfer Torque (STT)—MRAM has been described. Here, we have also addressed the issue of perpendicular STT-MRAM, application of STT-MRAM, the latest application of STT-MRAM, etc. We have carried out discussion on spintronics sensors that include *anisotropic magnetoresistance sensor, some GMR-based sensors applications and some magnetic tunnel junction-based TMR sensors applications.* Some very intriguing topics on Spin FET, Race Track Memory, Quantum Computing have also been highlighted and described in this chapter.

10.11 Exercises

1. What is the typical multilayer structure used in spin-valve devices?
2. What is called synthetic antiferromagnet?
3. How to use spintronic devices for the detection of magnetic field?
4. How could the non-volatile '*Selective bit programming*' could be achieved in MTJ-based Magnetoresistive Random Access Memories (MRAM)?
5. Write a short note on spin valve sensor for read head.
6. Write a short note on MTJ-based Magnetoresistive Random Access Memories (MRAM).

References

M.N. Baibich, J.M. Broto, A. Fert, et al., Giant magnetoresistance of (001)Fe/(001)Cr magnetic super lattices. Phys. Rev. Lett. **61** 2472–2475 (1988)

M. Bennati, K. Németh, P.R. Surjań, M. Mehring, Zero-field-splitting and ∏-electron spin densities in thelowest excited triplet state of oligothiophenes. J. Chem. Phys. **105**, 4441–4447 (1996)

L. Berger et al., J. Appl. Phys. **64**, 5459 (1988)

G. Binasch, P. Grünberg, F. Saurenbach, W. Zinn, Enhanced magnetoresistance in layered magnetic structures with antiferromagnetic interlayer exchange. Phys. Rev. B **39**, 4828(R) – 4830 (1989)

J.H. Burroughes, D.D.C. Bradley, A.R. Brown, Light-emitting diodes based on conjugated polymers. Nature (London) **347**, 539–541(1990)

P. Chuang, S.-C. Ho, L.W. Smith et al., All-electric and all-semiconductor spin field effect transistors. Nat. Nanotechnol. **10**, 35–39 (2015)

S. Datta, How we proposed the spin transistor. Nat. Electron. **1**, 604 (2018)

S. Datta, B. Das, Electronic analog of the electro-optic modulator. Appl. Phys. Lett. **56**, 665–667 (1990)

B. Dieny et al., Phys. Rev. B **43**, 1297 (1991)

B. Dieny, V.S. Speriosu, S. Metin, S.S.P. Parkin, B.A. Gurney, P. Baumgart, D.R. Wilhoit, Magnetotransport properties of magnetically soft spin-valve structures. J. Appl. Phys. **69**, 4774–4779 (1991)

B. Dieny, I.L. Prejbeanu, P. Bortolotti, Opportunities and challenges for spintronics in the microelectronics industry. Nat. Electron. **3**, 446–459 (2020)

T.L. Francis, O. Mermer, G. Veeraraghavan, M. Wohlgenannt, Large magneto resistance at room-temperature in small-molecular weight organic semiconductor sandwich devices. New J. Phys. **6**, 185e192 (2004)

G.B. Granley, T. Hurst, in *Sixth Biennial IEEE International Non-Volatile Memory Technology Conference Proceedings* (1996), p. 138

P. Grunberg, S. Demokritov, A. Fuss, R. Schreiber, J.A. Wolf, S.T. Purcell, J. Magn. Magn. Mater. **104–107**, 1734 (1992)

A.J. Heeger, Semiconducting and metallic polymers: the fourth generation of polymeric materials. Nobel Lecture, 8 Dec 2000

J. Heidecker, MRAM technology and status, in *Lecture from NEPP Electronic Technology Workshop*, 11–13 June 2012

A. Hirohata, K. Takanashi, Future perspectives for spintronic devices. J. Phys. D Appl. Phys. **47**, 193001 (2014)

A. Hirohataa, K. Yamadab et al., Review on spintronics: Principles and device applications. J. Magn. Magn. Mater. **509**, 166711 (2020)

J.S. Jiang, J.E. Pearson, S.D. Bader, Absence of spin transport in the organic semiconductor Alq_3. Phys. Rev. B **77**, 035303-1–035303-7 (2008)

T. Koga, J. Nitta, H. Takayanagi, S. Datta, Spin-filter device based on the Rashba effect using a nonmagnetic resonant tunneling diode. Phys. Rev. Lett. **88**, 126601–126604 (2002)

J. Kumar, R.K. Singh, P.K. Siwach, H.K. Singh, R. Singh, R.C. Rastogi, O.N. Srivastava, Enhanced magnetoresistance in $La_{0.82}Sr_{0.18}MnO_3$-$\prod$ conjugated semiconducting polymer heterostructure. Solid State Commun. **138**, 422–425 (2006)

M. Lederman, IEEE Trans. Magn. **35**, 794 (1999)

K. Liu, K. Nagodawithana, P.C. Searson, C.L. Chien, Phys. Rev. B **51**, 738 (1985)

X. Liu, K.H. Lam, K. Zhu, C. Zheng, X. Li, Y. Du, Chunhua Liu, P.W.T. Pong, Overview of spintronic sensors Internet of Things and smart living (2016), https://arxiv.org/abs/1611.00317

W.J.M. Naber, S. Faez, W.G. van der Wiel, Organic spintronics. J. Phys. D Appl. Phys. **40** R205eR228 (2007)

T.D. Nguyen, E. Ehrenfreund, Z.V. Vardeny, Spin-polarized light-emitting diode based on an organic bipolar spin valve. Science **337**, 204e209 (2012)

S.S.P. Parkin et al., J. Appl. Phys. **85**, 5828 (1999)

R.I. Potter, IEEE Trans. Magn. **10**, 502 (1974)

E.I. Rashba, Spintronics: sources and challenge. J. Supercond. **15**, 13–17 (2002)

S. Ravi, A. Karthikeyan, N.A. Nesakumari, K. Pugazhvadivu, K. Tamilarasan, Magnetoresistance in silicon based ferrite magnetic tunnel junction. Curr. Appl. Phys. **14**, 259–263 (2014)

N.J. Rolfe, M. Heeney, P.B. Wyatt, A.J. Drew, T. Kreouzis, W.P. Gillin, The effect of deuteration on organic magnetoresistance. Synth. Met. **161**, 608e612 (2011)

H. Sakakima et al., J. Magn. Magn. Mater. **210**, L20 (2000)

C.W. Tang, S.A. Van Slyke, Organic electroluminiscent diodes. Appl. Phys. Lett. **51**, 913 (1987)

C.W. Tang, S.A. VanSlyke, C.H. Chen, Electroluminiscence of doped organic thin films. J. Appl. Phys. **65**, 3610 (1989)

R. Tomasello, E. Martinez et al., A strategy for the design of skyrmion racetrack memories. Sci. Rep. **4**, 6784 (2014). Springer Nature

H. van den Berg et al., IEEE Trans. Magn. **32**, 4624 (1996)

A. Veloso et al., Appl. Phys. Lett. **77**, 1020 (2000)

T.W. Wlodarczyk, C. Rong, D. Waage, Challenges in healthcare and welfare intercloud, in *The 7th International Conference on Digital Content, Multimedia Technology and its Applications* (2011)
M. Ziese, M.J. Thornton (eds.), *Spin Electronics* (Springer, Berlin, 2001)
I. Zutic, J. Fabian, S.D. Sarma, Spintronics: fundamentals and applications. Rev. Modern Phys. **76**, 323–410 (2004)

Subject Index

P. Dey and J. N. Roy, *Spintronics*,
https://doi.org/10.1007/978-981-16-0069-2

T